Gel Electrophoresis of Nucleic Acids

The Practical Approach Series

SERIES EDITORS

D. RICKWOOD

Department of Biology, University of Essex
Wivenhoe Park, Colchester, Essex CO4 3SQ, UK

B. D. HAMES

Department of Biochemistry, University of Leeds
Leeds LS2 9JT, UK

Affinity Chromatography
Animal Cell Culture
Animal Virus Pathogenesis
Antibodies I and II
Biochemical Toxicology
Biological Membranes
Biosensors
Carbohydrate Analysis
Cell Growth and Division
Centrifugation (2nd edition)
Clinical Immunology
Computers in Microbiology
Directed Mutagenesis
DNA Cloning I, II, and III
Drosophila
Electron Microscopy in Molecular Biology
Fermentation
Flow Cytometry
Gel Electrophoresis of Nucleic Acids (2nd edition)
Gel Electrophoresis of Proteins (2nd edition)

Genome Analysis
HPLC of Small Molecules
HPLC of Macromolecules
Human Cytogenetics
Human Genetic Diseases
Immobilised Cells and Enzymes
Iodinated Density Gradient Media
Light Microscopy in Biology
Liposomes
Lymphocytes
Lymphokines and Interferons
Mammalian Development
Medical Bacteriology
Medical Mycology
Microcomputers in Biology
Microcomputers in Physiology
Mitochondria
Mutagenicity Testing
Neurochemistry
Nucleic Acid and Protein Sequence Analysis

Gel Electrophoresis of Nucleic Acids

A Practical Approach

SECOND EDITION

Edited by

D. RICKWOOD

Department of Biology, University of Essex

and

B. D. HAMES

Department of Biochemistry, University of Leeds

OXFORD UNIVERSITY PRESS
Oxford New York Tokyo

Oxford University Press
Walton Street, Oxford OX2 6DP

Oxford is a trade mark of Oxford University Press

Published in the United States
by Oxford University Press, New York

British Library Cataloguing in Publication Data
Gel electrophoresis of nucleic acids.—2nd ed.
1. Nucleic acids. Chemical analysis. Gel electrophoresis
I. Rickwood, D. David 1945–   II. Hames, B. D. (B. David)
III. Series
547.79046
ISBN 0-19-963082-8
ISBN 0-19-963083-6 pbk

Library of Congress Cataloging in Publication Data
Gel electrophoresis of nucleic acids: a practical approach / edited by
D. Rickwood and B. D. Hames.—2nd ed.
(Practical approach series)
Includes bibliographical references.
1. Nucleic acids—Analysis.   2. Gel electrophoresis.
I. Rickwood, D. (David) II. Hames, B. D. III. Series.
QP620.G45 1990   574.87'328'028—dc20   90-4348

ISBN 0 19 963082 8
ISBN 0 19 963083 6 (pbk)

Typeset and printed by Information Press Ltd, Oxford, England

Preface

SINCE the first edition of this book in 1982, some methods used for the electrophoretic analysis of nucleic acids have changed only marginally while in other cases entirely new techniques have been devised. The contents of this second edition reflect these developments. Nucleic acid sequencing is no longer included since there has been such an enormous increase in the diversity of techniques and extent of our knowledge in this area that the original chapters covering this topic have now been expanded into a separate companion volume in the Practical Approach series. One of the major new techniques now covered is pulsed field electrophoresis which allows the separation of exceptionally large molecules of DNA. Another key area described in detail are the ways that electrophoretic techniques can be used to study the specific interaction of proteins with DNA sequences by footprinting and gel retardation analyses. These techniques have proved invaluable in pinpointing the exact nature of the interaction of proteins with regulatory DNA sequences. In addition, the book now also describes the use of electrophoresis for the purification of synthetic oligonucleotides which are proving to be an essential tool in many areas of molecular biology.

In editing the second edition it has been necessary to balance the problem of the repetition of topics with the convenience of making each chapter as self-contained as possible. In all cases we have tried to make the book as comprehensive as possible while retaining the preferences of individual authors for particular methods. We wish to thank the authors for their help and forebearance of our editing their manuscripts which has allowed us to unite individual chapters into a complete text.

DAVID RICKWOOD
B. DAVID HAMES

Contents

2 Gel electrophoresis of DNA

Paul G. Sealey and Ed M. Southern

3 Pulsed field gel electrophoresis

Rakesh Anand and Ed M. Southern

4 The electrophoresis of synthetic oligonucleotides

J. William Efcavitch

5 Two-dimensional gel electrophoresis of nucleic acids

Rupert De Wachter, Jack Maniloff, and Walter Fiers

6 Gel retardation analysis of nucleic acid – protein interactions

Mark M. Garner and Arnold Revzin

Contents

7 The analysis of sequence-specific DNA-binding proteins in cell extracts

Graham H. Goodwin

8 Electrophoresis of nucleosomes

Robert H. Nicolas

9 Gel electrophoresis of ribonucleoproteins

Albert E. Dahlberg and Paula J. Grabowski

Contents

Appendices

Contributors

R. ANAND

Department of Biochemistry, Oxford University, South Parks Road, Oxford, UK.

A. E. DAHLBERG

Biology and Medicine Division, Brown University, Providence, RI 02912, USA.

R. DE WACHTER

Department of Biochemistry, Universiteit Antwerpen (UIA) Universiteitsplein 1 B-2610 Antwerpen, Belgium.

J. W. EFCAVITCH

Research and Development, Applied Biosystems Inc., 850 Lincoln Centre Drive, Foster City, CA 94404, USA.

W. FIERS

Laboratory of Molecular Biology, Rijksuniversiteit-Gent, B-9000 Gent, Belgium.

M. M. GARNER

Department of Chemistry, University of Wisconsin, Madison, WI 53706, USA.

G. H. GOODWIN

Chester Beatty Research Institute, Royal Cancer Hospital, Fulham Road, London SW3 6JB, UK.

P. GRABOWSKI

Biology and Medicine Division, Brown University, Providence, RI 02912, USA.

D. GRIERSON

Department of Physiological and Environmental Science, University of Nottingham School of Agriculture, Sutton Bonington, Loughborough, UK.

J. MANILOFF

Department of Microbiology and Immunology, University of Rochester School of Medicine and Dentistry, Rochester, NY 14642 USA.

R. H. NICOLAS

Chester Beatty Research Institute, Royal Cancer Hospital, Fulham Road, London SW3 6JB, UK.

A. REVZIN

Biochemistry Department, Michigan State University, East Lansing, MI 48824, USA.

P. G. SEALEY

Department of Zoology, Edinburgh University, West Mains Road, Edinburgh, UK.

E. M. SOUTHERN

Department of Biochemistry, Oxford University, South Parks Road, Oxford, UK.

IMPORTANT NOTICE
Health hazards of gel electrophoresis

1. A number of chemicals commonly used for gel electrophoresis are toxic whilst the status of others remains unknown. It is very important that experimenters acquaint themselves with the precautions required for handling all chemicals mentioned in this text. Particular care should be taken when handling acrylamide since this is a known potent neurotoxin. Polyacrylamide gel is not toxic unless it contains unpolymerised monomer.

2. Care should be taken when using gel electrophoresis apparatus that no electrical safety hazard exists. Particular care should be taken when using apparatus not obtained from commercial sources since this may not meet the usual required safety standards. It is recommended that all apparatus is checked by a competent electrician before use.

Abbreviations

A_{540}	absorption at 540 nm
A	adenine (only used as part of a sequence)
A	amp(s), ampere(s)
ABM-paper	aminobenzyloxymethyl-paper
AC	alternating current
% acrylamide	polyacrylamide gel concentration expressed in terms of *total* monomer (i.e. acrylamide and crosslinker)
AraCTP	arabinosyl CTP
BAC	*N,N'* bisacrylylcystamine
Bisacrylamide	*N,N'*-methylene bisacrylamide
bp	base pairs
Bq	Becquerel (1 disintegration/sec)
BSA	bovine serum albumin
C	cytosine
CAP	catabolite activator protein
CHEF	contour-clamped homogeneous electric field
Ci	Curie (3.7×10^{10} Bq)
cpm	counts per minute (the exact amount of radioactivity depends on the isotope and the method of measurement used)
% crosslinker	expressed in terms of *total* monomer (i.e. acrylamide and crosslinker)
CTAB	cetyltrimethylammonium bromide
DATD	*N,N'* diallyltartardiamide
DBM-paper	diazobenzyloxymethyl-paper
DC	direct current
ddH_2O	double distilled water
DEAE	diethylaminoethyl
DMAPN	3-dimethylamino-propionitrile
DMS	dimethyl sulphate
DMSO	dimethyl sulphoxide
DNase	deoxyribonuclease
dNTP	deoxynucleoside triphosphate
dpm	disintegrations per minute (60 Bq)
DPT	diazophenylthioester
DTT	dithiothreitol
EDTA	ethylenediaminetetra-acetate
EGTA	ethylene glycol-bis-(2-aminoethylether)tetracetic acid
G	guanine
g	gram(me)
g	centrifugal force ($\times$ unit gravitational field)
h	hour(s)
HEPES	*N*-2-hydroxyethylpiperazine-*N'*-2-ethanesulphonic acid
HMG	high mobility group
HPLC	high pressure liquid chromatography

HTAB	hexadecyl-trimethyl ammonium bromide
i.d.	internal diameter
IPTG	isopropylthio-β-D-galactoside
kb	kilobases
kBq	kilobecquerels (Bq $\times$ 10^3) see Bq
kV	kilovolts
LGT	low gelling temperature
M	molarity
M	mobility
mA	milliamp(s)
Mb	megabases (bp $\times$ 10^6)
MBq	megabecquerels (Bq $\times$ 10^6) see Bq
nm	nanometres
o.d.	outside diameter
PBS	phosphate buffered saline
PMSF	phenylmethylsulphonyl fluoride
PPO	2,5-diphenyloxazole
PVC	polyvinyl chloride
RNase	ribonuclease
S	Svedberg
SDS	sodium dodecyl sulphate
SSC	saline sodium citrate (0.15 M sodium chloride, 15 mM sodium citrate, pH 7.0)
TAE	Tris-acetate-EDTA
TBE	Tris-borate-EDTA
TCA	trichloroacetic acid
TEMED	*N,N,N',N',* tetramethylethylenediamine
UV	ultraviolet
V	volts

1

Gel electrophoresis of RNA

DON GRIERSON

1. Introduction

The term electrophoresis is generally applied to the movement of small ions and charged macromolecules in solution under the influence of an electric field. The rate of migration depends on the size and shape of the molecule, the charge carried, the applied current and the resistance of the medium. Zone electrophoresis is the separation of charged molecules in a supporting medium, resulting in the migration of charged species in distinct zones and is distinguished from boundary electrophoresis which is carried out in free solution.

Successful fractionation of RNA by electrophoresis was achieved in the middle of the 1960s and very good separations became routine by the end of that decade. The chief reasons for success were improvements in RNA preparation and handling techniques and the introduction of supporting gels for electrophoresis. The basic gel technology arose by trial and error without complex theory. In an electric field, at moderate pH, negatively-charged RNA migrates towards the anode. A fractionation is achieved because large molecules move more slowly through the gel than small molecules and separation of RNA within a given size range is obtained by selecting a gel of appropriate pore size.

Electrolytes used in electrophoresis generally consist of an aqueous buffer, containing a chelating agent such as ethylenediaminetetracetate (EDTA) and a nuclease inhibitor such as sodium dodecyl sulphate (SDS). Nuclease inhibitors may be omitted from the buffers, provided all apparatus and reagents are kept scrupulously clean or are sterilized. For some purposes it is important to reduce the effects of secondary structure and a denaturing agent may be added or used in place of water as solvent. A number of factors affect the fractionation of RNA. Increasing the current leads to higher rates of migration, but the flow of current also results in the production of heat, which, if excessive, adversely affects the separation by causing trailing and broadening of the zones. Increasing the ionic strength results in lower mobilities. If the current is increased to compensate, this again may result in adverse effects due to heating. On the other hand, if the ionic strength is too low, this may seriously reduce the buffering capacity of the solution and lead to undesirable pH changes. This problem can be overcome either by recirculating the buffer between the reservoirs or by changing it at intervals during the run.

Electrophoresis is carried out in the gels cast either in tubes or as slabs. A number

1

of gel materials have been used successfully, including agar, agarose, and polyacrylamide. Agar and agarose gels are made by heating the granular material in the appropriate electrolyte buffer, casting the gels and allowing them to set on cooling. The resolving power of these gels depends on the concentration of dissolved agar or agarose; dilute gels are used for very large RNA molecules and more concentrated gels for low molecular weight RNA.

Polyacrylamide gels are made from acrylamide and *N,N'*-methylene bisacrylamide (bisacrylamide) mixtures dissolved in electrolyte and polymerised by the addition of chemical catalyst. The physical properties and resolving power of the gels are determined by the acrylamide concentration and the proportion of bisacrylamide added as crosslinker. Details of the polymerization reaction and the effects of varying the ratio of acrylamide to bisacrylamide have been discussed in another volume of the Practical Approach series (32). Examples of the resolution obtainable with different concentrations of polyacrylamide in tubes and slab gels are given later in this Chapter (*Figures 2, 10, 11,* and *16*). The wide range of controlled pore size which can be obtained reproducibly with polyacrylamide makes this a very popular choice and gels can be made for most purposes. However, some researchers find very dilute polyacrylamide gels, used for fractionating very large RNAs, difficult to handle and so either strengthen them with agarose or use agarose on its own.

In many cases, the distance moved by an RNA molecule during electrophoresis is inversely related to the $\log_{10}$ molecular weight. However, base composition, secondary structure and other factors can affect this relationship. For accurate measurement of molecular weights it is necessary to carry out electrophoresis under

---

Table 1. Equipment required for polyacrylamide gel electrophoresis in tubes.

Electrophoresis

Electrophoresis apparatus made from Perspex (see *Figure 1*).

Gel tubes, preferably made from Perspex. These can be of any convenient dimensions, such as 0.6 cm or 0.9 cm internal diameter tubing cut to 8 – 12-cm lengths. With larger diameter tubes, heat dissipation during electrophoresis is less efficient and this may cause problems.

A good vacuum pump.

A power pack providing up to 500 V and 100 mA and the facility for constant voltage. It is not necessary to provide constant current.

Scanning of gels

Gel scanning equipment. Several types are available, such as a 'Polyfrac' made by Joyce-Loebl Ltd or a spectrophotometer fitted with a gel-scanning attachment. These normally require a parallel-sided quartz glass cuvette or a tube to which the gels are transferred after electrophoresis. Alternatively, the gels may be run in quartz glass tubes and scanned *in situ* but these tubes are expensive.

Measurement of radioactivity

Gel slicer. Accurate slicing of dilute gels (2.2% or 2.4%) is difficult unless they are first frozen and then sliced using an automated gel slicer (e.g. from Joyce-Loebl Ltd). More concentrated gels can be sliced unfrozen, but with less precision, using razor blades.

Scintillation counter or planchette counter.

2

completely denaturing conditions, for example by the inclusion of methylmercuric hydroxide in the gel system, by electrophoresis in the presence of formamide, or by reacting the RNA with glyoxal or formamide prior to loading it on to the gel. Each of these techniques is discussed later in this chapter.

It is not possible to review all the individual variations that are used in the gel electrophoresis of RNA. Therefore this chapter describes some of the most frequently used methods and their limitations. The choice between tube and slab gels, and the selection of the gel medium, will need to be determined by individual requirements. Polyacrylamide tube gels (*Table 2* and *Protocols 1* and *2*) are best if the RNA is to be detected and quantified after electrophoresis by UV scanning (*Figure 2*), since this avoids having to cut and manipulate strips from slab gels which are usually very fragile. On the other hand, if the RNA is to be detected by staining or autoradio-

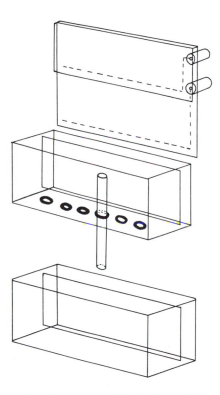

Figure 1. A tube gel electrophoresis apparatus. This design, which is used in many laboratories, can be constructed from 3-mm-thick Perspex sheet. It consists of two rectangular buffer reservoirs plus electrodes. The upper reservoir rests on the lower reservoir; gel tubes pass through holes drilled in the base of the upper reservoir and are held in position by rubber grommets. The one-piece electrode assembly, made from Perspex sheet and platinum wire, has two limbs which fit into the upper and lower reservoirs. The dimensions of the electrophoresis apparatus depend on the size and number of gel tubes. The gel tubes are shielded from the electrodes to localize pH changes and the products of electrolysis.

graphy, slab gels (either vertical or horizontal) are quicker and easier, especially if one wishes to compare many samples. Agarose is now most frequently used as the gel medium, because of its ease of use, although polyacrylamide is still best for high resolution or the separation of very small RNA molecules. At the present time the most popular method is to separate RNA by submarine electrophoresis in horizontal slabs of agarose (see Section 7), since the gels are easy to make and run, and can be used subsequently for northern blotting. In addition, the RNA bands can be detected by a variety of very sensitive staining procedures or by autoradiography of the gel before or after it has been dried.

2. Polyacrylamide tube gel electrophoresis

2.1 Apparatus

Polyacrylamide gel electrophoresis in tubes is convenient for the analysis of relatively small numbers of samples or where small quantities of RNA are to be fractionated,

---

Table 2. Stock solutions for the preparation of polyacrylamide gels.

Stock acrylamide solution

- Acrylamide   15 g
- Bisacrylamide   0.75 g
- Distilled water to 100 ml

For gels above 5% final acrylamide concentration, the bisacrylamide can be reduced to 0.375 g per 100 ml. Specially purified acrylamide and bisacrylamide may be purchased for electrophoresis. Alternatively, laboratory grade reagents can be recrystallized as follows:

Dissolve 70 g acrylamide in 1 litre of chloroform at 50°C using a heating mantle in a fume cupboard. Filter the solution hot under suction and obtain crystals by transferring to a deep freeze at -20°C. Filter the crystals using a chilled filter funnel and wash with chilled chloroform. Thoroughly dry the crystals before use.

Dissolve 10 g bisacrylamide in 1 litre of acetone at $40-50$°C using a heating mantle in a fume cupboard. Filter the solution hot and then cool to -20°C. Filter off the crystals and wash with chilled acetone. Thoroughly dry the crystals before use.

Caution: Do not inhale, swallow, or allow acrylamide to contact the skin; its toxic effects are cumulative.

Gel buffer[a]

- Tris base   21.7 g (180 mM)
- $NaH_2PO_4.2H_2O$ 23.8 g (150 mM)
- $EDTA.Na_2.2H_2O$ 1.85 g (5 mM)
- Distilled water to   1 litre

The buffer should be pH 7.7 at room temperature.

Catalysts

N,N,N',N'-tetramethylethylenediamine (TEMED).
Freshly dissolved 10% ammonium persulphate.

Reservoir buffer

This is gel buffer, prepared as described above, diluted fivefold and containing 0.2% SDS.

[a] The concentrations of reagents given in brackets are final concentrations in the gel buffer.

Table 3. Recipes for the preparation of polyacrylamide gels

	Final polyacrylamide concentration (%)									
	2.0	2.2	2.4	2.5	2.6	3.0	4.0	5.0	7.7	10.0
Stock acrylamide solution (ml)	5.0	5.0	5.0	5.0	5.0	5.0	5.0	10.0	10.0	10.0
Gel buffer (ml)	7.5	6.8	6.25	6.0	5.8	5.0	3.75	6.0	4.0	3.0
Water (ml)	24.7	22.0	19.7	18.7	17.8	14.7	9.7	13.4	5.4	2.0
Final volume (ml)	37.5	34.5	31.25	30.0	28.9	25.0	18.75	29.7	19.4	15.0

After mixing add 25 μl of TEMED, followed by 250 μl of freshly dissolved 10% ammonium persulphate.

although it can be scaled up as a preparative procedure. *Table 1* lists the equipment that is required. Suitable electrophoresis apparatus and gel tubes (*Figure 1*) can be easily and inexpensively constructed in a workshop or purchased as complete units from many suppliers. Perspex sheets can either be glued or fused together with chloroform. The electrodes should be made of platinum wire. Perspex tubing cut to the required length is the best material for gel tubes; glass tubes are not really suitable because the gels stick to the glass and so are difficult to recover for analysis.

2.2 Preparation of gels

Stock solutions for the preparation of polyacrylamide gels are set out in *Table 2*, recipes for making gels from 2−10% polyacrylamide are given in *Protocol 1* and *Table 3* and illustrations of the resolving power of 2.4%, 3%, and 10% gels are shown in *Figure 2*. Acrylamide is toxic and should not be inhaled, swallowed or allowed to contact the skin. Its effects are cumulative.

Protocol 1. Preparation of polyacrylamide tube gels

1. Before preparation of the gels, seal the tubes temporarily at the base and hold them vertically in a rack or in the electrophoresis apparatus. Parafilm, rubber bungs or caps can be used to seal the base of the gel tubes. Another method is to cut a transverse section of flexible plastic tubing to fit the inside of the tube and then to plug the hole with a short piece of glass rod.

2. Prepare the gel mixture by diluting the stock solution of acrylamide and bisacrylamide to the appropriate concentration with gel buffer and water (*Table 3*).

3. Remove dissolved oxygen, which inhibits polymerization, using a vacuum pump for approximately 30 sec. The vacuum should be good enough to cause the solution to bubble vigorously. Gels will polymerize even if oxygen is not removed, but with low concentrations of acrylamide this often results in weak

Protocol 1. *continued*

or uneven gels. It is also difficult to polymerize gels with similar properties from day to day without degassing.

4. After degassing add the tetramethylethylenediamine (TEMED) and ammonium persulphate (the polymerization catalysts). Gently mix the solution and immediately pipette the gel mixture into the tubes. A gel tube 10 cm long and 0.6 cm internal diameter holds about 3 ml.

5. Carefully overlayer the gel mixture with water to a depth of about 1 cm using a Pasteur pipette with a curved tip or a syringe fitted with a bent needle. This produces a flat surface at the top of the gel and also prevents the gel from drying out before use.

6. Leave the gel mixture to polymerize on a vibration-free surface. Usually, 2.4% gels will show signs of polymerization after approximately 15 min, whereas 7.5% gels will polymerize more rapidly. In order to retard polymerization and allow time for pipetting concentrated mixtures into the gel tubes and overlayering with water, the gel mixture can be either cooled in ice before adding the catalysts or the amount of TEMED added to the gel mixture can be reduced.

Although most gels are ready to use within two hours they continue to polymerize slowly if left longer. They can be stored for a few days at room temperature or in a refrigerator. Gels below approximately 2.6% polyacrylamide are almost transparent whereas more concentrated gels are progressively more translucent to visible light.

2.3 Setting up the electrophoresis apparatus

It takes about half an hour to get the gels ready for loading the samples, and this is described in *Protocol 2*.

Protocol 2. Setting up the tube gel apparatus

1. Arrange the gel tubes vertically by pushing them through the rubber grommets in the upper electrophoresis reservoir.

2. Remove the water overlay from the tops of the gels using a Pasteur pipette or by a flick of the wrist, and then remove the stoppers from the bottom of the gel tubes. Very dilute gels may slide out of the tubes unless they are supported by a small piece of muslin held in place over the bottom of the tube with a rubber band. A plastic ring placed inside the bottom of the gel tube will also prevent all but the most dilute gels from slipping out.

3. Fill the upper and lower electrophoresis reservoirs with sufficient reservoir buffer (*Table 2*) to cover the tops of the gels, and place the upper reservoir on top of the lower compartment. It is important to remove any air bubbles trapped at the bottom of each gel tube. This is easily done by directing a jet

Protocol 2. *continued*

of buffer at the base of the gel with the curved tip of a Pasteur pipette. This technique can also be used to remove air bubbles from the top of gels.

4. Place the electrodes in position (lower electrode positive) and switch on the power pack. Electrophoresis can be carried out at room temperature or in a cold room (5°C) if the SDS concentration is reduced (Section 3.3). It is normal practice to pre-run the gels at approximately 5 V/cm for 15−30 min to remove catalysts and UV-absorbing materials before loading the sample.

2.4 Sample preparation and loading of gels

RNA frequently tends to aggregate at the top of the gel if it is contaminated with proteins but normally this presents no problems since most procedures for RNA extraction and purification involve the use of proteolytic enzymes or deproteinizing mixtures such as phenol-cresol or phenol-chloroform (22). A typical mixture contains 1000 g redistilled phenol, 140 ml *m*-cresol and 1.0 g of 8-hydroxyquinoline and then this mixture should be saturated with water before use. Some people prefer to add an equal volume of chloroform to this mixture to improve the efficiency of deproteinization. Whenever possible, the purified RNA should be precipitated with ethanol or with a high concentration of salt both to concentrate the RNA and to remove any impurities which may otherwise interfere with electrophoresis or subsequent UV-scanning of the gels. Traces of phenol and detergents can be removed by dissolving the RNA in 0.15 M sodium acetate, adjusted to pH 6.0 with acetic acid, and re-precipitating it with two volumes of ethanol at −20°C for a few hours. It is a useful routine precaution to add 0.5% SDS to the sodium acetate solution to guard against any nuclease contamination. After ethanol precipitation, the RNA can be pelleted by centrifugation, drained of ethanol and, if necessary, partially dried in a vacuum dessicator for a few minutes (if there is more than a trace of alcohol in a sample it will not sink on to the surface of the gel during sample loading).

For safety reasons, the power pack should be switched off while the sample is loaded. Ideally, the RNA should be dissolved in small volume of electrophoresis buffer (36 mM Tris base, 30 mM NaH_2PO_4, 1 mM EDTA, 0.2% SDS, pH 7.7) containing 5−15% sucrose to increase its density and hence facilitate layering of the sample on to the gel. A good concentration to aim for is 1 μg RNA per microlitre. After the reservoirs have been filled with reservoir buffer, load the samples on to the gels using a micropipette with the tip held just above the surface of the gel. It is also possible to load samples dissolved in buffers of different composition to that used for making the gels. The main considerations are that the salt concentration should be low enough so that it does not interfere with the electrophoresis and that sufficient RNA can be loaded to allow subsequent detection. Bromophenol blue may be added as a marker if required.

The main consideration in deciding how much RNA to load is that sufficient RNA must be loaded to allow it to be detected after electrophoresis without overloading

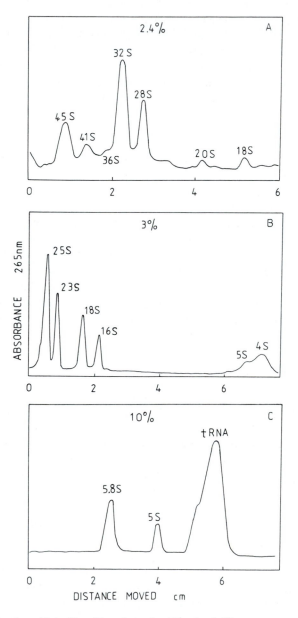

Figure 2. Separation obtainable with polyacrylamide gels of different concentrations. (a) HeLa cell nucleolar RNA separated on a 2.4% gel electrophoresed for 4 h at 50 V at room temperature. (b) RNA from spinach cotyledons separated on a 3% gel at 5°C for 4 h at 50 V. (c) RNA from mung bean cytoplasm separated on a 10% gel at 5°C for 5 h at 50 V. The RNA was heated at 65°C and rapidly cooled before loading. All gels were scanned at 265 nm after electrophoresis. The data in (a), (b) and (c) are redrawn from refs. 26, 27, and 28, respectively.

the gel and hence causing unnecessary broadening of the RNA bands. Good separations can be obtained in a 0.6-cm diameter gels by loading $10-50$ μl samples of solution containing $10-100$ μg RNA. Slight band-broadening may occur with volumes above 30 μl or with more than 50 μg RNA per gel. For a trial run, try 30 μl at 1 μg RNA/μl. In the case of 0.9-cm diameter gels, one can load $10-200$ μl of solution containing $10-250$ μg RNA although some band-broadening may occur with volumes above 150 μl or with more than 100 μg RNA per gel.

2.5 Electrophoresis of gels

Electrophoresis may be carried out quite successfully at room temperature. For 9 cm $\times$ 0.6 cm gels, 5 V/cm of gel and a current of approximately 6 mA per gel, gives very good separations. Under these conditions, a satisfactory fractionation of precursor rRNAs and rRNAs can be obtained in 2.2% or 2.4% gels in $3-4$ h (*Figure 2a*). Low molecular weight RNA migrates off the end of the gel under these conditions but can be retained on the gel by increasing the gel concentration as shown in *Figures 2b* and *2c*. The gels do warm up during the run but some cooling is provided by the electrophoresis buffer in which the gel tubes are immersed. The running time can be shortened by using up to 10 V/cm without much loss of resolution, but at higher voltages bands may broaden and trail, probably due to excessive heating. Gels can also be run overnight provided that the voltage is reduced to take account of the increased running time.

2.6 Scanning the gels

It is not essential to scan the gels immediately at the end of the run and they can be left in the gel tubes for a few hours if required; there is very little band-broadening due to diffusion. A simple way of removing gels from the tubes is to fill a rubber teat with water and use this to squeeze the gel gently out of the tube into a dish or a tray containing distilled water. Concentrated gels may prove a little difficult to remove; they are looser if the SDS from the upper buffer tank has migrated all the way through the gel. This means that it is better to pre-run gels for an hour or more before loading the samples. Very dilute gels are difficult to handle but they can easily be picked up from the dish using a syringe attached to a long piece of the gel tubing. Stronger gels can be manipulated by hand but care should be taken to wear disposable plastic gloves since fingerprints can be detected by UV-scanning.

Frequently, gels can be scanned immediately after running but occasionally impurities are present near the top of the gel which absorb UV light. If present, these impurities normally wash out into distilled water within $1-2$ h. There is very little loss of RNA during such washing, except of low molecular weight RNA which diffuses quite rapidly. Beginners often find that washing gels is necessary to reduce background and hence improve the scans but the problem seems to disappear with practice. Provided that a suitable instrument is available, UV-scanning is relatively simple. Most systems use open troughs of parallel-sided quartz glass or quartz cylinders to hold the gel. Care should be taken to keep the gel unscratched and free

9

Table 4. Staining procedures for detecting RNA.

Wash and fix the gels in 1 M acetic acid or 10% trichloroacetic acid for 30 – 60 min prior to staining of RNA bands by one of the following methods:

Method 1[a]

Soak the gel overnight in 0.2% methylene blue dissolved in 0.2 M sodium acetate buffer (pH 4.7) and then destain by frequent washes with water or in continuous running water. Toluidine blue O, thionin and Azure A can be used in place of methylene blue.

Method 2

Stain with 0.1% pyronin Y dissolved in 0.5% acetic acid, 1 mM citric acid and destain in 0.5% acetic acid.

Method 3[a]

Stain with 5 μg/ml ethidium bromide dissolved in 0.5 M ammonium acetate for 30 – 60 min and then destain by washing the gel in water. The RNA is visualized by its fluorescence when the gel is viewed under light. (See also recipes in Section 7.1).

[a] Gels stained by Method 1 can be photographed using a yellow filter and visible white light transillumination. For gels stained with ethidium bromide, use red and yellow (or orange) filters, with Polaroid film at f 4.5 for ½ sec, with short wavelength (254 nm) UV transillumination.

---

of dust. If these precautions are taken it is extremely easy to see a 0.1 μg peak of RNA above the background. Scanning gels by UV is a quick and reliable method for analysing samples and it has the advantage of providing quantitative as well as qualitative data, since, at low RNA loadings, the peak area is approximately proportional to the amount of RNA present. If this type of analysis is not possible, gels can be stained to localize the bands (see *Table 4*).

2.7 Slicing of gels

After scanning it may be necessary to slice the gels, either to measure radioactivity in the slices, or to elute and recover certain fractions. Concentrated gels can be sliced using a linear arrangement of razor blades (*Figure 3*) but in the case of dilute gels, or where more precision is required, a mechanical slicer such as the Mickle gel slicer (Joyce-Loebl Ltd) should be used. Chop the gels frozen. With a little practice, slices 0.5 mm thick or less can easily be cut. An example of such a fractionation is shown in *Figure 4*. Place the gels in separate aluminium foil troughs, moulded to the approximate dimensions of the gel, with a rubber stopper at each end to hold the gel at the same length as when it was scanned. Freeze the gel slowly. The best method is to place the trough containing the gel on powdered dry ice (solid CO_2) so that it freezes without lateral displacement. Some vertical displacement may occur so that the gel becomes pear-shaped in cross-section. In the absence of dry ice, a freezing aerosol spray can be used, provided that the spray is directed to the lower side of the aluminium trough. Alternatively, if care is taken, it is possible to use liquid nitrogen. If necessary, gels can be stored frozen in a deep freeze before slicing.

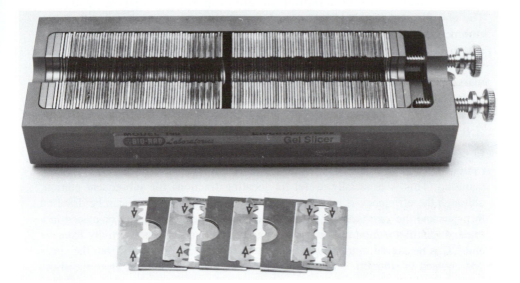

Figure 3. A manual gel slicer (Bio-Rad model). The gel slicer consists of a series of razor blades separated by 0.1-cm-thick metal spacers. In this photograph four razor blades and three spacers have been removed and are displayed in front of the slicer.

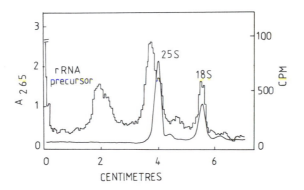

Figure 4. Detection of radioactive RNA after gel electrophoresis. Total leaf RNA was extracted from primary leaves of mung beans germinated for 60 h in darkness and incubated in ^{32}P-phosphate for the last hour. DNA was removed by deoxyribonuclease digestion and the RNA fractionated in a 2.4% polyacrylamide gel electrophoresed for 4.5 h at 50 V at room temperature. The gel was scanned with UV, frozen, sliced into 0.5 mm slices and the radioactivity in each slice measured (29).

When using the Mickle gel slicer, a strip of sticking plaster, or some similar material, is attached to the platform first, with a strip of wet filter paper on top; this enhances the adhesion of the gel to the platform. The frozen gel is then placed on the filter paper and frozen to the platform with an aerosol spray or with powdered dry ice. The chopping blade should be inclined at a slight angle towards the gel. At the commencement of slicing, the gel must not be too cold, otherwise the slices may shatter or the gel may be displaced from the platform. If difficulty is experienced when slicing frozen gels, the inclusion of 10% glycerol in the gel solutions (*Table 3*) makes the gels easier to slice. Alternatively, the gels can be equilibrated with glycerol during the staining and destaining procedures.

2.8 Measurement of radioactivity of gel slices

In the case of nucleic acids labelled with ^{32}P-phosphate, it is possible to measure radioactivity in wet gel slices, either alone or covered with buffer, by Cerenkov counting. This is especially useful for locating fractions for subsequent elution, although the efficiency is relatively low. Alternatively, the slices can be dried on to paper and the radioactivity measured using a planchette counter or placed in a toluene-scintillator mixture for liquid scintillation counting. Radioactivity from ^{3}H and ^{14}C is measured more efficiently if the RNA can be dissolved from the slices and counted in solution. To do this degrade the RNA and elute if from the slices as described in *Protocol 3*.

Protocol 3. Elution of radioactively labelled RNA from gels

1. Place each slice in scintillation vial and add 1 ml of '880' ammonia solution or 10% piperidine containing 1 mM EDTA.

2. Seal the vials and heat them for a few hours at 60°C, or simply leave them in a fume cupboard.

3. Evaporate off the liquid either at 60°C in an oven fitted with an extractor fan and fume outlet or overnight at room temperature in a fume cupboard.

4. Add a standard volume of water (0.5 ml) to each vial to swell each slice and to dissolve the hydrolysed RNA.

5. Add 10 ml of a scintillation mixture which is miscible with water to each vial. Suitable mixtures are toluene/Triton scintillator which contains 5.0 g of 2,5-diphenyloxazole (PPO) dissolved in 1 litre of toluene and 500 ml of Triton X-100, or toluene/methoxyethanol scintillator which contains 0.5 g of PPO, 0.1 g *p*-bis-(2-(5-phenyloxazoyl))benzene (POPOP), dissolved in 600 ml of toluene and 400 ml of methoxyethanol.

A simpler but more expensive method is to add to the wet gel slices one of the commercially-available solubilizing agents designed for liquid scintillation counting.

2.9 Elution and recovery of RNA

If the RNA fractions to be recovered can be detected by UV-scanning, it is best to place markers in the gels before they are frozen. The most convenient way to

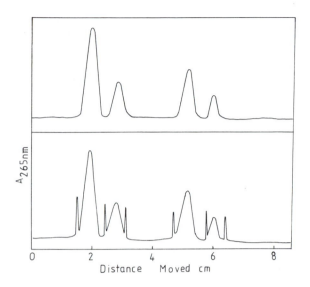

Figure 5. Localization of RNA fractions by inserting wires into the gel. Higher plant leaf RNA was fractionated in a 2.4% gel at 5°C for 3.5 h. (a) Gel scanned immediately after electrophoresis; (b) gel re-scanned after inserting wire markers into the gel.

do this is to scan the gel first and calculate the positions of the required RNAs. Then insert short pieces of thin wire into the gel in the appropriate positions at right angles to the long axis and scan the gel again. If the wire markers are not exactly in the right place they can be removed with forceps and replaced in the correct position. When the markers are in the correct positions (*Figure 5*) freeze the gel and slice it. Keep the slices between the wire markers. If the appropriate slices cannot be located in this way they can be detected by Cerenkov counting if ^{32}P-phosphate label is used, or samples of RNA can be eluted from each slice and measured by liquid scintillation counting.

RNA can be eluted intact from the slices into a variety of solutions including 2 × SSC and 6 × SSC with or without 0.2% SDS (1 × SSC is 0.15 M NaCl, 0.015 M sodium citrate, pH 7.0); 0.15 M sodium acetate adjusted to pH 6.0 with acetic acid and containing 0.5% SDS; 0.6 M or 0.8 M lithium acetate (pH 6.0) containing 0.5% SDS. The kinetics of elution of rRNA at 30°C from 0.5 mm slices of a 2.4% gel into 0.5% SDS, 0.6 M lithium acetate (pH 6.0), are shown in *Figure 6*. Generally speaking, 0.5 – 1.0 ml of solution should be used per gel slice. If the volume is too small then a substantial proportion of the RNA will remain in the slices, if it is too large there may be difficulties in recovering the RNA since it may be too dilute. Between 60 and 80% of the RNA can be recovered within a few hours using a one-step washing and the yield can be increased by re-extracting the slices. Shaking slices overnight is often the most convenient protocol since it fits in with most laboratory routines.

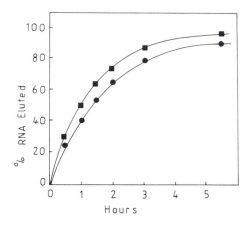

Figure 6. Elution of RNA from gel slices. Higher plant RNA, labelled with ^{3}H-uridine, was fractionated in a 2.4% gel for 3 h and the RNA peaks located by UV-scanning. The gel was frozen and sliced into 0.5 mm slices. Pairs of adjacent slices corresponding to the 25S and 18S rRNAs were placed in vials containing 1 ml of 0.5% SDS, 0.6 M lithium acetate (pH 6.0), and shaken at 30°C. The eluate was replaced with fresh solution at regular intervals. The cumulative total ^{3}H radioactivity eluted with time is plotted as a percentage of the total radioactivity in the slices. ■———■, 18S rRNA; ●———●, 25S rRNA. (From ref. 29.)

Some polyacrylamide elutes from the slices together with the RNA. Soluble (presumably non-crosslinked) polyacrylamide does not appear to interfere with DNA−RNA hybridization so that it is possible to elute RNA into 2 × SSC or 6 × SSC and to use the solution directly for hybridization to DNA attached to nitrocellulose filters. For other purposes it may be necessary to recover the RNA from solution. In this case, first remove some polyacrylamide and small fragments of gel by filtration. The best method is to squeeze the solution through a glass-fibre filter disc then a membrane filter using a syringe. Recover the RNA by precipitation with two volumes of ethanol. Soluble polyacrylamide always contaminates the RNA precipitates. In fact, this is a great advantage for RNA recovery since exceedingly small amounts of RNA precipitate from alcohol in its presence. It is possible to remove the precipitated polyacrylamide solubilized by this procedure by applying the RNA to a small lysine-Sepharose 4B column in 20 mM Tris-HCl (pH 7.5). The column should be washed with the same buffer, to remove the polyacrylamide, and the RNA can then be eluted with 0.5 M NaCl, 20 mM Tris-HCl (pH 7.5) and then recovered by ethanol precipitation. As an alternative to precipitation with ethanol, RNA eluted from the slices can be pelleted by centrifugation at 100 000 g overnight from 0.5% SDS, 0.8 M lithium acetate (pH 6.0). RNA recovered by any of these methods should be undegraded (*Figure 7*). An alternative approach for recovering RNA from gels is to homogenize the slices in buffer, then shake the aqueous phase with phenol/cresol, separate the phases by centrifugation (10 000 g for 10 min) and finally to precipitate the RNA from the aqueous buffer layer with two volumes of ethanol.

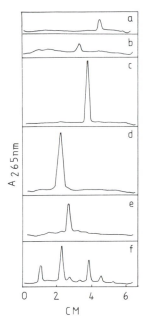

Figure 7. Re-electrophoresis of RNA recovered from polyacrylamide gels. Higher plant leaf RNA was separated in 2.6% gel and recovered by elution with 0.5% SDS 0.6 M lithium acetate (pH 6.0) at room temperature for 6 h. After adjusting the lithium acetate concentration to 0.8 M, the RNA fractions were centrifuged overnight at 100 000 *g* at 15°C. The RNA pellets were drained, dissolved in electrophoresis buffer containing sucrose and electrophoresed for 3 h in a 2.4% gel at room temperature. (a) 16S chloroplast rRNA; (b) $M_r = 0.9 \times 10^6$ fragment of chloroplast 23S rRNA; (c) 18S cytoplasmic rRNA; (d) 25S cytoplasmic rRNA; (e) 23S chloroplast rRNA; (f) sample of the original RNA run as a marker. The peak at about 1 cm from the top of the gel is DNA. (From ref. 29.)

3. Modifications to the basic tube gel method

3.1 Removal of DNA prior to electrophoresis

If DNA is not removed during nucleic acid extraction and purification it may be digested before electrophoresis as follows. Dissolve the total nucleic acid in a small volume of sterile buffer, without sucrose and SDS, but containing magnesium ions (e.g. 36 mM Tris base, 30 mM NaH_2PO_4, 2 mM $MgCl_2$). Digest the DNA by adding ribonuclease-free deoxyribonuclease to a final concentration of 10 μg/ml and incubate the mixture for 15 min. At the end of the incubation, stop the reaction and prepare the sample for electrophoresis by adding sucrose and SDS to final concentrations of 10% and 0.2%, respectively.

3.2 Denaturation of RNA before loading

This may be necessary if the RNA has a tendency to aggregate, or if it is desired to reveal 'hidden breaks' in the RNA, as discussed in Section 7. Either dissolve the RNA in normal loading buffer and heat at 60−90°C for a few minutes or dissolve it in half-strength buffer containing 50% (v/v) formamide or 8 M urea and heat the sample to 75°C before loading. These methods are normally sufficient to disrupt aggregates and hydrogen-bonded structures. However, intact RNA is only temporarily unfolded and will refold immediately it cools or as it enters the gel in the absence of a denaturing agent. In fact, RNA with a high degree of secondary structure may not be completely denatured by these procedures. To denature these molecules it is necessary to use completely denaturing gels as discussed in Section 4.

3.3 Running gels at non-ambient temperatures

Low temperatures prevent the separation of RNA fragments held together by only short hydrogen-bonded regions (see Section 8). If gels are to be run at low temperature, reduce the SDS concentration in the electrophoresis buffer to $0.05-0.10\%$ to prevent its precipitation, and carry out electrophoresis in a cold room. There are also separate effects of temperature on the mobility of RNAs in polyacrylamide gels. Different RNA molecules of similar size can behave differently when the temperature is changed, presumably due to specific effects of secondary or tertiary structure and base composition. Two types of RNA may be resolved by electrophoresis at one temperature but run in exactly the same position at a different temperature. Therefore it is well worthwhile examining temperature effects when attempting to fractionate RNAs of particular interest.

3.4 Gradient gels

It is sometimes advantageous to use discontinuous gels of two concentrations if one wishes to fractionate high and low molecular weight RNA on a single gel. The best concentrations to use depend upon the individual circumstances but, for example, a tube with a 6% gel polymerized in the lower half, and a 2.4% gel subsequently polymerized on top is often useful. One disadvantage is that some RNA may accumulate at the interface between the two gels forming an artefactual band. To avoid this, the problem can be approached in a different way by making a continuous gradient of acrylamide (see *Tables 7* and *8* and Section 6). However, when RNA is plentiful it is often simpler to separate the RNA sample on two or three individual gels of different concentrations.

3.5 Solubilizable polyacrylamide gels

Polyacrylamide gel containing bisacrylamide crosslinker can be solubilized by incubating the gel in about 10 volumes of 30% hydrogen peroxide at 50°C. However, if it is wished to solubilize the gels, other crosslinkers are generally used. Polyacrylamide gels that are soluble in alkali can be constructed using ethylene diacrylate as the crosslinker instead of bisacrylamide (1). However, RNA is rapidly degraded in alkaline conditions. Several other crosslinkers are available which can be cleaved under acidic or neutral conditions, for example, diallyltartardiamide (DATD) or N,N'-bisacrylylcystamine (BAC). Gels polymerized using these crosslinkers instead of bisacrylamide are solubilized by 2% periodic acid and 2-mercaptoethanol, respectively (23,24).

3.6 Agarose – acrylamide composite gels

For separating high molecular weight RNA, large-pore polyacrylamide gels are required. Although low concentration polyacrylamide gels are suitable for this purpose they are difficult to handle because they lack mechanical strength. The addition of agarose to 0.5% overcomes this problem. The preparation of mixed agarose−acrylamide composite gels, as originally described by Peacock and Dingman (2), is described in detail in Chapter 9.

Table 5. Alternative buffers for electrophoresis.[a]

(a) Tris base   3.63 g (30 mM)
    EDTA.Na$_2$.2H$_2$O   0.04 g (0.1 mM)
    SDS   1.00 g (0.1%)
    1.0 M HCl   16 ml
    Distilled water to 1 litre

    The buffer should be pH 8.0 at room temperature.
    Changes in conformation/mobility may be observed if 2 mM magnesium acetate is added
    to this buffer. The gels are electrophoresed at 25°C to prevent the SDS precipitating.

(b) Tris base   0.44 g (3.6 mM)
    NaH$_2$PO$_4$.2H$_2$O   0.48 g (3.0 mM)
    EDTA.Na$_2$.2H$_2$O   0.04 g (0.1 mM)
    SDS   1.00 g (0.1%)
    Distilled water to 1 litre

    The buffer should be pH 7.7 at room temperature.

(c) Tris base   10.80 g (89 mM)
    Boric acid   5.50 g (89 mM)
    EDTA.Na$_2$.2H$_2$O   0.93 g (2.5 mM)
    SDS   1.00 g (0.1%)
    Distilled water to 1 litre

    The buffer should be pH 8.3 at room temperature.
The corresponding buffer used to make the gels (*Table 2*) should be five times the above
concentrations and without SDS. When using buffer (b), it is best to circulate the solution
between the two buffer reservoirs during electrophoresis, to minimize pH changes.

[a] The concentrations of reagent given in brackets are the final concentrations in the buffer.

3.7 Other electrophoresis buffers

The main advantage to be gained in using buffers other than those already described
(*Table 2*) is that a lowering of the ionic strength may alter the conformation of the
RNAs being fractionated differentially and hence increase the chance of their being
separated. Useful alternative buffers are given in *Table 5*. Note that if different buffers
are used for electrophoresis, corresponding changes should be made to the buffer
used to make the gels. The running times may also be different from those when
using the standard buffer given in *Table 2*. In addition, pH changes during electrophoresis may be more pronounced and it may be necessary to circulate the buffer
between the buffer reservoirs during electrophoresis. This can be done easily by
setting an empty gel tube a few centimetres above the level of the gels as an overflow,
and pumping buffer from the lower reservoir to the upper reservoir using a peristaltic
pump.

4. Denaturing gels

For many experimental purposes, unambiguous data can only be obtained if all the
secondary structure of the RNA is abolished during electrophoresis (see Section 3.2).
This is particularly important for measuring the molecular weights of RNA accurately.

Several denaturing gel systems have been developed and the most important ones are described in this section.

4.1 Formamide gels

The procedure outlined in *Protocol 4* for making formamide gels is a modification of the method of Staynov *et al.* (9). The gels contain approximately 98% formamide, which unfolds the RNA completely. The formamide should be deionized before use by stirring with Amberlite Monobed MB-1 resin (3 g/100 ml) for 2 h and then filtering.

Protocol 4. Recipes for the preparation of denaturing formamide polyacrylamide gels

	Final acrylamide concentration (%)		
	3.5	4.0	10.0
Acrylamide (g)	0.75	0.91	2.4
Bisacrylamide (g)	0.13	0.09	0.10
Diethylbarbituric acid (g) (Final concentration 20 mM)	0.092	0.092	0.092
TEMED (ml)	0.06	0.06	0.06
Formamide (ml) (deionized)	20	20	20

1. After mixing all the ingredients, adjust the gel mixture to pH 9.0 with concentrated HCl and the total volume to 25 ml with deionized formamide. Degassing of the mixture is not necessary.

2. Add 0.2 ml of freshly dissolved 18% ammonium persulphate, mix and pipette into the gel tubes.

3. Overlayer the gels with 70% buffered formamide.

4. After approximately 30 min, replace the 70% buffered formamide with 100% buffered formamide. 100% buffered formamide is 0.092 g diethylbarbituric acid in 25 ml deionized formamide adjusted to pH 9.0. 70% buffered formamide is made by diluting this with water.

Note that the relative bisacrylamide concentration varies for different concentration polyacrylamide gels.

Variations: in place of diethylbarbituric acid, one can use 0.25 ml of 1 M Na_2HPO_4 (final concentration 10 mM, final pH adjusted to pH 9.0). Alternatively, use 0.25 ml of 1 M NaH_2PO_4 to give a final pH of pH 6.0. In each case, compensate for the added water by using 0.1 ml of 36% ammonium persulphate catalyst and concentrated HCl for pH adjustment. This gives gels 98% formamide.

Caution: Formamide and diethylbarbituric acid are both toxic.

The apparatus and method used for formamide gels are essentially similar to those described for non-denaturing polyacrylamide tube gels with the following slight modifications. The gels are not pre-run. The RNA sample should be dried *in vacuo* and dissolved in 100% buffered formamide containing 10% sucrose. The sample may be heated to ensure denaturation of the RNA. After cooling, layer the sample beneath the 100% formamide gel overlay and on to the gel surface before adding electrophoresis buffer to the top reservoir. The best results are obtained if only $20-30$ μg of RNA are loaded in a volume of $10-20$ μl. After loading, fill the top reservoir with electrophoresis buffer without disturbing the 100% formamide overlay. This helps to prevent the transfer of water from the electrolyte to the RNA. With diethylbarbituric acid in the gels (*Protocol 4*), the usual electrophoresis medium is 20 mM NaCl in water. When diethylbarbiturate is replaced with sodium phosphate (*Protocol 4*), the buffer is 10 mM sodium phosphate in water adjusted to pH 6.0 or pH 9.0. Some workers prefer to use 98% formamide in the electrophoresis buffer reservoirs in place of water. Electrophoresis is carried out at $50-100$ V for several hours. The electrolyte must be circulated between the compartments to neutralise pH changes at the electrodes. Both formamide and diethylbarbituric acid are very toxic and must be handled with great care.

Gels can be scanned in the UV immediately after the run is finished but results are often better if the gels are first washed with several changes of water over a period of $1-2$ h to remove the formamide. The gels are also easier to slice after the formamide has been washed out. The bands of RNA can be stained by placing the gels in 0.5% acetic acid, 0.1% pyronin Y, 1 mM citric acid overnight, followed by destaining in 10% acetic acid.

4.2 Urea gels

Electrophoresis in 8 M urea at 60°C can be used in place of formamide. The urea solution should be deionized with Amberlite mixed-bed resin (MB-1).

Protocol 5. Preparation of denaturing urea gels

1. Prepare the gels according to *Table 3* except that all solutions used to make the gels (i.e. stock acrylamide, gel buffer and water in *Table 3*) should be made up containing a final concentration of 8 M urea.

2. For these gels dissolve the RNA sample in buffer containing 8 M urea, 10% sucrose and heat to at least 60°C before loading. Any of the buffer systems in Sections 2.2 and 3.7 may be used, but since the object is to denature the RNA it makes sense to use a low ionic strength buffer.

3. Overlayer the gels with 8 M urea, load the RNA sample on to the gel and then carefully add aqueous buffer at 60°C to the top buffer reservoir. Alternatively, urea may be added to the electrophoresis buffer to a final concentration of 8 M.

4. Carry out electrophoresis in an incubator at 60°C at 50 V for several hours.

4.3 Methylmercuric hydroxide gels

Methylmercuric hydroxide acts as a denaturing agent by reacting with the NH groups of purine and pyrimidine residues involved in Watson−Crick base-pairing. It is soluble in aqueous solutions and concentrations within the range 3−5 mM are sufficient to achieve complete denaturation of RNA so that mobility during electrophoresis is inversely related to $\log_{10}$ of the molecular weight (17). Since methylmercuric hydroxide reacts with free radicals, which are required for the polymerization of polyacrylamide gels, it is normally used only with agarose gels. The reaction of methylmercuric hydroxide with NH groups is reversible so that un-complexed nucleic acids can be recovered after electrophoresis by soaking the gel in 0.5 M ammonium acetate, 10 mM dithiothreitol (17) or 2-mercaptoethanol (*Protocol 12*). Methylmercuric hydroxide is very poisonous and slightly volatile and hence great care must be taken when handling this compound; storage and all manipulations should be carried out using a fume cupboard.

The gel buffer used contains:

- Boric acid   3.1 g (50 mM)
- Disodium tetraborate (decahydrate)   1.9 g (5 mM)
- Na_2SO_4 1.4 g (10 mM)
- EDTA.Na_2.$2H_2O$ 0.4 g (1 mM)
- Distilled water to 1 litre

The buffer should be pH 8.19 at room temperature.

To prepare the gel dissolve 1.0 g of agarose in 10 ml of hot buffer. Cool to 60°C and add methylmercuric hydroxide to 5 mM final concentration. Pour the solution into tubes or cast as a slab gel and allow to cool.

It is not necessary to add methylmercuric hydroxide to the buffer in the electrophoresis reservoirs; it is uncharged and does not migrate out of the gels rapidly. Conditions for electrophoresis are similar to those described elsewhere in this chapter for non-denaturing tube or slab gels. Nitrogen-containing bases, EDTA and Cl⁻ complex with methylmercuric hydroxide and so high concentrations of these should be avoided. RNA molecules, when denatured, may migrate more slowly that in non-denaturing gels. In addition, empirical observations suggest that the RNA bands may be slightly broader than in non-denaturing gel systems. After electrophoresis, the gels may be scanned in the UV or prepared for fluorography (*Protocols 7* and *8*), or stained with ethidium bromide (*Table 4*).

5. Molecular weight estimations

5.1 Non-denaturing gels

The original studies carried out using polyacrylamide gel electrophoresis suggested that in most cases the mobility of RNA molecules was inversely related to the $\log_{10}$ of the molecular weight (1,3−6). Plotting the $\log_{10}$ molecular weight as a function of distance moved by various RNA components fractionated in the same gel produced a straight line, irrespective of gel concentration or time of electrophoresis. This provided a very easy method for determining the size of an unknown RNA, by co-electrophoresing it with marker RNAs whose molecular weight was known, plotting the results graphically and estimating the $\log_{10}$ molecular weight of the unknown RNA from its mobility. Molecules which differed in apparent molecular weight by as little as 4% could be easily distinguished. It was realized at the time that some anomalies did arise, for example when comparing molecules differing in their degree of secondary structure or base composition, but initial experiments suggested that the assumption that mobility is inversely proportional to $\log_{10}$ molecular weight resulted in relatively small errors. A major increase in activity in the field of RNA analysis followed, leading to the identification of specific RNA molecules in the ribosomes of chloroplasts and mitochondria, and the recognition of major differences between the rRNAs of different groups of animals and plants. In addition, the great resolving power of gels led to very detailed studies of RNA metabolism and the identification of previously undetected metabolic intermediates and minor RNA components.

One interesting discovery was that a single, apparently Gaussian, RNA peak, representing a single homogeneous RNA species, actually contains a range of metastable conformations. The more unfolded molecules migrate more slowly than the more compact molecules, producing a single peak slightly broader than one might predict. Radioactive molecules eluted from slices taken from the leading edge or the trailing edge of a peak, and fractionated on another gel with an unlabelled RNA sample, run as a sharp band on either the leading or the trailing edge of the main peak. If they are denatured before electrophoresis, for example by heating, they seem to adopt the complete range of possible conformations and now co-migrate exactly with the main peak.

As work progressed it became clear that substantial changes in relative mobility could occur when some RNAs are subjected to changes in temperature, gel concentration, or ionic strength. Different types of RNA are not equally affected. This means that widely different molecular weight estimates can be obtained for a single RNA depending on the conditions used (7,8). While such phenomena can sometimes be exploited for fractionation purposes, they cast some doubt on molecular weight measurements of non-denatured RNA. The solution to this problem is to carry out the electrophoresis under conditions where all secondary structure is abolished so that essentially the only variable affecting mobility is molecular weight. This condition is met by using a denaturing gel system.

5.2 Denaturing gels

Detailed procedures for the preparation and electrophoresis of denaturing gels are given in Sections 4 and 7. To measure molecular weights accurately, several standard RNAs of known size should be included in the gel with the unknown RNAs. Plots of $\log_{10}$ of the molecular weight against distance moved should be linear with no anomalies caused by secondary structure or base composition. The only exception to this general rule is if the RNA sample contains non-linear RNA molecules such as lariat-splicing intermediates or other RNA molecules with branch points. RNA with hidden breaks will, of course, produce fragments. Certain experimental conditions, such as long exposure to urea at 60°C, may cause partial hydrolysis of the originally intact RNA and so should be avoided. RNAs of known sequence such as tRNA, *Escherichia coli* 16S rRNA (1542 nucleotides), maize chloroplast 16S rRNA (1491 nucleotides) or MS2 RNA (3569 nucleotides) make good markers. A more comprehensive list of molecular weight markers is given in Appendix I of this book.

6. Electrophoresis of slab gels

It is frequently more convenient to fractionate RNA samples by electrophoresis in slab gels, as opposed to tube gels. Eight or more samples can be fractionated under identical conditions in separate tracks of a single slab gel, which can be stained after electrophoresis to reveal the RNA bands for visual comparison and photography. RNA labelled with ^{32}P-phosphate can be detected by autoradiography of the wet slab, or the gel can be dried for the detection of ^{14}C or ^{3}H by autoradiography or fluorography. Thicker slab gels can also be used for preparative electrophoresis of a large quantity of RNA.

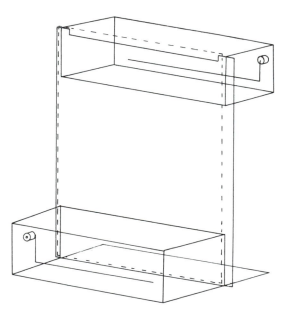

Figure 8. A Studier-type apparatus for slab gel electrophoresis. The apparatus is made from Perspex sheet. The lower reservoir is attached to a vertical plate which supports the upper reservoir. The apparatus is stabilized by an extended base. The gel may be cast either between the vertical plate clamped to a second glass plate or between two plass plates as in *Figure 9*. In the latter case, the top of one of the glass plates should be notched so as to allow communication with the buffer in the upper reservoir when the gel is clamped in position. Further details are given in ref. 11.

6.1 **Apparatus**

The apparatus for slab gel electrophoresis can quite easily be constructed in a workshop. Alternatively, several commercial systems are available. A design similar to that of Studier (11) is shown in *Figure 8*. This apparatus is very simple to use, as outlined in the legend to *Figure 8*. The gels are cast between two plates approximately 20 cm wide and 20 cm high, normally made of glass, held $1-3$ mm apart by lightly-greased glass or plastic strips each about 20 cm long, 1 cm wide, and of the required thickness. Soft PVC (polyvinyl chloride) or Perspex can be used as spacers, although Perspex and some other plastics may inhibit polymerization. The thickness of the gel affects the amount of sample that can be loaded, and the subsequent drying of the gel. Initially, try gels 2 mm thick, although polyacrylamide gels 1 mm thick are not too difficult to handle. Before use, thoroughly clean the glass plates with a laboratory detergent, rinse with distilled water and finally with ethanol $-$ ether (1:1 v/v) before drying with paper tissue. Alternatively, instead of detergent, some workers prefer to clean the plates with chromic acid. Clamp the cleaned plates vertically between four clamps, or in the electrophoresis apparatus itself. When assembled, the plates stand vertically, supported by the lower pair of clamps (*Figure 9*). The base can be sealed by firmly embedding it in Blu Tack.

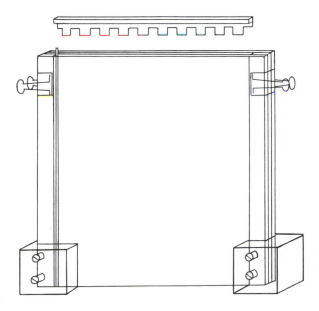

Figure 9. Gel mould for casting slab gels. The plates are held between two sets of clamps and have spacers at each side. The base of the plates is sealed, acrylamide or agarose solution is pumped or poured into the mould and the comb is inserted into the gel mixture to form loading wells.

Table 6. Recipes for polyacrylamide slab gels.

4% polyacrylamide slab gel

Stock solution: 15% acrylamide, 0.75% bisacrylamide	10.7 ml
Gel buffer: 180 mM Tris base, 150 mM NaH_2PO_4, 5 mM EDTA, pH 7.7 (see Section 2.2)	8.0 ml
Distilled water	21.0 ml

Degas the solution then add:

TEMED	30 μl
Ammonium persulphate (10% solution, freshly dissolved)	300 μl

The approximate final volume is 40 ml

15% polyacrylamide slab gel

Stock solution: 30% acrylamide, 0.37% bisacrylamide	20 ml
Gel buffer: 180 mM Tris base, 150 mM NaH_2PO_4, 5 mM EDTA, pH 7.7	8 ml
Distilled water	12 ml

Degas the solution then add:

TEMED	30 μl
Ammonium persulphate (10% solution, freshly dissolved)	300 μl

The approximate final volume is 40 ml.

4% stacking gel for the top of the 15% slab

Stock solution: 15% acrylamide, 0.75% bisacrylamide	2.5 ml
Gel buffer: 180 mM Tris base, 150 mM NaH_2PO_4, 5 mM EDTA, pH 7.7	0.38 ml
Distilled water	6.35 ml

Degas the solution then add:

TEMED	30 μl
Ammonium persulphate (10% solution, freshly dissolved)	300 μl

The approximate final volume is 9.5 ml.

Note that the final buffer concentration in the stacking gel is one-fifth of that of the lower gel.

Table 7. Recipe for a 2.4 – 5.0% polyacrylamide gradient slab gel.

Solution A: 5.0% acrylamide, 0.125% bisacrylamide, 20% surcrose

30% acrylamide solution	4.16 ml
2% bisacrylamide solution	1.56 ml
40% sucrose solution	12.5 ml
Buffer: 180 mM Tris base, 150 mM NaH_2PO_4, 5 mM EDTA, pH 7.7	5.0 ml
Distilled water	1.8 ml

Degas the solution then add:

10% ammonium persulphate solution	62.5 μl
10% TEMED solution	25 μl

The approximate final volume is 25 ml.

Solution B: 2.4% acrylamide, 0.12 bisacrylamide, 5% sucrose

30% acrylamide solution	2.0 ml
2% bisacrylamide solution	1.56 ml
40% sucrose solution	3.1 ml
Buffer: 180 mM Tris base, 150 mM NaH_2PO_4, 5 mM EDTA, pH 7.7	5.0 ml
Distilled water	13.3 ml

Degas the solution then add:

10% ammonium persulphate solution	62.5 μl
10% TEMED solution	93.8 μl

The approximate final volume is 25 ml.

Table 8. Recipe for a 4–15% polyacrylamide gradient slab gel.

Solution A: 15.0% acrylamide, 0.35% bisacrylamide, 20% sucrose

30% acrylamide solution	10 ml
2% bisacrylamide solution	3.51 ml
60% sucrose dissolved in 108 mM Tris base, 90 mM NaH$_2$PO$_4$, 3 mM EDTA	6.67 ml

Degas the solution then add:

10% ammonium persulphate solution	62.5 μl
10% TEMED solution	25 μl

The approximate final volume is 20 ml.

Solution B: 4% acrylamide, 0.15% bisacrylamide, 5% sucrose

30% acrylamide solution	2.67 ml
2% bisacrylamide solution	1.5 ml
60% sucrose dissolved in 108 mM Tris base, 90 mM NaH$_2$PO$_4$, 3 mM EDTA	1.67 ml
Buffer: 180 mM Tris base, 150 mM NaH$_2$PO$_4$, 5 mM EDTA	3.0 ml
Distilled water	11.0 ml

Degas the solution then add:

10% ammonium persulphate solution	62.5 μl
10% TEMED solution	93.8 μl

The approximate final volume is 20 ml.

Alternatively, this can be done with a strip of soft polyvinyl chloride. After the gel has polymerized (see Section 6.2) transfer the glass plate assembly to the electrophoresis reservoirs which are made from Perspex sheet with platinum electrodes (*Figure 8*). In most designs one of the glass plates used to form the gel mould has a notch cut in the top, to allow continuity of the buffer between the gel and the upper tank during electrophoresis.

6.2 Recipes for gels

Homogeneous slab gels can be made from polyacrylamide or agarose, using normal or denaturing conditions. Alternatively, gels consisting of a gradient of polyacrylamide can be prepared. Sample recipes for non-denaturing 4% and 15% polyacrylamide slab gels are given in *Table 6* and for 2.4–5% and 4–15% non-denaturing polyacrylamide gradient gels in *Tables 7* and *8*. Examples of the resolution obtainable are shown in *Figures 10, 11*, and *14b*. Some other possible buffers are given in Section 3.7. Fully denaturing (98% formamide, *Protocol 4*) or partially denaturing (6 M urea or 50% formamide) polyacrylamide slab gels may also prove useful. For non-denaturing agarose gels, add the agarose powder to the buffer and heat to boiling-point while stirring. Cool the solution to a few degrees above the setting temperature and cast the slab. First, try 1.5% agarose dissolved in the buffer given in Section 2 or buffer (b) in *Table 5*. For partially or completely denaturing buffer systems used with agarose gels, try 1.5% agarose with 50% formamide or with 6 M urea (10) or with 4 mM methylmercuric hydroxide (17). When formamide is present in the gels they should be allowed to set at 5–10°C. Sea Plaque agarose

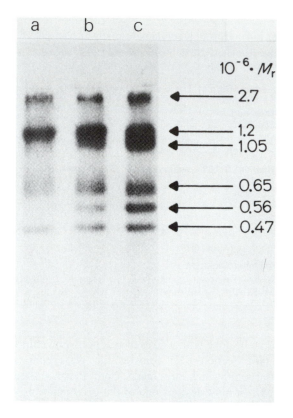

Figure 10. Detection of radioactive RNA in polyacrylamide slab gels by autoradiography. Spinach chloroplast RNA labelled with ^{32}P was prepared from chloroplasts isolated from leaves incubated in ^{32}P-orthophosphate for (a) 2 h, (b) 4 h, or (c) 6 h. Samples were fractionated in a 2.4–5% linear gradient gel (*Table 7*) at 50 V for about 18 h at room temperature. The radioactive RNA lane (a) 12 500 cpm, (b) 20 000 cpm, (c) 20 000 cpm, was detected by autoradiography of the wet slab. The RNAs are referred to by their molecular weights × 10^{-6}. The M$_r$ = 1.05 × 10^6 and 0.56 × 10^6 RNAs are the 23S and 16S chloroplast rRNAs. The M$_r$ = 1.2 × 10^6 and 0.65 × 10^6 RNAs are the immediate precursors to the 23S and 16S rRNAs, respectively, and the M$_r$ = 2.7 × 10^6 RNA is a polycistronic precursor of chloroplast rRNA. Reproduced from ref. 15 with permission.

(Marine Colloids, Inc.) and other similar products have the useful property of remelting at a moderate temperature which may facilitate the recovery of RNA from the gel after electrophoresis (*See Protocol 6*).

6.3 Casting the gels

For homogeneous slab gels, simply pour, or pump the gel mixture into the gel mould to the desired height. Insert a plastic sample well-former (comb) the same thickness

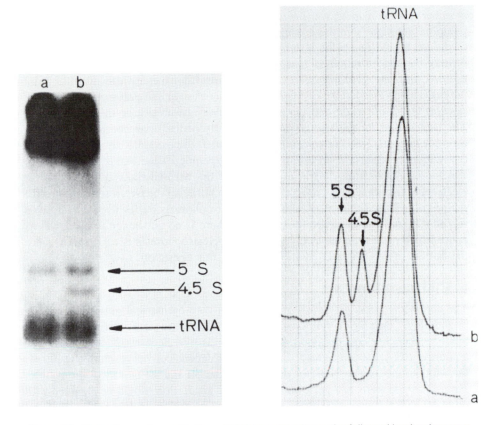

Figure 11. Detection and quantitation of RNA by autoradiography followed by densitometry. Spinach chloroplast RNA, labelled with ^{32}P-orthophosphate, was fractionated in a 15% polyacrylamide slab gel overlayered with a 4% stacking gel (*Table 6*). Electrophoresis was for 18 h at 50 V at room temperature. ^{32}P-labelled RNA was located by autoradiography of the wet slab and the developed film was photographed and scanned with a densitometer. The chloroplast rRNAs and rRNA precursors are at the top of the gel; the 5S, 4.5S and tRNA have been fractionated in the lower half of the gel. Lane (a) shows RNA from plants labelled in the presence of chloramphenicol, which inhibits 4.5S RNA production, whereas lane (b) shows RNA labelled in the absence of chloramphenicol. Reproduced from ref. 15.

as the gel, into the mixture to form the loading wells, about 1 cm wide and 1 cm deep. Alternatively, overlayer the top of the gel with water or buffer and allow to polymerize. Remove the liquid overlay and polymerize a separate 'stacking' gel (*Table 6*), with loading wells on top.

For gradient gels, place a glass capillary, made by drawing out a Pasteur pipette, or some narrow-bore plastic tubing, next to one of the glass spacers, extending to the bottom of the gel mould. Add a few millilitres of water or buffer to the space between the glass plates, to serve as an overlay for the gel. Add equal volumes of the two acrylamide solutions (*Tables 7* and *8*) to the reservoirs of a simple two-chamber gradient marker, with the more dilute solution in the mixing chamber. Slowly pump the solution from the mixing chamber through the capillary into the bottom of the gel mould under the water overlay using a peristaltic pump.

Since the concentrations of catalyst are different in the two acrylamide mixtures used to form the gradient, the gel polymerizes from the top downwards. After polymerization remove the overlay on the top of the gel and polymerize a separate stacking gel (*Table 6*) above the gradient gel, using a comb to form the loading wells. The buffer used for the stacking gel is normally of the same composition as that used for the main gel, but sometimes it is diluted, for example, to one-fifth of the concentration because it encourages zone sharpening of the bands.

After polymerization, unseal the bottom of the gel and transfer the gel to the electrophoresis apparatus. Add electrophoresis buffer to the two reservoirs (for the recipes of gels in *Tables 6, 7,* and *8*, the electrophoresis buffer used is that given in *Table 2*) and pre-electrophorese the gel. There is no advantage in using a low-concentration buffer in the stacking gel if pre-electrophoresis is carried out, since the buffers in the two gels equilibrate during this operation.

6.4 Electrophoresis

The comments made previously in connection with cylindrical gels about sample preparation, running conditions, artefacts, problems and molecular weight measurements are equally applicable to slab gels (see Sections 2, 3, and 4). The amount of RNA that can be loaded onto the gels depends on the resolution required, the width of the loading well, the thickness of the gel and the quality of the sample. For 1-cm-wide wells in a gel 2 mm thick, good separation should be obtained with up to 50 μg RNA per well loaded in up to 50 μl of buffer. If gels are cast without loading wells, milligram amounts of RNA may be loaded across the entire width of the gel.

After electrophoresis, gently separate the plates using a metal spatula and carefully remove the gel for staining and autoradiography.

6.5 Detection and recovery of RNA in slab gels

Non-radioactive RNA bands in analytical gels can be visualized by staining the gel (*Table 4*). For preparative procedures where the sample is spread across the whole width of the gel, it is better to cut strips from each side of the gel and stain only these in order to locate the positions of the required bands in the remaining gel. The sections of the gel containing the unstained RNA can then be cut out with a razor blade and the RNA eluted (see Section 2.9). Another method of detecting RNA is by UV shadowing. To do this place the gel on a clear polythene sheet above a Polygram Cel 300 UV$_{234}$ thin-layer chromatography sheet (Macherey-Nagel). The RNA bands can then be located under UV light, gel segments cut out and the RNA eluted. The methods described in Section 2.9 for the elution and recovery of RNA from tube polyacrylamide gels can also be used with slab gel segments. In the case of agarose gels containing 6 M urea, gel slices will melt at 75°C and remain liquid at 20°C. Sea Plaque agarose without urea melts at 70°C and remains liquid at 37°C. RNA can then be purified from the solutions using hexadecyl-trimethyl ammonium bromide and butanol (*Protocol 6*, ref. 18).

Protocol 6. Recovery of RNA from agarose gels

Melting the gel slices
Sea Plaque agarose (Marine Colloids, Inc.) gels will melt at 70°C and subsequently remain liquid at 37°C. Ordinary agarose gels containing 6 M urea melt at 75°C and remain liquid at 20°C.

Recovering the RNA
RNA in the liquid agarose is extracted into water-saturated butanol containing hexadecyl-trimethyl ammonium bromide (HTAB), either at 20°C or at 37°C depending on the gel system (see above), and recovered from this fraction with NaCl solution, from which the HTAB is removed by precipitation with chloroform (18).

Solutions
Equilibrate 150 ml 1-butanol with 150 ml distilled water by shaking in a separating funnel. Separate the phases and add 5.5 g HTAB (previously recrystallized from water and dried) to the butanol fraction. Add 100 ml of the previously equilibrated aqueous fraction to the butanol-HTAB together with 50 μl Antifoam A (Sigma Chemical Co.) and equilibrate by shaking. Allow the phases to separate and store the butanol-HTAB and the aqueous fractions separately.

Procedure
1. Measure the volume of molten gel containing the RNA and add equal volumes of the butanol-HTAB and aqueous fractions described above.
2. Shake the contents of the tube thoroughly to mix the reagents and allow the phases to separate on standing or after brief centrifugation (for agarose-urea gels the operations can be at 20°C whereas, for Sea Plaque agar alone, the temperature should be 37°C).
3. Remove the butanol layer and store. Re-extract the aqueous phase twice more by shaking with an equal volume of the equilibrated butanol-HTAB solution.
4. Combine the butanol extracts and add one-quarter volume of 0.2 M NaCl. Shake the tube contents thoroughly and allow the phases to separate. Take the *lower* (aqueous) phase and store.
5. Re-extract the butanol-HTAB phase with one quarter volume of 0.2 M NaCl.
6. Combine the aqueous extracts and add an equal volume of chloroform, drop by drop.
7. Place the tube in ice to precipitate the HTAB.
8. Remove the aqueous layer from the chloroform and evaporate residual

Protocol 6. *continued*

chloroform using a stream of air. The RNA can be desalted by gel filtration (e.g. Sephadex G-25) if necessary or precipitated with ethanol. Ethidium bromide, used to locate the RNA in the gel, remains in the butanol phase during the extraction.

RNA labelled with ^{32}P or ^{125}I can be detected by autoradiography of wet slab gels. These should be covered with plastic-wrap film (e.g. Saran Wrap) and autoradiographed by exposure to X-ray film (e.g. Kodak X-omat R or Fuji RX) for from several hours up to a few days (*Figure 10*). Better resolution on the autoradiograph may be obtained if the gel is dried before placing it in contact with the X-ray film.

Stained or unstained gels can be dried for autoradiography using a commercial gel drier, most of which include a heating element that allows gels to be dried down in 1−2 h. Alternatively one can use an apparatus consisting of a shallow, rectangular, hollow box made of plywood fitted with a porous polythene sheet on the top and an evacuating tube on one side. To use this apparatus place the gel on a sheet of filter paper and transfer it, gel uppermost, to the porous polythene sheet. Connect the evacuating tube to a vacuum line and place the drying apparatus inside a polythene bag. Secure the neck of the bag around the evacuating tube with sticky tape and turn on the vacuum. As the air is withdrawn, carefully smooth the polythene bag over the top of the gel. After a few hours the gel can be warmed by placing a 250 W reflecting infra-red bulb 50−80 cm above the gel. However, this is not absolutely necessary, and if applied too soon may crack the gel. Leave the gel under vacuum for about 24 hours or until it has dried down to a thin film. For autoradiography, place the dried gel, stuck to the filter paper, in contact with X-ray film and expose at −70°C. Sensitivity to ^{32}P and ^{125}I can be increased tenfold or more by pre-fogging the film to an A_{540} of 0.1−0.2. Then place the film over the dry gel, with the fogged side uppermost, and place a calcium tungstate intensifying screen on top (12). Clamp the gel film and screen together, wrapped in black plastic sheeting to exclude light, or place in a special cassette, and leave at −70°C. After exposure, it is best if the film is unwrapped before it warms up to reduce the possibility of physical fogging as a result of condensation, and then develop the film according to the manufacturer's instructions.

Fluorography, outlined in *Protocols 7* and *8*, is another method of increasing sensitivity. In the original protocols the gel was soaked in dimethyl sulphoxide (13,14) or acetic acid (16). It was then transferred to the same solvent containing diphenyloxazole (PPO) which is allowed to permeate the gel, and finally washed in water to precipitate PPO in the gel (see *Protocol 7*). The gel was then dried down and exposed to prefogged X-ray film (*Figures 16* and *17*). An alternative and simpler procedure which has been devised utilizes sodium salicylate (21) and is able to produce results of a similar quality to those obtained using PPO. *Protocol 8* outlines this method of fluorography. A number of commercial products are also available for the fluorography of gels.

Protocol 7. Fluorography using PPO as scintillator

1. Soak the stained or unstained polyacrylamide gel in about 20 times its volume of dimethyl sulphoxide for 30 min. Transfer to fresh dimethyl sulphoxide for 30 min.

2. Soak the polyacrylamide gel in four volumes of 20% PPO (w/w) dissolved in dimethyl sulphoxide for 3 h.

3. Wash the gel in 20 volumes of water for at least 1 h. This removes the dimethyl sulphoxide and precipitates the PPO in the gel. A longer washing period may be necessary if artificial blackening of the film by residual dimethyl sulphoxide occurs.

4. Dry the gel under vacuum.

5. Place the dried gel in contact with preflashed Kodak X-Omat R film at −70°C.

6. Develop the film using Ilford PQX-1 or Kodak DX-80. See refs 13 and 14 for further details.

Caution: all solutions of dimethyl sulphoxide should be handled with gloves and used in a fume cupboard.

Variations: Probst *et al.* (16) have suggested the use of 100% acetic acid in place of dimethyl sulphoxide as the PPO solvent since this can be used for both agarose and polyacrylamide gels. For agarose gels, or 2% (or less) agarose−acrylamide composite gels which do not change volume in methanol, 100% methanol can also be used in place of dimethyl sulphoxide (13).

Protocol 8. Fluorography using sodium salicylate as scintillator

1. Fix the gel, if required, in 5% TCA or methanol:acetic acid:water (5:1:5 by vol).

2. If the gel has been fixed, the acid must be removed by soaking the gel in 20 volumes of distilled water for 30 min; this step is necessary to prevent the sodium salicylate precipitating.

3. Soak the gel in 10 volumes of 1.0 M sodium salicylate (pH 7) for 20 min.

4. Lay the gel on to wetted Whatman 3 MM paper and dry under vacuum.

5. Carry out autoradiography as in *Protocol 7*.

Variations: some people prefer to add 5% methanol and 1% glycerol to the sodium salicylate solution to facilitate subsequent drying of the gel.

Developed films can be scanned using a densitometer (see *Figure 11*). Within certain limits, the area under a peak is proportional to the amount of radioactivity present but it is best to check the linearity of the film response, using radioactive standards.

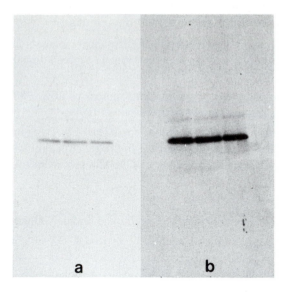

Figure 12. Intensification of autoradiographs. Autoradiograph films exposed for 16 h (a) were activated and re-exposed for 16 h, (b) as described in *Protocol 10*. Quantitative densitometric methods revealed up to tenfold enhancement of bands. Data kindly provided by Amersham International Ltd.

One problem frequently encountered, particularly when using isotopes with short half-lives (e.g. ^{32}P-phosphate) is that at the end of the autoradiography or fluorography procedure the film remains under-exposed. In the case of long-lived isotopes it is possible to repeat the procedure using a longer exposure time. However, an alternative solution is to intensify the image by treating the film with a solution of radioactive thiourea which binds to the silver grains (31). The now radioactive film is then re-exposed to fresh unfogged film. The full procedure is given in *Protocol 9*. A typical result of this intensification process is shown in *Figure 12*. A non-radioactive method that works well is to enhance the image on the X-ray film by binding chromium to the silver grains (33).

Protocol 9. Intensification of X-ray film images using ^{35}S-thiourea

This protocol describes a method[a] to allow intensification of X-ray film images which have been obtained by autoradiography or fluorography. It can be used effectively on films up to 2 years old.

1. *Pre-washes*

   The following series of washes should be performed using large volumes of solution (e.g., 1 litre for a 18 cm × 24 cm sheet of X-ray film).

   (a) Distilled water   2 min

   (b) Distilled water   2 min

Protocol 9. *continued*
    (c) 20% (v/v) methanol   2 min
    (d) 50% (v/v) methanol   2 min
    (e) 20% (v/v) methanol   2 min
    (f) Distilled water   10 min
    (g) Distilled water   10 min
    Dry the film.

2. *Activation[b,c,d]*
   Make a solution of ^{35}S-thiourea in 0.1 M ammonia (pH 11) and filter it carefully. The radioactive concentration of the solution should be approximately (74 kBq/ml, 2 μCi/ml).
   Shake the film in the solution of ^{35}S-thiourea for 6 min at room temperature. (Allow 37 kBq, 1 μCi, ^{35}S-thiourea per square centimetre of film). This activating solution can be used at least three times although there will be a slight loss of activity after each use.

3. *Washes*
   Wash the activated film successively in a large volume (1 litre) of each of the following solutions:
       (a) Distilled water   2 min
       (b) Distilled water   2 min
       (c) 20% (v/v) methanol   5 min
       (d) 50% (v/v) methanol   2 min
       (e) 20% (v/v) methanol   5 min
       (f) Distilled water   5 min
       (g) Distilled water   5 min
       Dry the film.
   It is essential that these washing stages are performed thoroughly.

4. *Exposure*
   Expose the activated film to a sheet of unflashed X-ray film at −70°C. Do not put anything between the two sheets of film and monitor the activated film using a thin-walled Geiger tube to determine which side of the film has been activated and to give a guide to the exposure time required.

[a] Covered by US Patent No. 4,101,780.
[b] It is not possible to use ^{14}C-thiourea for intensification.
[c] For significantly lower backgrounds, dissolve the thiourea in ammonia solution adjusted to pH 8 with HCl or in dilute NaOH at pH 8. The lower pH solution activates the film more slowly, so the film should be shaken for 2 h.
[d] Thiourea should be handled with care as it has been found to be a weak thyroid carcinogen in animals.

7. Electrophoresis of RNA in horizontal slab gels after denaturation with glyoxal or formamide

7.1 Preparation of gels and electrophoresis conditions

The method most frequently used for fractionating RNA is by electrophoresis in horizontal slabs of agarose (submarine electrophoresis), following denaturation of

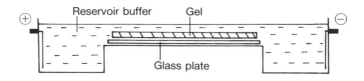

Figure 13. Apparatus for electrophoresis of RNA in horizontal slab gels. After the gel (shown hatched in the diagram) has set it is placed on a platform connecting the two buffer reservoirs and submerged in electrophoresis buffer. For many applications it is necessary to recirculate the buffer, with a pump, or to change it frequently, to minimize pH changes that occur during the run.

the RNA at high temperatures in the presence of glyoxal or formamide. The gels can be made quickly and the apparatus is easy to use. Provided that appropriate precautions are taken to maintain the correct pH during electrophoresis, the RNA remains fully denatured and its mobility through the gel is inversely related to the $\log_{10}$ of its molecular weight. After fractionation, very small amounts of RNA can be detected by staining with ethidium bromide and photographing the gel, or the gel may be used for northern blotting (see Section 7.3). A diagram of a submarine slab gel electrophoresis apparatus is shown in *Figure 13*. Cast the gel on a horizontal glass plate by pouring agarose melted in the electrophoresis buffer (*Protocol 10*) into a rectangular gel mould. Form the sample wells by inserting a comb vertically near one end and leave the gel to set for about one hour. The teeth of the comb should not touch the glass plate. Remove the gel mould and comb, then place the gel in the tank and cover with electrophoresis buffer. Prior to loading, denature the RNA with either 1 M glyoxal in 50% (v/v) aqueous dimethyl sulphoxide or 2.2 M formaldehyde in 50% (v/v) aqueous formamide. The procedures and buffers used in these two methods are given in *Protocols 10* and *11*. For RNA up to 1 kb in size use 1.4% or 1.5% agarose and for longer RNA molecules try $0.8-1.0\%$ agarose. The use of methylmercuric hydroxide as a denaturing agent is outlined in Section 4.3. It is important to emphasize that denaturing the RNA by any of these methods allows accurate determination of molecular weights of linear RNA molecules to be made, provided that appropriate size-markers are run in the same gel. A number of RNA 'ladders' are available commerically and a list of molecular weight markers is given in Appendix I. It is also important to realize that denaturation not only disrupts aggregates, but reveals hidden breaks in the polynucleotide chain (see Section 8.2.5 and *Figures 15* and *16*).

Protocol 10. Recipes for horizontal slab agarose gel electrophoresis of RNA denatured with glyoxal and dimethylsulphoxide

1. Denature the RNA in an autoclaved microcentrifuge tube by mixing:
   - RNA (up to 20 μg)   3.7 μl
   - 6 M glyoxal[a]   2.7 μl

Protocol 10. *continued*

- dimethylsulphoxide (DMSO)   8.0 μl
- 0.1 M sodium phosphate (pH 7.0)[b]   1.6 μl

   Incubate the RNA solution at 50°C for 60 min in a tightly closed tube. If size markers are used, they should also be denatured before electrophoresis.

2. For RNA up to 1 kb in size use 1.4% or 1.5% agarose gels, if the RNA is larger try using 0.8−1.0% agarose gels. For electrophoresis in a 11 × 16 cm 1% agarose gel, mix 0.5 g agarose, 5 ml of 0.1 M sodium phosphate (pH 7.0), and 45 ml of water.[b] Heat the mixture to dissolve the agarose and mix thoroughly. Adjust the volume if necessary, allow to cool to 60°C and pour the gel.

3. Insert the comb to form the loading wells. The teeth of the comb should be 1−2 mm above the glass plate. Allow the gel to set for at least 60 min at room temperature.

4. Cool the glyoxylated RNA sample solution (16 μl) to 20°C and add 4 μl of sterile loading buffer: 50% glycerol, 0.01 M sodium phosphate (pH 7.0), 0.4% bromophenol blue.

5. Fill the buffer reservoirs and cover the gel with 10 mM sodium phosphate (pH 7.0).[c]

6. Load the samples into the wells and run the gel at 100 V until the bromophenol blue has moved approximately half-way through the gel. Glyoxal dissociates from the RNA at pH greater than pH 8.0, so it is necessary to recirculate the buffer constantly, to maintain the pH. Alternatively, the buffer may be changed every 30 min during the run.

---

[a] Glyoxal is usually obtained as a 6 M (40%) solution. It oxidizes in air and must be deionized before use by passing it through a mixed-bed resin (Bio-Rad AG 501-X8) until the pH is neutral.
[b] The 0.1 M sodium phosphate buffer should be autoclaved.
[c] Do not add ethidium bromide because glyoxal reacts with it.

Protocol 11. Recipes for horizontal slab gel electrophoresis of RNA denatured with formaldehyde and formamide

1. Denature the RNA in an autoclaved microcentrifuge tube by mixing:
   - RNA (up to 20 μg)   4.5 μl
   - 5 × Mops running buffer[a]   2.0 μl
   - formaldehyde (37%)[b]   3.5 μl
   - formamide[c]   10.0 μl

   Incubate the RNA solution at 55°C for 15 min in a fume hood and cool in ice.

2. For electrophoresis in a 11 × 16 cm 1% agarose gel, mix 0.5 g agarose, 10 ml 5 × Mops running buffer (see Notes) and 31 ml water. Heat the mixture to dissolve the agarose. Adjust the volume if necessary, allow to cool to 60°C

Protocol 11. *continued*

and add 9 ml of 37% (12.3 M) formaldehyde (final concentration 2.2 M). Mix thoroughly, pour the gel, and let it set for a least 60 min at room temperature.

3. After cooling the denatured RNA solution (18 μl) in ice, add 2 μl of sterile loading buffer: 50% glycerol, 1 mM EDTA, 0.4% Bromophenol blue.

4. Fill the buffer reservoirs and cover the gel with 1 $\times$ Mops/EDTA running buffer, pH 7.0.

5. Pre-electrophorese the gel at 100 V for 10 min.

6. Load the samples into the wells and run the gel at 100 V until the bromophenol blue has moved approximately half-way through the gel. Constant buffer recirculation is not necessary but if this is not done it is advisable to change the buffer several times during the run.

[a] 5 $\times$ Mops gel running buffer is: 0.2 M morpholinopropanesulphonic acid (pH 7.0), 50 mM sodium acetate, 5 mM EDTA (pH 7.0).
[b] Formaldehyde should be handled in a fume hood. It is usually supplied as a 12.3 M (37%) solution in water. Check that the pH is greater than pH 4.
[c] Formamide oxidizes in air and should be deionized by passing it through a mixed-bed resin (Bio-Rad AG 501-X8) until the pH is neutral.

7.2 Detection of RNA after electrophoresis in denaturing slab gels

RNA labelled with ^{32}P-phosphate can readily be detected by autoradiography (see Section 6.5), or by a variety of staining methods (*Table 4*). The procedure most commonly used to locate the RNA bands is by staining with ethidium bromide. This binds to the RNA and can be visualized by UV-induced fluorescence (*Table 9*). In some electrophoresis protocols, ethidium bromide is added to the RNA sample before loading it on to the gel. However, this should not be done when using glyoxal, as

---

Table 9. Staining denaturing gels with ethidium bromide.[a]

Glyoxal gels
Stain the gel for 15 min in the dark in 0.5 M ammonium acetate containing 5 μg/ml ethidium bromide. Destain for 15 min in the dark in 0.5 M ammonium acetate.
Alternatively, stain in 0.5 μg/ml ethidium bromide in water and destain in the dark.

Formaldehyde gels
Stain the gel for 5 min in the dark with 5 μg/ml ethidium bromide and destain for 2 h in the dark. *Alternatively*, wash the gel for 60 min with two changes of 0.1 M ammonium acetate. Stain in the dark with 0.5 μg/ml ethidium bromide in 0.1 M ammonium acetate, containing 0.1 M 2-mercaptoethanol. Destain in the dark in 0.1 M ammonium acetate containing 0.01 M 2-mercaptoethanol. RNA bands stained with ethidium bromide are visualized by trans-illumination with short wavelength (254 nm) UV light[b] and photographed with Polaroid film using a red plus yellow (or orange) filter at *f* 4.5 for ½ − 1 sec. Prolonged exposure to UV light can damage the RNA.

[a] **Caution:** ethidium bromide is toxic and carcinogenic and should be handled with care.
[b] Wear protective goggles when working with UV.

---

ethidium bromide reacts with it. It is not satisfactory for formaldehyde gels either, due to the high background fluorescence of the denaturing agents, although ethidium bromide (0.033 μg/μl) can be added if the formaldehyde concentration is lowered to 0.66 M (35). Using this method, no staining or destaining of the gel is necessary. However, high concentrations of ethidium bromide are believed to reduce the transfer efficiency of the RNA during northern blotting. Thus, most protocols involve staining of the gel after electrophoresis and, where the gel is to be blotted afterwards, marker tracks run at the edges of the gel are cut off and stained. RNA can also be detected in polyacrylamide gels (40), denaturing agarose gels (41) and after blotting on to nitrocellulose or Nylon membranes (42).

7.3 Northern blotting RNA on to diazobenzyloxymethyl-paper

Alwine *et al.* (19) have developed a method for transferring RNA from slab gels to diazobenzyloxymethyl-paper (DBM-paper), preserving the original fractionation pattern. The method is analogous to the transfer of DNA from gels to nitrocellulose paper developed by Southern (20), frequently referred to as 'Southern blotting'. In most cases this method has now been superseded by northern blotting to nitrocellulose (25) or Nylon membranes (see Section 7.4).

The procedure for making aminobenzyloxymethyl (ABM)-paper which is converted to DBM-paper just before use, is described in ref. 19. ABM-paper is also available commercially. The method for transferring the RNA, which is similar to that used for DNA is outlined in *Protocol 12*. RNA transferred from gels to DBM-paper remains attached to the paper by covalent linkage and can be used for hybridization by soaking the paper in solutions of radioactive probes (*Protocol 13*). The hybridized radioactivity can then be located on the paper by autoradiography.

Protocol 12. Transfer of RNA to DBM-paper

The procedure described, Alwine *et al.* (19), is similar to Southern's method for DNA (20).

1. Prepare DBM paper from ABM paper just before use, as follows: Soak the paper for 30 min at 4°C in 120 ml of 1.2 M HCl pre-mixed with 3.2 ml of freshly dissolved NaNO$_2$ (10 mg/ml). (Check the solution for free HNO$_2$, which turns starch−iodide paper black.) After 30 min, wash the paper five times (5 min each time) with 100 ml cold water, followed by two washes with ice-cold 50 mM sodium borate buffer (pH 8.0) for 10 min. The paper should turn bright yellow, and is kept cold until used for transfer within 15 min. For further details see ref. 19.

2. To remove methylmercuric hydroxide from the gel,[a,b] and to hydrolyse the RNA partially to assist the transfer of high molecular weight RNA,[c] soak the gel in:

   (a) 20 ml of 50 mM NaOH, 5 mM 2-mercaptoethanol with gentle shaking for 40 min at room temperature.

37

Protocol 12. *continued*

    (b) Transfer to 200 ml of 200 mM sodium borate (pH 8.0), containing 7 mM iodoacetic acid for 10 min at room temperature. Repeat once.

    (c) Wash twice with 200 ml of 50 mM sodium borate (pH 8.0) for 5 min at room temperature.

3. Place the gel on two sheets of Whatman 3MM paper saturated with 50 mM sodium borate buffer (pH 8.0). Cover the gel with a sheet of freshly-prepared DBM-paper saturated in the same buffer. Use Perspex strips to support the DBM-paper at the edges and prevent it from contacting the filter paper or buffer beneath the gel. Cover the DBM-paper with two layers of dry Whatman 3MM paper, several layers of paper towels and a glass plate.

4. Allow buffer to soak through the gel and DBM paper and into the paper towels transferring the RNA to the DBM-paper where it becomes fixed. Replace the wet towels with dry ones at intervals and allow the transfer to continue overnight.

  a If the gel does not contain methylmercuric hydroxide, but the RNA is high molecular weight, omit the 2-mercaptoethanol and iodoacetic acid in solutions (a) and (b) respectively.

  b For transfer of RNA from gels lacking methylmercuric hydroxide and without hydrolysis of the RNA, simply soak the gel in solution (c).

  c In the case of low molecular weight RNA, NaOH is not required and can be replaced in solution (a) with electrophoresis buffer containing 2-mercaptoethanol.

Protocol 13. Hybridization of radioactive DNA to RNA attached to DBM-paper

1. Incubate the DBM-paper, to which the RNA has been bound (see *Protocol 12*), in 50% formamide, 0.75 M NaCl, 75 mM sodium citrate containing 0.02% bovine serum albumin, 0.02% Ficoll, 0.02% polyvinylpyrrolidone, 1−2 mg of sonicated denatured heterologous DNA per ml and 1% (w/v) glycine, for 4−24 h at 42°C. This treatment hydrolyses remaining diazo groups and blocks non-specific sites on the paper to which DNA may adsorb.

2. Place the DBM-paper strips in plastic bags with 50−100 μl per cm^2 paper of the above solution, minus glycine, containing the radioactive single-stranded DNA probe. Remove the air and then seal the bags using a heat sealer and incubate for 36 h at 42°C with gentle rocking.

3. After hybridization, wash the paper for several hours at 42°C in 50% formamide, 0.75 mM sodium citrate. Change the solution several times. After blotting, the paper can be covered with plastic wrap (e.g. Saran Wrap) and exposed to X-ray film for autoradiography or, alternatively, the paper can be completely dried before autoradiography.

7.4 Northern blotting RNA on to nitrocellulose or Nylon membranes.

Blotting of RNA samples from the gel to nitrocellulose or Nylon membranes is easier and quicker than using DBM-paper (*Protocol 14*). Transfer is normally effected by capillary blotting. Capillary blotting overnight is usually sufficient, but the efficiency of transfer depends on the molecular weight and the amount of RNA in each band, the thickness of the gel, and the volume of buffer that moves through the gel. One modification of the original method is vacuum blotting (39) in which the transfer is speeded up by applying a vacuum across the gel. It is claimed that this gives a better transfer and resolution of bands. Another modification which speeds up the transfer is electroblotting, in which case the RNA is electrophoresed out of the gel and on to the membrane; for an example of the protocol used see Section 5 of Chapter 8. Apparatus for the vacuum blotting and electroblotting of gels is now commercially available from several sources.

Protocol 14. Capillary blotting of RNA from slab gels to nitrocellulose or Nylon membranes

1. Soak formamide gels for 5 min in several changes of distilled water, followed by the blotting solution (20 × SSC: 3 M NaCl, 0.3 M trisodium citrate, pH 7.0) for 30−60 min. No washing is necessary for glyoxal gels.

2. Cut the nitrocellulose or Nylon membrane to the exact size of the gel and pre-wet it by soaking it first in distilled water for 5 min and then the blotting solution (20 × SSC) for 5 min.

3. Set up the capillary blot (*Figure 14*) by making a platform and covering it with three layers of Whatman 3MM filter paper, moistened with 20 × SSC. Place the gel face down on the 3MM filter paper, taking care to exclude all air bubbles. A small-pore sponge placed under the paper wick will enhance the flow of 20 × SSC through the gel. Lay the membrane carefully on top of the gel, taking care to exclude air bubbles. Cover the membrane with three layers of 3 MM filter paper and a large thickness of absorbent towels cut to the size of the membrane. Apply a weight to the top (0.5−1 kg) and allow to transfer for 8−24 h. Change the absorbent towels as necessary.

4. Carefully disassemble the blot and cut off one corner of the membrane to permit subsequent orientation. Rinse membranes blotted from formaldehyde gels in 3 × SSC. Do *not* rinse membranes blotted from glyoxal gels.

5. Allow membranes to air-dry and then bake in a vacuum oven at 80°C for 2−4 h.

6. The RNA may be crosslinked to Nylon membranes by placing them, RNA side downwards, on a UV transilluminator and irradiating them for 3−4 min. *Do not* expose nitrocellulose membranes to UV light, due to the risk of fire.

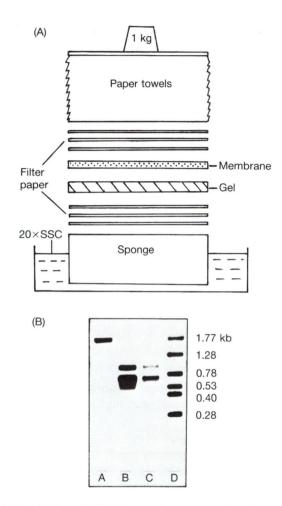

Figure 14. Northern blotting of RNA to membranes and detection of RNA bands by hybridization and autoradiography. An exploded view of the components of the blotting 'sandwich' is shown in (A). A small-pore, firm, sponge aids transfer of blotting solution (20 × SSC) but is not essential. If a solid support is used instead of a sponge, the filter-paper wicks need to be placed so that they are in contact with the filter-paper sheets and dip into the blotting solution. Specific RNA bands detected by hybridization to ^{32}P-labelled nucleic acid probes are shown in (B). Track A shows polygalacturonase mRNA from ripening tomatoes, tracks B and C are *antisense* polygalacturonase mRNA in unripe and ripening fruit of transgenic tomato plants expressing an antisense gene. Track D contains RNA markers. Sizes are shown in kilobases. Data in (B) from ref. 38.

It is desirable to check the efficiency of transfer by staining the gel with ethidium bromide *after* blotting (*Table 9*). For high molecular weight RNA, the transfer may be improved by soaking the gel in 50 mM NaOH for 40 min (*Protocol 12*). RNA is hydrolysed at alkaline pH and this treatment breaks the polynucleotide chain into shorter fragments that transfer more efficiently. The transfer of RNA is also improved if the gel is not stained with ethidium bromide before blotting. The RNA transferred to the membrane during blotting may be visualized by staining with histones and Indian ink (34), or directly by UV fluorescence, if the RNA is blotted from a gel containing 0.66 M formaldehyde and a low concentration of ethidium bromide (35).

As shown in *Figure 14*, specific RNA bands may be detected by hybridization of the membrane with radioactive DNA or RNA probes (36). *Protocol 13* is typical, but the exact conditions will depend on the nature of the RNA. Nitrocellulose membranes give good results, but Nylon membranes are often preferred because, provided that the membrane is not dried, the probe can be removed after autoradiography (see Section 6.5) and the membrane challenged with a second probe. A variety of membranes are now available commercially. Specific instructions for use of each are supplied by the manufacturers.

Several methods are now available for the production of sensitive non-radioactive probes. A number of these are based on the use of biotin to 'label' the nucleic acid probe, by incorporating biotin-tagged nucleotides enzymically. Biotinylated hybridization probes can also be prepared chemically, by tagging them with 'photo-biotin' (N-(4-azido-2-nitrophenyl)-N-(N-D-biotinyl-3-aminopropyl)-N'-methyl-1,3-propanediamine)(37). Following hybridization, the probe can be detected and quantified by binding to avidin or streptavidin, linked to an enzyme such as peroxidase or alkaline phosphatase, which, when incubated with an appropriate substrate, gives a coloured product. In another system, the steroid hapten digoxigenin is incorporated into the probe, linked to UTP, and the probe is detected using an antibody–enzyme conjugate.

8. Problems of polyacrylamide gel electrophoresis and their remedies

8.1 Problems with gels

8.1.1 Gels do not polymerize or polymerize unevenly

This may be due to any one of a number of causes:

(a) traces of organic solvents, used to recrystallize the acrylamide or bisacrylamide, may be present and these inhibit polymerization. After recrystallizing acrylamide, thoroughly air-dry or leave the crystals in a vacuum dessicator for 1−2 days before making the stock solution. Alternatively, purchase reagents specially purified for electrophoresis;

(b) dissolved oxygen may be present in the gel mixture. This inhibits polymerization. Check the joints on the vacuum system or use a better vacuum pump to improve degassing;

(c) faulty ammonium persulphate. The solution must be freshly dissolved. Occasionally the ammonium persulphate crystals are too old and a fresh supply must be used.

8.1.2 Gels polymerize too quickly

This can happen either before pipetting is complete or before water is layered on top, especially with high concentration polyacrylamide gels. Possible remedies are:

(a) work faster;

(b) cool the gel mixture before adding the catalysts, hence slowing the rate of polymerization;

(c) reduce the concentrations of TEMED or ammonium persulphate.

8.1.3 Gels slide out of the tubes after polymerization

This occurs most frequently with low concentration polyacrylamide gels and can be avoided by either:

(a) covering the bottom of each tube with gauze or dialysis tubing secured with a rubber band, or

(b) inserting a disc of porous polythene into the bottom of each gel.

8.2 Problems with RNA samples

8.2.1 No RNA detected in the gel after electrophoresis

The electrodes may be connected with the incorrect (i.e. reversed) polarity. However, the usual causes are:

(a) too little RNA loaded on to the gel;

(b) contamination of the RNA with denatured proteins may cause the RNA to 'stick' at the top of the gel, although low molecular weight RNA should enter the gel. Try an alternative or additional deproteinizing step, that is, use phenol/*m*-cresol instead of phenol/chloroform or use proteinase K (Boehringer Mannheim GmbH) to digest contaminating protein. Typically, use 200 μg/ml proteinase K in 0.1 M KCl, 10 mM magnesium acetate, 0.05 M Tris-HCl (pH 7.5) incubated for 30 min at 10°C;

(c) the RNA may have been degraded to small fragments which pass straight through the gel. Try a more concentrated gel or run for a shorter time and use bromophenol blue as a marker for the buffer front. If the RNA is degraded, check the starting material or the extraction and purification steps. Alternatively, ribonuclease may be present in the electrophoresis reservoir or buffer. Clean the apparatus thoroughly and autoclave the electrophoresis buffer and keep sterile until electrophoresis of the sample;

(d) the sample may have floated off the gel during loading. This is easy to see when loading the samples if a marker dye, such as bromophenol blue, is present. The problem can be prevented by evaporating alcohol from the RNA during sample preparation or by adding more sucrose before loading. A few minutes after starting electrophoresis, refractile bands of RNA should be seen entering the gel;

(e) the gel concentration may be too high. See *Figure 2a−2c* as a rough guide as to the correct gel concentration to use.

8.2.2 RNA bands very broad or trailing

The fault normally lies in the sample preparation or loading;

(a) check the composition of the electrophoresis buffer used. If this is correct, check that the salt concentration in the loading solution is not too high and the pH of the solution is as expected;

(b) load a smaller volume of RNA sample;

(c) use a lower concentration of RNA for loading;

(d) try an alternative deproteinizing procedure (see 8.2.1b).

8.2.3 Small amounts of RNA present in the gel are obscured by background 'noise' when scanning

This is caused either by impurities in the gel which absorb UV light or by dirt and dust introduced into the gel reagents, or stuck to the gels, or by air bubbles and scratches on the cuvette or the gels. The remedies are:

(a) to remove high UV absorbance in the gels, try washing the gels in distilled water for 1−2 h. Alternatively, remove impurities from the sample before electrophoresis by re-precipitation or by washing RNA precipitates with ethanol and from the gel reagents by recrystallization (*Table 2*);

(b) filter all solutions used for electrophoresis and keep all apparatus scrupulously clean. With very small amounts of RNA, use membrane filters to remove debris from solutions and exclude dust and air bubbles from the cuvette when scanning.

8.2.4 Aggregation of RNA

RNA extracted from certain organisms has a great tendency to aggregate causing problems if the RNA is analysed on non-denaturing gels. In extreme cases, where most of the RNA is aggregated, this is quite easy to spot and may be remedied by improving the method of deproteinization (see 8.2.1b) or by denaturing the RNA before loading on to the gel (see Section 3.2). A greater problem is the production of specific aggregates of RNA in small quantities which might be interpreted as authentic RNA components. Sometimes these can be formed during RNA extraction at elevated temperatures or where RNA is dissolved at high concentrations. For example, specific aggregates of rRNA are formed by association between molecules of 25S (28S) and 18S rRNA, or by two molecules of 25S (28S) RNA. These aggregates migrate into the gel and produce peaks of apparently high molecular weight RNA which resemble precursor rRNA molecules (genuine precursor rRNAs are shown in *Figure 2*). When in doubt about the authenticity of even minor components it is always a good idea to see if they disappear when the sample is denatured before it is loaded on to the gel. An even more rigorous criterion is to electrophorese the sample under denaturing conditions (see Section 4).

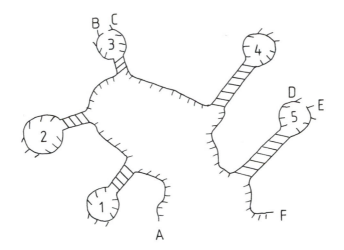

Figure 15. Origin of hidden breaks. A hypothetical single-stranded RNA molecule (A – F), with 5 hydrogen-bonded regions (1 – 5) is shown. The exposed loops at positions 3 and 5 have been cleaved by nuclease action. Under completely denaturing conditions three RNA fragments will be formed (A – B, C – D, E – F). Under milder conditions only two fragments (A – B and C – F) will form. At low temperatures or in the presence of divalent cations fragmentation may be completely prevented.

It is also possible for relatively small amounts of RNA to remain stuck at the top of the gel. This frequently happens if traces of protein are present in the sample or if too high a concentration (or volume) of RNA is applied to the gel. In the case of tube gels a sign of possible trouble is shrinking of the top of the gel a short time after loading the sample. If this happens, staining or detection of radioactive RNA will reveal RNA stuck at the top of the gel. If RNA is no longer found at the top of the gel after the sample has been denatured or run in denaturing conditions, the simplest and most probable interpretation is that it was an aggregate.

8.2.5 Fragments and 'hidden breaks' in RNA

Many single-stranded RNA molecules contain 'hairpin loops' or regions where the RNA chain folds back upon itself and forms short helical sections held together by hydrogen bonds. Sometimes the loops at the end of the hairpins are exposed and open to attack by ribonuclease (*Figure 15*). This can occur during RNA extraction but it also seems to happen *in vivo*, particularly with rRNA. Many rRNA molecules start off as continuous chains but may be 'nicked' in certain positions during their lifetime in the cell. Frequently the fragments are held together by hydrogen bonds between base pairs in the stem (*Figure 15*), where the phosphodiester bonds are less sensitive to nuclease attack, so that these 'hidden breaks' may go undetected in non-denaturing gels. They are immediately apparent, however, if the RNA is denatured before electrophoresis (Section 3.2) or if the RNA is electrophoresed under denaturing conditions (Section 4). Under these conditions the fragments separate, there is a

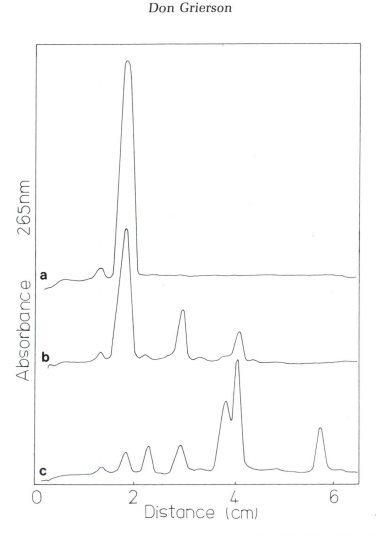

Figure 16. Temperature-induced fragmentation of chloroplast 23S rRNA. 23S chloroplast rRNA from spinach was purified by sucrose gradient centrifugation at 4°C and dissolved in electrophoresis buffer at 0°C. Separate aliquots were incubated at (a) 0°C, (b) 25°C and (c) 40°C for 10 min and then fractionated in 3% polyacrylamide tube gels for 5 h at 5°C. The small peak to the left of the 23S RNA is a 25S cytoplasmic rRNA contaminant. (Redrawn from ref. 30.)

reduction in the amount of one of the original components and 'new' RNA molecules are generated. These can be unequivocally identified by isolating the parent molecule and subsequently denaturing it to generate fragments.

Sometimes the fragments dissociate even at room temperature so that it is not necessary to heat the RNA to reveal the breaks. A good example of this is found with the 23S rRNA from chloroplast ribosomes of higher plants. *Figure 16* shows that if the 23S rRNA is brought to room temperature it produces several fragments, whereas below 4°C the molecule remains intact. In some cases it may be useful to prevent fragmentation when studying RNA molecules of similar mobility to the fragments. This is most easily achieved by carrying out sample preparation and gel electrophoresis at low temperatures (see Section 3.3).

8.2.6 Identification of radioactive bands in gels

The position of an RNA band in the gel, relative to other peaks, is frequently used to identify the RNA or to estimate its molecular weight. Therefore, when using tube gels, it is important that UV scans are correctly synchronized with the plots of radioactivity measured in the gel slices. Low concentration polyacrylamide gels are liable to expand or contract and so it is important that they are frozen for slicing while held at exactly the same length as when scanned. It is also important to freeze gels when they are horizontal. If frozen when slightly inclined the gel will stretch at one end and be compressed at the other, giving rise to an apparent shift in electrophoretic mobility. With very dilute gels it may be necessary to insert short

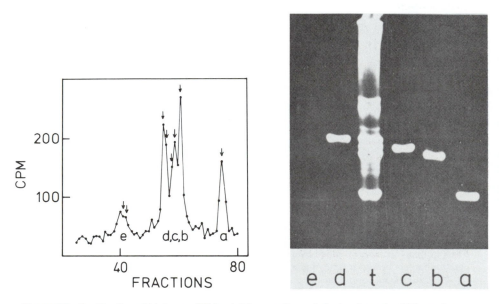

Figure 17. Purification of histone mRNAs. (a) Preparative gel electrophoresis. 140 μg of sea-urchin polyribosomal RNA plus 3×10^5 cpm ^{3}H-uridine-labelled RNA as tracer, were fractionated on a 6% polyacrylamide tube gel. The gel was sliced and the RNA was eluted as described in the text. 25 μl of each 1 ml sample of eluate was counted to locate the RNA. (a – e) Refer to the subfractions of histone mRNA. (b) Electrophoresis of purified mRNAs. The individual subfractions (a – e in *Figure 17a*) were re-purified as described in the text. To examine the final purity of each fraction, approximately 600 cpm of each subfraction were fractionated in a slab gel, together with 15 000 cpm of total polyribosomal RNA (lane t). The photograph shows the result of fluorographic exposure of the gel for 30 days. Reproduced from ref. 10 with permission.

pieces of wire at right angles to the long axis of the gel to help synchronize the UV scan with slice number (see *Figure 5*).

The presence of radioactivity at a particular position in a gel can sometimes be misleading. For example, DNA normally migrates as a single peak of polyacrylamide gels (*Figure 2*) and may fortuitously coincide with an RNA molecule such as a precursor rRNA molecule. It is impossible to be certain whether the DNA or the RNA, or both, are radioactive without treating the sample with deoxyribonuclease or ribonuclease before electrophoresis. Furthermore, carbohydrate molecules may also become labelled. These frequently migrate slowly in the gel but can be distinguished from authentic nucleic acids because they survive nuclease digestion of the sample. Many unpolymerized radioactive contaminants move very rapidly through the gel and may coincide with low molecular weight RNA running close to the buffer front. Such material tends to diffuse out during staining and destaining procedures. Alternatively, low molecular weight contaminants can be removed from the sample either by passing the sample down a Sephadex G-25 column before electrophoresis or by choosing a gel of higher acrylamide concentration for the fractionation, in which case the RNA often separates from the contaminants during electrophoresis.

9. Purification of histone messenger RNA: a case study

Preparative electrophoresis of sea-urchin histone mRNAs in tube gels has been described by Gross *et al.* (10). *Figure 17a* shows the initial preparative gel, the purified mRNAs run in slab gels are shown in *Figure 17b* and the protein translation products produced when purified mRNAs were added to a wheat-germ cell-free protein synthesis system are shown in *Figure 18*.

Tubes, 0.7 cm internal diameter and 10 cm long, were filled to within 2.5 cm of the top with 6% acrylamide gel mixture. After polymerisation, the water overlay was removed and a 1.5 cm length of 3% acrylamide gel was cast over the 6% gel. Buffer (b) in *Table 5* was used in the gels and the buffer reservoirs. Approximately 140 µg of sea-urchin total polyribosomal RNA were fractionated on each gel. Electrophoresis was at 34°C for 10−15 h at 60 V with a current of about 1.5 mA per gel. The electrophoresis buffer was recirculated between the reservoirs during the run. After electrophoresis the gels were sliced into 0.5 mm fractions using a Mickle gel slicer. The RNA was eluted at room temperature from pairs of adjacent slices into 1 ml of 0.6 M lithium acetate (pH 5.8) containing 0.2% SDS and 7.5 µg of purified *E.coli* tRNA to act as carrier. After elution, the fractions were passed through 0.45 µ Millipore filters and 25 µl of each fraction was counted in a liquid scintillation counter. The profile obtained is shown in *Figure 17a*. The peak fractions were pooled and the RNA was precipitated with 2.5 volumes of ethanol at −20°C overnight. The RNA precipitates were pelleted by centrifugation (20 000 *g* for 15 min), washed several times with 95% ethanol and dried. The RNA was then dissolved in sterile distilled water and insoluble material removed by centrifugation.

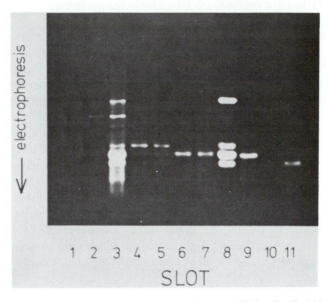

Figure 18. *In vitro* translation products of purified histone mRNAs. Purified histone mRNAs (equivalent to subfractions a – e in *Figure 17a*) were used to direct protein synthesis in a wheat-germ system. The ^{35}S-methionine-labelled protein products were fractionated in a slab gel together with authentic sea-urchin histones labelled with ^{3}H-lysine *in vivo*. Lanes (1) and (2) RNA e; (3) total polyribosomal RNA; (4) and (5) RNA d; (6) and (7) RNA c; (8) *in vivo* labelled sea-urchin histones; (9) RNA b; (10) endogenous products of wheat-germ system; (11) RNA a. Reproduced from ref. 10 with permission.

Each RNA fraction was then further purified by re-electrophoresis before being used for hybridization experiments. The final purified histone mRNA fractions were examined by slab gel electrophoresis as shown in *Figure 17b*.

Histone mRNAs fractionated by the method described were also used for *in vitro* protein synthesis in a wheat-germ cell-free system. In this case, the gels were washed in several changes of distilled water for 24 hours at 4°C before slicing to remove components present in the gel (e.g. SDS) which might inhibit subsequent *in vitro* translation of the RNA. Re-purified yeast tRNA was used as carrier and the eluted RNAs were recovered, subjected to a second round of electrophoresis and elution and subsequently used to direct histone protein synthesis *in vitro* (*Figure 18*).

References

1. Choules, G. L. and Zimm, B. H. (1965). *Anal. Biochem.*, **13**, 366.
2. Peacock, A. C. and Dingman, C. W. (1968). *Biochemistry*, **7**, 668.
3. Loening, U. E. (1967). *Biochem. J.*, **102**, 251.
4. Loening, U. E. (1968). *J. Mol. Biol.*, **38**, 355.
5. Loening, U. E. (1969). *Biochem. J.*, **113**, 131.

6. Bishop, D. H. L., Claybrook, J. R., and Spiegelman, S. (1967). *J. Mol. Biol.*, **26**, 373.
7. Fisher, M. P. and Digman, C. W. (1971). *Biochemistry*, **10**, 1895.
8. Reijnders, L., Sloof, P., Sival, J., and Borst, P. (1973). *Biochim. Biophys. Acta*, **324**, 320.
9. Gould, H. J. and Hamlyn, P. H. (1973). *FEBS Lett.*, **30**, 301.
10. Gross, K., Probst, E., Schaffner, N., and Birnstiel, M. L. (1976). *Cell*, **8**, 455.
11. Studier, F. W. (1973). *J. Mol. Biol.*, **79**, 237.
12. Laskey, R. A. and Mills, A. D. (1977). *FEBS Lett.*, **82**, 314.
13. Laskey, R. A. and Mills, A. D. (1975). *Eur. J. Biochem.*, **56**, 355.
14. Bonner, W. M. and Laskey, R. A. (1974). *Eur. J. Biochem.*, **46**, 83.
15. Hartley, M. R. (1979). *Eur. J. Biochem.*, **96**, 311.
16. Probst, E., Dressman, A., and Birnstiel, M. L. (1979). *J. Mol. Biol.*, **135**, 709.
17. Bailey, J. M. and Davidson, N. (1976). *Anal. Biochem.*, **70**, 75.
18. Langridge, J., Langridge, P., and Bergquist, P. L. (1980). *Anal. Biochem.*, **103**, 264.
19. Alwine, J. C., Kemp, D. J., and Stork, G. R. (1977). *Proc. Natl. Acad. Sci. (USA)*, **74**, 5350.
20. Southern, E. M. (1975). *J. Mol. Biol.*, **98**, 503.
21. Chamberlain, J. P. (1979). *Anal. Biochem.*, **98**, 132.
22. Kirby, K. S. (1968). In *Methods in Enzymology* (ed. L.Grossman and K.Moldave) Vol. XIIB, p. 87. Academic Press, New York and London.
23. Spath, P. J. and Koblet, H. (1979). *Anal. Biochem.*, **93**, 275.
24. Hansen, J. N., Pheiffer, P. H., and Boehnert, J. A. (1980). *Anal. Biochem.*, **105**, 192.
25. Thomas, P. S. (1980). *Proc. Natl. Acad. Sci. (USA)*, **77**, 5201.
26. Loening, U. E. (1970). *Symp. Soc. Gen. Microbiol.*, **20**, 77.
27. Grierson, D. (1979). *Z. Pflanzenphysiologie*, **95**, 171.
28. Leaver, C. J. and Pope, P. K. (1980). In *Nucleic Acids and Protein Synthesis in Plants* (ed. L. Bogorad and J. H. Weil) p. 213. Plenum Press, New York.
29. Grierson, D. (1972). Ph. D. Thesis, University of Edinburgh.
30. Speirs, J. and Grierson, D. (1978). *Biochim. Biophys. Acta.*, **521**, 619.
31. Vachon, D. Y., Owunwanne, A., Carroll, B. H., O'Mara, R. E., and Griffiths, H. J. L. (1981). *Invest. Radiol.*, **16**, 221.
32. Hames, B. D. (1990). *Gel Electrophoresis of Proteins: A Practical Approach*, 2nd Edition (ed. B. D. Hames and D. Rickwood). IRL Press at OUP, Oxford.
33. Mitchell, R. L. (1986). *Focus*, **8**, 10.
34. Bülow, S. and Link, G. (1986). *Nucleic Acids Res.*, **14**, 3973.
35. Fourney, R. M., Mikatoshi, J., Day, R. S. III, and Paterson, M. C. (1988). *Focus*, **10**, 5.
36. Hames, B. D. and Higgins, S. J. (eds) (1985). *Nucleic Acid Hybridisation: A Practical Approach*. IRL Press, Oxford.
37. McInnes, J. L., Vise, P. D., Habili, N., and Symons, R. H. (1987). *Focus*, **9**, 1.
38. Smith, C. J. S., Watson, C., Ray, J., Bird, C. R., Morris, P. C., Schuch, W., and Grierson, D. (1988). *Nature*, **334**, 724.
39. Olszewska, E. and Jones, K. (1988). *Trends in Genetics*, **4**, 92.
40. Lomholt, B. and Frederiksen, S. (1987). *Anal. Biochem.*, **164**, 146.
41. Skopp, R. N. and Lane, L. C. (1988). *Anal. Biochem.*, **169**, 132.
42. Su, X. (1987). *Anal. Biochem.*, **163**, 535.

2

Gel electrophoresis of DNA

PAUL G. SEALEY and ED M. SOUTHERN

1. Introduction

The applications of gel electrophoresis for the analysis of DNA range from determining base sequence to separating the DNAs of whole chromosomes and it would clearly be difficult to include all applications in a single chapter. In this initial chapter, the treatment is divided into two major sections. The first is devoted to the basic applications which are in most common usage, whilst the second part describes the construction and use of two devices for preparative fractionations. Descriptions of other more specialized techniques, such as those used for footprinting, and those involving varying the relative orientation of the electric field and the gel (pulsed field gel electrophoresis), are described in other chapters. The theory of gel electrophoresis as it applies to DNA is outside the scope of this article and in any case is not well understood. This chapter does not attempt any theoretical treatment of the subject but does include a few background notes on the various parts of the process where it is felt that an understanding of the background may help in carrying out the process. Empirical methods are given for deriving DNA size from electrophoretic mobility.

2. Basic techniques and their application

Electrophoresis of DNA may be carried out in agarose, polyacrylamide, or agarose−acrylamide composite gels. The most common application is the separation of duplex DNA fragments at neutral pH values; under these conditions the DNA is negatively charged so that fragments loaded into a sample well at the cathode (−) end of a gel move through the gel towards the anode (+). The electrophoretic mobility of DNA fragments larger than oligonucleotides is dependent on fragment size and fairly independent of base composition or sequence.

Agarose gels can be used to analyse double-stranded DNA fragments from 70-base-pairs (bp) (3% agarose gel) to 800 000 bp (0.1% agarose gel) (1). The fact that for any set of conditions there is an upper fragment size at which resolution deteriorates sharply, coupled with the difficulties of handling gels below 0.5% agarose, means that for separating fragments above 50 000 bp one of the techniques of pulsed field gel electrophoresis (Chapter 3) will be the method of choice.

Polyacrylamide gels are used for fragments between 6 bp (20% acrylamide) (2) and 1000 bp (3% acrylamide). The smallest detectable amount of non-radioactive DNA in any single size class is less than a nanogram; resolution of one size class from another can be better than 0.5% of the fragment size.

Denaturing gel systems of several types are available for the analysis of single-stranded DNA fragments, and again the fragments migrate with a size-dependent mobility, though single strands of DNA moving in non-denaturing gels show some dependence of mobility on secondary structure. In the search for higher resolution, gels containing a gradient of increasing denaturing power have been used. In this case the mobility of a fragment drops sharply at the point in the gel where complementary strands begin separating. The final fragment position is therefore sequence-dependent and, in combination with size separation, this makes two-dimensional separation of DNA fragments possible (ref. 3 and Chapter 5).

In the eight years since the authors wrote this chapter for the first edition there has been little change in the basic range of procedures available. There has, however, been a very large increase in the number of commercial products available; where the authors have had direct experience of them these are mentioned in the text. Substantial changes have been made in the section dealing with the recovery of DNA fragments from gels to take into account improved procedures. The descriptions of two dedicated, high capacity, preparative devices for the electrophoretic separation of DNA have, however, been retained unchanged.

2.1 Equipment

2.1.1 Electrophoresis apparatus

The most versatile design consists of a buffer reservoir in the form of a square tank, one side of which forms the back plate of the gel mould (*Figure 1*). A removable front plate is separated from the back plate by spacers of any desired thickness (0.2 mm to 2.0 cm). Sample wells are cast in the gel which is formed between the front and back plates using a sample well-former. The whole apparatus is held together by metal clips. One electrode (the cathode) is located inside the buffer tank which forms the cathodic reservoir. Prior to electrophoresis, the tank is placed in a glass or plastic tray full of buffer in which the anode is located, thus forming the anodic reservoir. Both reservoirs are filled with buffer and electrophoresis takes place in a downward direction, through the vertical gel. Most of the detailed procedures for preparing and running gels, in subsequent parts of this chapter, refer to this type of apparatus.

The overall dimensions of the buffer tank can be varied depending on the size of the gels; common values for gel height and width are 10 × 10 cm, 10 × 20 cm, 20 × 10 cm and 40 × 20 cm. The best material for the front and back gel plates is glass (3 mm thick); the high thermal expansion of Perspex can cause problems with agarose, and it inhibits polymerization of acrylamide. Polycarbonate (Lexan) is somewhat better in this respect. When used with agarose, the very largest tanks may be improved by using plate glass, which bends less under the weight of gel

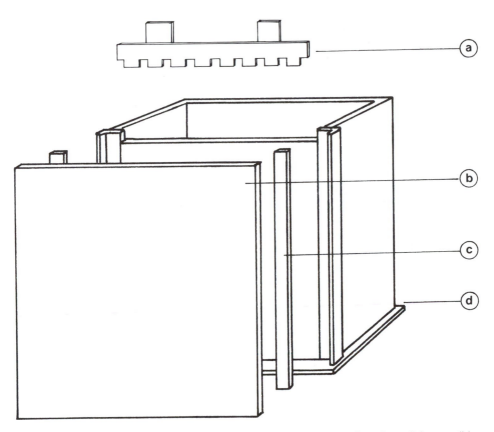

Figure 1. Vertical apparatus for gel electrophoresis of DNA. (a) Sample well-former; (b) removable front plate; (c) spacer; (d) buffer reservoir. For further details see the text.

and buffer. The rest of the tank can be constructed from glass or Perspex. Platinum wire (0.25 mm diameter) is used for the electrodes, and these can be built-in or removable as desired. Smaller tanks, up to 20 × 20 cm, if for use with agarose only, can be made with the front plates glued permanently in position. The voltages used can give a dangerous shock, so that it is advisable to incorporate a lid in the apparatus that will disconnect the power supply when removed, making a short circuit between the buffer reservoirs impossible.

The spacers for the glass plates may be fixed in position or removable. Typical thicknesses are 1 mm (polyacrylamide gels), 3 mm (agarose gels), or up to 2 cm (preparative agarose gels). Suitable materials are Neoprene rubber and Perspex, for agarose gels, and 'sticky' PVC, PTFE, or polystyrene ('Plastikard') for polyacrylamide. Some materials require a thin layer of silicone grease to give a good seal. Sample well (slot) formers can be made of similar materials, providing they can be cut to give a straight, flat bottom to the teeth. The tank itself can be glued

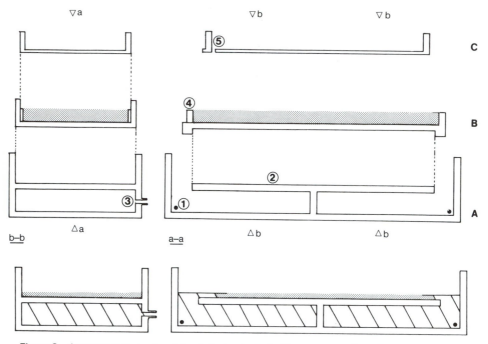

Figure 2. Apparatus for horizontal gel electrophoresis. The apparatus has three nesting parts. These are shown in *Figure 2a* as cross-sectional plans, on the left through the axis b — b and on the right through axis a — a. (A) is a box, divided into two buffer compartments fitted with platinum wire electrodes, (1); a gel platform, (2); and ports for buffer circulation, (3). (B) is a gel tray which fits onto the platform (2); one end (4) can be removable. (C) is a lid for this tray with a slit (5) for a slot former. The lid rests on removable side pieces which determine the gel thickness. The external dimensions of the box are approximately 32 cm long × 7 cm high × 14.5 cm wide. Construction is from Perspex sheet, mainly 6 mm thick.

In the simplest applications (*Figure 2b*) only box (A) is used. This is a 'submarine' technique similar to that developed by Schaffner (personal communication). The gel (stippled) area is cast in a separate mould (not shown) and laid directly onto the gel platform. The buffer level (hatched area) is adjusted until the gel is just flooded. For loading, the volume of the sample should be adjusted to fill the sample well; electrophoresis is as for vertical gels. In the presence of ethidium bromide the DNA can be viewed directly using overhead UV illumination, whilst the gel is running. Radioactive fragments may be detected with a hand monitor. Depending on their length, several gels can be laid in series along the platform (i.e. down the voltage gradient). Alternatively, the tray (B) and lid (C) may be used (*Figure 2a*). The gel is cast in the tray with the lid in place and the assembly fitted onto the platform. The buffer level is adjusted to overflow the ends of gel and electrophoresis carried out as described. The tray bottom, or the platform, can be made from black Perspex, or glass (clear or black); if the lid is of UV-transparent Perspex, the gel can be viewed without removing it. We have found this second procedure useful for dilute gels.

together using a solvent-based plastic glue (e.g. 'Tensol') for Perspex/Perspex joints and a silicone-rubber caulk for joints involving glass. Slot formers may be of the comb type (*Figure 1*) or, alternatively, individual formers may be made and used

---

Table 1. Tris-borate electrophoresis buffer.

1 × TBE is: 89 mM Tris-borate
             2.5 mM EDTA
             pH 8.3 at 20°C

To make up 5 litres of 5 × TBE, dissolve:
             272.5 g Tris base
             139.1 g boric acid
             23.3 g EDTA (MW 372.24)
          in: 4.5 litres distilled water

Check the pH and make up to a final volume of 5 litres. If available 125 ml of 0.5 M EDTA, pH 8 may be used instead of the solid.

---

in any desired combination of widths. The sample well thickness should be slightly less (0.02 mm) than the gel thickness for gels more than 2 mm thick. Very low percentage agarose gels (0.3%) are difficult to run in a vertical apparatus since they tend to collapse. This can be prevented by using a deep anodic reservoir which allows one to use similar buffer levels in both reservoirs.

A horizontal slab gel electrophoresis apparatus (*Figure 2*) will place less mechanical stress on the gel, but in the authors' experience better resolution is obtained using the vertical apparatus. In a vertical apparatus the sample spreads itself over the whole cross-sectional area of the gel slot, but in a horizontal gel this will only be achieved by careful adjustment of the sample volume for each slot size. However, if the highest resolution is not required, many workers find it is easier to use a well-designed horizontal apparatus routinely; and the authors use both the GNA series (Pharmacia-LKB) and the H series (Gibco-BRL), as well as designs of our own construction.

2.1.2 Power pack

A power supply delivering up to 200 mA DC at 200 V is suitable for nearly all applications, except sequencing gels.

2.1.3 Peristaltic pump

It may be necessary to circulate buffer between the reservoirs to prevent pH changes arising from electrolysis of the buffer; a pump with a flow rate of 1 ml/min is sufficient.

2.2 Buffers for non-denaturing gels

For polyacrylamide gels, 89 mM Tris-borate buffer (pH 8.3), 1 × TBE (*Table 1*), is a good and common choice, and this borate buffer has become increasingly popular for use with agarose gels also. However, with some types of agarose, borate−agarose complexes may form and cause high electroendosmosis and gel damage. This is not a problem with the types of agarose specified later in this chapter and 0.5 × TBE or 1 × TBE buffer may be used routinely. Acetate-based buffers, for example 50 mM Tris-acetate (pH 7.5−8.0) or 50 mM sodium acetate−acetic acid (pH 7.5−8.0), are also popular. Acetate is oxidized to carbonate during electrophoresis, raising the pH, but, especially where DNA is to be recovered, acetate buffer is a good choice when combined with buffer replacement or recirculation as necessary. Alternatively, it is possible to use phosphate-based buffers, either 50 mM Tris-NaH$_2$PO$_4$ (pH 7.5−8.0) or 50 mM Na$_2$HPO$_4$-NaH$_2$PO$_4$ (pH 7.5−8.0). The former has a higher buffering capacity; the latter is cheaper but during long runs it may be necessary to circulate the buffer between the reservoirs. These buffers can be prepared as five-fold or 25-fold stock solutions. Phosphate buffers should be avoided when DNA is to be recovered by ethanol precipitation, as the phosphate will also precipitate from solutions at concentrations of greater than about 10 mM. Tris base is oxidized to a UV-absorbing compound during electrophoresis, whilst chloride-containing buffers lead to the generation of chlorine. In all cases, 1-5 mM EDTA should be present to chelate divalent cations, and the pH of the buffer should be checked at intervals during electrophoresis until the characteristics of a particular protocol have been established.

2.3 Buffers for denaturing gels

There are many different denaturing buffer systems but those using urea, alkali, formamide, or methylmercuric hydroxide as denaturants are the most common. Many of the techniques are similar to those used for gel electrophoresis of RNA (see Section 4 of Chapter 1). The choice of denaturant is limited by the type of gel; for example, agarose will not set in the presence of chaotropic solvents such as formamide while alkali deaminates polyacrylamide and methylmercuric hydroxide inhibits its polymerization. Typical denaturing buffer systems for agarose gels are either 30 mM NaOH, 2 mM EDTA (4) or 5 mM methylmercuric hydroxide, 50 mM boric acid, 5 mM sodium borate, 25 mM Na$_2$SO$_4$, 1 mM EDTA (pH 8.2) (5). Typical denaturing buffers for polyacrylamide gels are either 98% formamide containing 20 mM diethylbarbituric acid (pH 9.0) (6) or 7 M urea, 2.5 mM EDTA, 89 mM Tris-borate (pH 8.3) (7).

The choice of buffer system for use with any given set of DNA fragments depends on several factors.

● If absolute molecular weight determination in the absence of secondary structure is required, then methylmercuric hydroxide is probably the best denaturant.

● Methylmercuric hydroxide cannot be used with polyacrylamide and therefore another system must be used to study small fragments. Methylmercuric hydroxide is also very poisonous; the risks in its use have been analysed and precautions must be taken against its toxic and potentially lethal effects (8).

● The alkali method is probably the simplest to use with agarose. The agarose is dissolved in distilled water, cooled to about 60°C, and stock NaOH-EDTA is added to give a final concentration of 30 mM NaOH, 2 mM EDTA at the desired agarose concentration. To use alkali with polyacrylamide is more difficult since deamination of polyacrylamide gel causes high electroendosmosis. However, in an apparatus where the sample is loaded in contact with the electrode buffer, the gel may be cast in 30 mM NaCl, 2 mM EDTA, and then 30 mM NaOH, 2 mM EDTA used as the electrode buffer. The hydroxyl ions migrate faster than the DNA, hence maintaining a denaturing environment for the sample.

- 7 M urea stops agarose from setting, but does not affect polyacrylamide and is a simpler alternative to formamide.
- Special methods have been developed for DNA-sequencing gels but these are outside the scope of this chapter.

2.4 Gel preparation

2.4.1 Agarose gels

Agarose gels are typically 3 mm thick; thinner gels are more difficult to cast and handle. The agarose should, preferably, be of the low/medium (EEO less than 2.0) electroendosmosis type (e.g. Sigma Chemical Co. Type II); at very low concentrations this feature may have to be sacrificed somewhat in favour of higher mechanical strength. Several specialist agaroses are available, offering properties such as low-gelling and low-melting temperatures (Sea Plaque agarose from FMC BioProducts, Easy Plaque agarose from Park Scientific Ltd). Nusieve (FMC BioProducts) can be used to cast gels up to 4% agarose that will resolve DNA fragments smaller than 20 bp, which can otherwise only be resolved by polyacrylamide gels, and whose low melting temperature is useful for preparative separations (see later). These specialist agaroses tend to be much more expensive than standard types.

Protocol 1. Preparation of agarose gels for the vertical apparatus

1. Assemble the apparatus as shown in *Figure 1*, with a 3 mm gap at the bottom of the front plate.

2. Dissolve the agarose in the chosen buffer to give the desired final concentration in an autoclavable bottle. This is done by heating in a boiling water bath (on a bunsen burner or hotplate, with stirring), or in a microwave oven. The latter of these methods is fast and avoids burning; but care must be taken to avoid metal caps on bottles, or sealed bottles. Superheating may occur leading to sudden boiling when the agarose solution is removed from the oven. Handle the bottle with care—a 'Hot Hand' (Bel-Art Products) is useful. When standardization is important, the bottle and contents should be weighed before and after dissolving the agarose, and losses in the course of boiling compensated for by adding distilled water.

3. Cool the solution to 45°C for less than 1% agarose or to 50°C for greater than 1% agarose.

4. Tilt the apparatus back to about 60° to the vertical and seal the bottom of the gel mould by pouring agarose into the gap at the bottom of the front plate so that it fills up about 10% of the gel mould. Avoid air bubbles in the agarose solution. For gel concentrations less than 1% use a 1% agarose plug, for those greater than 1% use an agarose plug of equal concentration.

Protocol 1. *continued*

5. Allow the agarose plug to set at room temperature for about 15 min, keeping the rest of the gel mixture above the setting temperature.
6. Pour the rest of the gel, filling the gel mould from the top.
7. Quickly place the sample well-former into the molten agarose; avoid trapping air bubbles at the bottom of the teeth.
8. The agarose will shrink as it sets and so add additional agarose as necessary.
9. Allow the gel to set and then allow it to cool for about 60 min.
10. Place the tank with its attached gel mould in a tray of buffer and then fill the tank with buffer.
11. Remove the sample well-former; slight tilting and twisting of the former is necessary to let the buffer into the partial vacuum created as the teeth are pulled up.

Thin agarose gels, less than 1 mm thick, require a different procedure as described in *Protocol 2*:

Protocol 2. Preparation of thin agarose gels for the vertical gel apparatus

1. Assemble the buffer tank, spacers and comb, keeping the top plate separate.
2. Heat the assembly to 60°C.
3. Lay the buffer tank down the back gel plate uppermost.
4. Pour a 'puddle' of agarose on to the back plate, then quickly and carefully lower the top plate into position.
5. Clip the gel plates together, insert the sample well-former and allow the gel to cool and set as described previously for 3-mm-thick gels.

For the horizontal apparatus the gel is prepared as follows:

Protocol 3. Preparation of agarose gels for a horizontal apparatus

1. Make the agarose solution as described in *Protocol 1*.
2. Cast the gel in any conveniently shaped tray or dish. The sample well-former can be suspended between two fold-back clips resting on the sides of the tray with the teeth held about 1 mm above the bottom.
3. After cooling the gel can be transferred from the mould on to the platform for electrophoresis by the submarine technique.
4. Commercial apparatus should be used according to manufacturer's instructions.

Table 2. Stock solutions for the preparation of polyacrylamide gels.

A — 40% acrylamide

B — 2% bisacrylamide

C — 10 × TBE (0.89 M Tris-borate, 25 mM EDTA, pH 8.3)

D — TEMED

E — 10% ammonium persulphate (freshly prepared)

Table 3. Recipes for the preparation of non-denaturing polyacrylamide gels.

Stock solution	Final polyacrylamide gel concentrations (%)		
	5	12	20
A (ml)	12.5	30.0	50.0
B (ml)	12.5	20.0	33.0
C (ml)	10.0	10.0	10.0
D (ml)	0.05	0.05	0.05
E (ml)	1.0	1.0	1.0
Distilled water (ml)	63.95	38.95	5.95

The final volume in each case is 100 ml.

2.4.2 Polyacrylamide gels

Normally these gels are not more than 1 mm thick. The compositions of the stock solutions are given in *Table 2* and recipes for non-denaturing and denaturing polyacrylamide gels are given in *Table 3* and *Protocol 5*, respectively. Gels of differing pore size are prepared by varying the amount of acrylamide (stock Solution A) used. Varying the amount of bisacrylamide crosslinker (stock Solution B) also affects the gel pore size but it also affects the brittleness and clarity of the gel. Typically, a non-denaturing polyacrylamide gel of 12% acrylamide, 0.4% bisacrylamide is used to fractionate nucleic acids 10 – 100 nucleotides long (the gels are electrophoresed at a constant voltage of 10 V/cm for 18 h); further details of the electrophoresis of oligodeoxynucleotides are given in Chapter 4. Pre-mixed electrophoresis grade acrylamide and bisacrylamide solids, in a range of ratios, are available (Electran range, BDH). These may be used to make stock solutions without any weighing out of the dangerous acrylamide monomers.

Slab polyacrylamide gels are prepared as follows:

Protocol 4. Preparation of non-denaturing polyacrylamide gels for the vertical apparatus

1. Clean the front and back gel plates with ethanol and dry them using a clean tissue.

Protocol 4. *continued*

2. Siliconize the back plate only.

3. Assemble the apparatus and seal the bottom of gel mould with 1% agarose (Section 2.4.1).

4. Mix the stock acrylamide solutions with the stock buffer and degas, to remove oxygen which inhibits polymerization (for this reason polyacrylamide gels cannot be cast in horizontal open top trays like agarose). Add the polymerization initiator and catalyst (Solutions D and E) and mix well.

5. Pour the gel with the buffer tank tilted forward on one corner so that the gel mould fills from that corner.

6. Position the sample well former. A close-fitting former reduces the diffusion of oxygen into the gel and leads to more even polymerization.

The gel will polymerize within 60 min. The polymerization rate of polyacrylamide gels is temperature dependent. Adding larger amounts of initiator and catalyst, usually TEMED and ammonium persulphate, also causes quicker polymerization.

Protocol 5. Preparation of denaturing polyacrylamide gels

Urea-containing gels

Prepare these as for non-denaturing gels (*Table 3*) except that the volume of distilled water is reduced to allow 42 g of urea to be added to give a final volume of 99 ml. Then add 0.05 ml of solution D and 1.0 ml of solution E immediately before pouring the gel. The final gel mixture contains 7 M urea.

Formamide-containing gels

1. Prepare buffered deionized formamide by adding 20 ml of deionized formamide to 92 mg of diethylbarbituric acid and adjusting the apparent pH to pH 9.0 with 1 M NaOH. Then adjust the volume to 25 ml with deionized formamide.

2. Dissolve the samples in buffered deionized formamide and add one-tenth volume of 20% Ficoll, 0.25% bromophenol blue.

3. Prepare the gel mixture using the following stock solutions.

Stock solution	Final polyacrylamide concentration	
	4%	10%
A (ml)	10.0	25.0
B (ml)	0.1	0.1
diethylbarbituric acid (g)	0.37	0.37
deionized formamide (ml)	80.0	80.0
D (ml)	0.24	0.24
E (ml)	1.50	1.50

Adjust the apparent pH to pH 9.0 with 1 M HCl and add deionized formamide to a final volume of 100 ml.

Protocol 5. *continued*

4. Pour the gel (degassing is unnecessary) and place the sample well former in position.
5. When the gel has set, remove the sample well-former and fill the wells with buffered deionized formamide.
6. Load the samples into the wells under the buffered deionized formamide.
7. Fill the buffer reservoirs with 20 mM NaCl.
8. Electrophorese for the required time with continuous buffer recirculation.

2.4.3 Agarose – acrylamide composite gels

For nearly all purposes, either pure agarose or pure polyacrylamide gels will give equally good results. However, low concentration polyacrylamide gels as used in some separations are very weak and these must be strengthened by the addition of agarose. This is also necessary when polyacrylamide is used in the preparative gel apparatus and so this type of gel is described later (Section 3). The use of Nusieve (FMC) extends the size range for fractionation using agarose well into the range in which acrylamide has been the only choice previously.

2.5 Sample preparation and loading of the gels

The maximum amount of DNA that can be loaded on to a gel (that is, the capacity of a gel) depends on several factors.

- Well size: the capacity of the gel is proportional to the area of the base of the sample well, in a vertical apparatus.
- Fragment size: the capacity drops sharply as the fragment size increases, especially over a few thousand base pairs.
- Distribution of DNA fragment size: the highest capacity is when the DNA fragments have a continuous, even distribution across a wide size range. Roughly speaking, 50 μg/cm^2 of the area of the sample well base is the upper limit for an *Eco*RI or *Hin*dIII digest of a complex DNA on an agarose gel. The limit for a digest of a viral DNA yielding few fragments of different sizes might be only 10 μg/cm^2.
- Voltage gradient: higher voltage gradients generally lead to poorer resolution and hence lower capacity, especially for larger DNA fragments.

Most DNA preparations in buffers of low ionic strength and at concentrations of less than about 1 μg/μl can be loaded directly on to gels by adding one-tenth volume of the following 'stop mix': 30% Ficoll (Pharmacia-LKB Ltd.), 0.25% Orange G, 250 mM EDTA. Ficoll is better than low molecular weight compounds (e.g. sucrose or glycerol) for increasing the density of the sample for loading in that it does not 'stream' up the sides of the sample well and so the resultant DNA bands have a less significant 'trailing edge'. Orange G should be added to make the sample easily visible and to act as a marker close to the migrating front of smaller-size DNA fragments. Its position relative to the DNA fragments depends very much on the gel concentration. Other tracking dyes, for example bromophenol blue (0.025%), can also be used. For alkaline gels, bromocresol green (0.025%) and xylene cyanol F.F. (0.025%) are suitable. EDTA is added in order to complex divalent cations. When this mixture is added to restriction digests it stops the reaction and the digest can be loaded directly on to the gel without further treatment. For some DNA fragments it is desirable to denature any 'sticky ends' present and this can be done by heating at 65°C for 5 min followed by rapid cooling in an ice−water mixture. Sometimes it is necessary to estimate the amount of DNA loaded on to each track. In the absence of careful initial spectrophotometry and dispensation, this may be achieved by using the dye Hoechst 33258 (bis-benzimidazole) which binds to DNA, leading to a DNA specific enhancement of UV-induced fluorescence (source 365 nm, emission 458 nm); contaminating proteins, or RNA do not cause fluorescence. The samples can be run out on a small gel quickly and the amount of DNA in each track estimated by scanning densitometry of the gel photograph (9). Alternatively, a device specifically designed to measure this fluorescence may be used (e.g. TKO 100, Hoefer Scientific Instruments).

If the sample contains a high concentration of salt, or needs to be concentrated, or is contaminated, the procedures described in *Protocols 6, 7* and *8* may be useful.

Protocol 6. Removal of salt from DNA samples

This procedure is useful for fractions from CsCl gradients, although not essential if some retardation and distortion of DNA bands is acceptable.

1. Add 3 volumes of 70% ethanol to the sample, mix and leave at −20°C for at least 60 min. For very small amounts of DNA (<1 μg), it may be useful to add *E. coli* tRNA as a carrier.

2. Centrifuge the solution at 10 000 *g* for 5 min (e.g. using a microcentrifuge) and discard the supernatant.

3. Wash the pellet with 2 volumes of 70% ethanol and recover the pellet by centrifugation as in Step 2.

4. Dry the pellet under vacuum since traces of ethanol in a sample can reduce its density to the extent that the sample floats out of the well during loading.

5. Dissolve the pellet in the required volume of 10 times diluted 'stop mix' (see Section 2.5).

Protocol 7. Methods for the concentration of DNA samples

A. *Standard procedure*

1. Add NaCl to a final concentration of 0.3 M.

2. Precipitate the DNA by the addition of 2 volumes of 95% ethanol.

Protocol 7. *continued*

3. Wash the pelleted DNA with ethanol, dry it under vacuum and dissolve the DNA in buffer as described in *Protocol 6*.

B. *Procedure for large volumes of dilute DNA[a]*

1. Add 2−2.5 volumes of butan-2-ol to the DNA sample, shake and remove the lower, aqueous phase containing the DNA.

2. Re-extract the aqueous phase with 1 volume of chloroform. The upper (aqueous) phase contains the DNA now in a reduced volume.

3. Recover the DNA by ethanol precipitation as in the standard procedure. The degree of concentration depends on the initial salt concentration; distilled water is completely miscible with butan-2-ol, whereas if 0.001 M NaCl is extracted its volume is reduced about fivefold by the above procedure. Salts are concentrated along with the DNA.

[a] This procedure avoids the necessity of adding carrier to obtain efficient precipitation of DNA in dilute solutions.

Protocol 8. Removal of impurities from DNA samples

1. Extract the DNA sample twice, each time with an equal volume of buffered, redistilled phenol.

2. Extract twice with two volumes of diethyl ether. Allow traces of ether to evaporate.

3. If necessary precipitate the DNA with ethanol as described in *Protocol 7A*.

In some cases, an alternative to this method is to pellet the DNA directly by centrifuging it at about 40 000 g for a few hours, the exact time depending on the amount, size and concentration of the DNA.

To apply the sample to the gel it should be taken up in a narrow-tipped dispenser; for example, a 0−200 μl automatic pipettor or a glass capillary tube. The apparatus should be already filled with buffer. Carefully direct the dispenser tip into the open top of the sample well and slowly eject the sample so that it sinks to the bottom and forms a narrow, dense layer. Viscous samples are often difficult to load because trapped bubbles of air tend to lift the sample out of the well. This can be avoided by removing bubbles in the sample by centrifugation in a microcentrifuge prior to loading the sample and using a dispenser with a somewhat wider hole; that is, using a standard plastic tip with about 0.5 cm cut off.

2.6 Size markers for gels

Dyes such as orange G and bromophenol blue can be used as fairly accurate size markers for small DNA fragments. Their mobility relative to a given size fragment of DNA does, however, vary with the type of gel and running conditions (see *Table 1* of Chapter 4). For accurate size estimation it is necessary to have a range of DNA fragments of known size. These can easily be made by preparing restriction digests of viral or plasmid DNAs. The entire base sequence of several of these are now known; for example, the bacteriophage lambda cI ts 857 (10) and the plasmid pBR322 (11). Hence the size of various restriction fragments can be determined precisely. Care should be taken to use markers giving good coverage of the size range of interest; for example, an *Eco*RI/*Hind*III double digest of bacteriophage lambda cI ts 857 gives marker fragments ranging from 21 226 to 125 bp (see later, *Figure 14*) with a gap between 5148 and 21 226 bp. Other digests are better in this size range; for example, *Bgl*II gives fragments of 22 010, 13 286, 9688, and 2392 bp. It should be noted that different strains of bacteriophage lambda may generate different size fragments on digestion. Appendix I includes a list of DNA restriction fragments that can be used as markers.

Concatemers of plasmids or phage, generated by ligation, or by reassociation of sticky ends, can be used to extend the marker range to several hundred thousand base pairs. Thus dv21 (12), which has a monomeric size of 3153 bp, grows as an oligomer in an appropriate host. On partial digestion one obtains a series of seven fragments, each a multiple of the monomer in size up to a maximum of 22 071 bp.

It is useful to store stock marker solutions containing 0.1 $\mu g/\mu l$ of DNA in loading buffer (see 'stop mix', Section 2.5) at $-20°C$. A few microlitres of this will give easily visible bands after staining, although because all the fragments in digests of simple genomes are present in equimolar amounts, visualization of small fragments requires higher loadings of total DNA than visualization of large fragments.

2.7 Electrophoresis conditions

Most gels are electrophoresed at room temperature and no cooling is needed at the current and voltage normally used. If overheating occurs during electrophoresis, the DNA bands will appear distorted. Overheating is detectable by touching the front of the gel mould, after disconnecting the power supply. Immersing the tank in buffer with circulation helps to cool the gel and even out any temperature differences. A small temperature gradient across the thickness of the gel can produce tilted bands that are broadened in subsequent photographs and blots. The amount of heat generated depends on the dimensions of the gel and the current. Reducing the buffer concentration will reduce the current and hence the heating, at constant voltage. The optimum voltage gradient in turn depends on the degree of resolution required, fragment size, and time available. In general, large DNA fragments are best resolved by electrophoresing gels for long times at low voltage gradients. Smaller fragments diffuse faster and therefore band sharpness is increased by electrophoresing the gels using high voltage gradients, and a balance has to be struck between sharpness and separation. The voltage gradient normally used is in the range $1-10$ V/cm.

There are analogous situations with regard to gel length and gel concentration. The further that fragments migrate, the greater is the separation between them. However, increasing the migration distance requires either longer times or higher voltage gradients which may lead to increased diffusion of small fragments or less separation of large fragments, respectively. In general, separation is better on gels of higher concentration, for equivalent distances migrated. However, large DNA fragments move so slowly in concentrated gels that lower concentration gels must be used. In order to obtain the highest resolution of detail, sample wells up to 2 cm wide are advisable.

2.8 Analysis of gels after electrophoresis

2.8.1 Staining the gels

i. Vertical apparatus

After electrophoresis, empty the buffer from the buffer tank and remove the metal clips from the gel mould. Place the tank with the gel mould resting face down and gently lever off the front plate so that the gel comes with it, peeling off from the back plate. You can then use the plate to transfer the gel to a suitable staining tray. Agarose gels of concentration greater than 1% can be handled without too much difficulty; those less than 1% are more fragile. It may help to use a Perspex tray with a black Perspex bottom, slightly larger than the gel; once in this tray the gel can be stained, drained, and photographed using incident illumination with a minimum of handling. Trays of UV-transparent plastic may be used with transilluminators.

Remove the gels from gel moulds with fixed front plates using the following procedure:

Protocol 9. Removal of gels from gel moulds with fixed front plates

1. Cut the gel along the bottom of the front plate with a scalpel and discard the strip of agarose from the bottom few millimetres of the gel.

2. Insert a thin (1−2 mm) metal rod down one side of the gel and press gently inwards so that the gel is teased away from the edges and contact with the glass plates is loosened.

3. Repeat Step 2 for the other edge of the gel. The gel will now, usually, slide out quite easily.

4. If difficulty is experienced, the gel can be pushed out, from the bottom, with a wide, flexible plastic strip (e.g. used X-ray film with a 2 mm or 3 mm strip bent over at one end).

Neutral pH gels can be stained directly by immersion in electrophoresis buffer containing 0.5 μg/ml ethidium bromide (equivalent to about 1 drop of a 10 mg/ml stock solution per litre of buffer). For a 3-mm-thick 1% agarose gel, staining for 30−60 min with occasional rocking is sufficient. For the ethidium bromide to penetrate the gel completely takes longer. If a low background is required, the gel should be rinsed in fresh buffer or destained in fresh buffer (30 min with occasional rocking). Prolonged destaining causes losses of DNA-bound dye. Alkaline gels must be washed in a neutral buffer (twice for 30 min) before staining.

Caution: ethidium bromide is a powerful mutagen and gloves should be worn when handling gels in the staining solution (13).

ii. Horizontal apparatus

Generally, the gel can be removed from the apparatus easily. Many workers like to add ethidium bromide to both the gel solution (0.5 μg/ml, achieved by adding 5 μl of 10 mg/ml stock solution per 100 ml), before it is poured, as well as to the buffer. The dye will have some effect on DNA mobility but the gel can be viewed directly during electrophoresis. This is especially easy if the gel tray, used for both running and casting, is made of UV-transparent plastic and can be used to move the gel to and from a transilluminator.

2.8.2 Viewing and photography of gels

The complex formed between ethidium bromide and DNA has a fluorescence excitation spectrum with a maximum at 302 nm; the fluorescence of DNA-bound ethidium bromide is at least ten times greater than that of free ethidium bromide and its maximum is in the red (590 nm). It is thought that UV light absorbed by the DNA is transferred to bound ethidium bromide and re-emitted at the longer wavelength. In addition to the fluorescence of the DNA-bound ethidium bromide, the ethidium bromide in the gel (agarose or polyacrylamide) also gives background fluorescence. Lamps are available with emission spectra maxima at 254 nm, 302 nm, or 366 nm. Illumination at 254 nm causes photo-nicking, dimerization and rapid bleaching of the complex; these effects are relatively absent if higher wavelength sources are used (14). However, the 366 nm source produces fluorescence much less efficiently than those at 302 nm and 254 nm. Therefore, for general use, illumination at 302 nm is to be preferred. Unfortunately, lamps with 302 nm maxima also emit light in the red end of the visible spectrum which must be removed by additional filters. Thus 302 nm lamps may not give the same sensitivity as 254 nm lamp for the detection of faint bands. If the DNA is high molecular weight and is to be transferred to a membrane filter, then the nicking resulting from illumination at 254 nm may improve the efficiency of the transfer.

Two types of illumination system are in common use, using incident and transmitted UV light, respectively. The incident system works well with unfiltered 254 nm lamps (each 1.5 kW), with the gel on a black glass or Perspex plate, and a red filter (Hoya 25A) on the camera. Exposure time is about 3 min at *f*5.6 with 12.5 cm × 10.0 cm panchromatic film (Kodak Plus-X Pan Professional) (*Figure 3a*).

The transillumination system (*Figure 3b*) works with a UV-pass, visible-blocking filter (Corning) attached to a 302 nm lamp. The gel can be placed on a UV-transparent plastic sheet (e.g. Plexiglass 218); this is useful if a gel slice is to be removed for fragment isolation because the blocking filter may be easily damaged by scalpel blades. A diffusing sheet between the UV-pass filter and the plastic sheet may be necessary. The camera set-up is similar to that used with incident illumination. An alternative is to use a less stringent UV-pass filter with an interference filter on the camera. With a commercial apparatus (UVP Inc.) the combination of an orange filter (Kodak 23A) and Kodak TMAX 100 professional film gives a good result after only

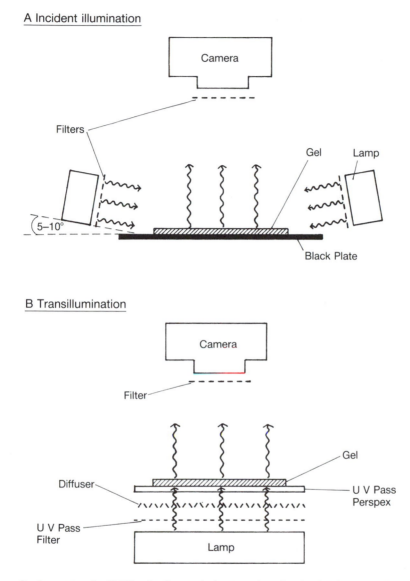

Figure 3. Apparatus for UV illumination and photography of stained gels. A, incident light illumination system; B, transillumination system.

20 sec exposure, although the quality of the negative is not as high as can be achieved with the overhead illumination system.

Faint bands detected by long exposure may show reciprocity failure, that is, the film response is less than proportional to the light emitted. Pre-fogging the film with an attenuated electronic flash (to give a fog level of $A_{540} = 0.05$) overcomes this problem; after pre-fogging, the film gives a linear response and shows enhanced sensitivity.

If difficulty in focusing the camera is experienced, oblique lighting with a desk or spot lamp, and a small strip of paper tissue laid on one edge of the gel, or a ruler laid alongside the gel, will help.

It is very important that the eyes are protected at all times from UV radiation by a suitable mask or goggles. In addition, prolonged exposure of the skin to UV light causes burns. Finally, ozone, which is toxic, is generated by UV lights and so effective ventilation is necessary.

2.8.3 Estimation of fragment size and quantity

Fragment sizes are estimated from their mobilities relative to those of fragments of known sizes. For rough measurements these values can be measured directly from the negative of the gel photograph. If the gel is photographed with a ruler alongside, the mobility values can be read off directly from the photograph. For more precise work a microdensitometer is necessary to obtain a density profile on an expanded scale, and then mobility values (distance from origin to peak) are measured with a ruler; some machines can do this automatically. Alternatively, the photograph can be projected or printed as an enlargement prior to making any measurements. A plot of mobility (M) against molecular weight or size (L) will give a curved line, which can be used for reading off size values for given mobilities. Interpolation is much more accurate if a function can be found which gives a straight line. A plot of $\log(L)$ against M is most commonly used, but in our experience a plot of L against $1/M$ generates a straight line over a wider range, for double-stranded and single-stranded DNA run at low voltage gradients (15). At higher voltage gradients, a plot of L against $1/M$ shows marked curvature. A comparison of two methods of dealing with this curvature, a least-squares analysis (16) using all the standard points available for each determination, and a local 2×3 point method (17), has revealed that the latter is more accurate (18). In this method the formula used is:

$$L = k_1/(M-M_0)+k_2$$

where:

$$M_0 = \frac{M_3-M_1\ ((L_1-L_2)/(L_2-L_3)\ \times\ (M_3-M_2)/(M_2-M_1))}{1-((L_1-L_2)/(L_2-L_3)\ \times\ (M_3-M_2)/(M_2-M_1))}$$

$$k_1 = \frac{L_1-L_2}{1/(M_1-M_0)-1/(M_2-M_0)}$$

$$k_2 = L_1-k_1/(M_1-M_0)$$

M_0 is an empirical correction factor determined by imposing the condition that the three lines joining the three points all have the same slope. L_1, L_2, and L_3 are the three molecular weight standards with mobilities M_1, M_2, and M_3 which most closely span the unknown molecular weight of mobility M. The estimation is performed twice for each value. The first calculation uses two markers that are larger, and one smaller, than the unknown; the second uses two markers that are smaller, and one larger. The final result is the average of these two values. These operations can be performed quickly on many electronic calculators, and a simple program can be used such that for any given set of standard mobilities only one step is required to obtain a size estimate from a mobility measurement.

The microdensitometer tracing can be used also to estimate the relative amounts of DNA in bands, by measuring the areas under the peaks using any one of several methods (cutting out and weighing, planimetry, etc.). To estimate the amount of DNA, calibration against known standards is necessary. Since the density of bands on a developed film does not vary linearly with DNA, the quantity standards used for calibration should span the range of intensities of the bands to be measured. Restriction digests of bacteriophage or plasmid DNA often yield a suitable series of bands whose DNA contents are directly proportional to their molecular weight, and thus these can be used to construct the required curve of film responses (peak area) to the amount of DNA. Caution is advisable, however, as reciprocity failure can lead to underestimation of a broad band as compared with a sharp one containing the same amount of DNA (19).

2.9 Recovery of DNA fragments from agarose gels

Techniques for the recovery of DNA fragments from agarose gels abound, but few are completely satisfactory. The difficulties arise because microgram (or even nanogram) quantities of DNA have to be recovered from a gel which may contain several milligrams of the gel material. The bands must be located accurately and a gel slice removed. Ideally, the recovered DNA must be free of contaminants which could inhibit subsequent enzyme-mediated reactions (e.g. kinase labelling) or single-strand nicks which would reduce molecular weight (e.g. during nick translation) and features such as single-stranded ends should be retained (e.g. for 'sticky end' ligation). The quality of the agarose used is important because those with a high sulphate content contain the largest amount of enzyme inhibitors. In some cases the quality of agarose may vary from batch to batch. An important development has been the observation that many enzyme reactions can be carried out in dilute solutions of low melting point agarose (20,21). This means that often, once a band has been cut out of a gel, no purification is needed. The simplest method of purification is to pass the dilute solution (usually 3 vol. of low salt buffer to 1 vol. of gel) through an Elutip-D (Schleicher & Schuell) mini-column (see *Protocol 3* of Chapter 7), the DNA binds to the column while the agarose solution is washed through. The DNA is subsequently eluted in a small volume and can then be ethanol precipitated. If these methods are not adequate, then the authors would recommend one of the more time-consuming procedures detailed below; alternatively, one can try using the 'crush and soak' method (Section 2.8 of Chapter 5). When monitoring the recovery of DNA, UV lamps with emission maxima of 302 nm or 366 nm must be used to minimize damage to the DNA.

Protocol 10. Isolation of DNA from low gelling temperature agarose (22)

1. Cast the desired type of gel using low gelling temperature agarose. Allow the gel to set at 4°C; load and electrophorese the DNA using acetate buffer, preferably at 4°C.

2. Stain the gel with ethidium bromide and identify the band of DNA to be recovered using a 302 nm or 366 nm lamp; cut the DNA band out of the gel. The resolution obtained may not be as good as on a normal agarose gel.

3. To the slice add an equal volume of 200 mM NaCl. Gently macerate the gel using a glass rod.

4. Heat the gel slurry to 65−70°C with occasional shaking until the agarose has melted.

5. Add an equal volume of buffered redistilled phenol at 20°C and then shake the mixture gently for 20 min; a white emulsion forms.

6. Centrifuge the mixture at 15 000 *g* for 10 min. A dense white layer of agarose forms at the interface.

7. Remove the upper aqueous layer and re-extract it with phenol, then remove residual phenol by extracting the aqueous layer twice with two volumes of diethyl ether. Precipitate the DNA with ethanol (see *Protocol 7*).

Protocol 11. Use of agarase for DNA isolation (23)[a]

1. Proceed as in *Protocol 10* until the gel slice has been melted.

2. Add 50 units agarase/ml (Calbiochem 121814, DNase free). Incubate overnight at 37°C.

3. Phenol extract, etc., as described in *Protocol 10*.

[a] This adaptation of the previous procedure produces higher yields with less risk of agarose contamination.

Protocol 12. Electroelution on to dialysis tubing (24)

This may be used with either agarose or polyacrylamide gels.

1. The gel can be run using any buffer except those based on phosphate.

2. Remove the gel slice as described previously.

3. Place the slice inside a piece of dialysis tubing, closed at one end with a tubing clip. Squeeze the piece of gel to one side of the tubing. Add one gel slice volume of 25 mM sodium acetate buffer (pH 8) and close the tubing with another clip, avoiding bubbles.

Protocol 12. *continued*

4. Place the assembly on the platform of a horizontal gel apparatus, with the tubing oriented across the electric field and held down by the two clips, and with the gel slice closest to the cathode.

5. Just immerse the tubing in 25 mM sodium acetate buffer (pH 8) and electrophorese at 10 V/cm for 1 h. The DNA migrates out of the gel and gathers on the inside of the tubing on the anode side. If ethidium bromide is present the progress of electroelution can be visualized by illumination with long-wave UV light.

6. Reverse the field polarity for 2 min to release the DNA from the inside of the membrane.

7. Unclip one end of the tubing and carefully remove the buffer; flush the tubing with one volume of buffer and pool the buffer solutions. The DNA can be precipitated with ethanol (*Protocol 7*). Alternatively, the material may be recovered in a small volume, using an Elutip-D, or equivalent, which has the advantage of filtering out any fragments of agarose recovered in the electroelution buffer.

2.10 Analysis of DNA fragments using hybridization

DNA can be transferred from agarose gels to sheets of nitrocellulose or chemically activated paper (e.g. DBM-paper), or more commonly nowadays a Nylon-based membrane, in such a way as to retain the original pattern. The filter so obtained can then be probed with radioactive DNA or RNA, resulting in the formation of a pattern of radioactive bands where the probe has re-associated with complementary filter-bound DNA. This pattern can then be detected by autoradiography. The following protocols describe general procedures for the transfer of DNA to nitrocellulose and subsequent hybridization. Examples of procedures for the use of DBM-paper and for autoradiography are given elsewhere in this volume (Section 7.3 of Chapter 1). Nylon membranes are represented by Hybond-N (Amersham International), a Nylon-66 membrane, and Zeta-Probe (Bio-Rad), a quaternary amine derivatized Nylon membrane, and specific reference is made to these, as well as a number of alternatives for each part of the process. Nylon membranes offer several advantages over nitrocellulose, including higher tensile strength, durability, higher DNA binding capacity and retention, the most important result of these features being the ability of these membranes to be re-probed several times. Another useful feature is that the low inherent fluorescence of Nylon membranes allows visualization of bands directly by UV illumination (35). The derivatization of Zeta-Probe allows blotting to take place in 0.4 M NaOH, considerably reducing the number of steps involved, as well as improving resolution, both by reducing diffusion and by preventing re-association during transfer. In the subsequent sections the description of the basic blotting and hybridization procedure, as used with nitrocellulose, has been retained, since this provides a reference point. Manufacturers all provide their

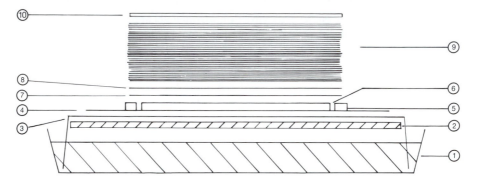

Figure 4. Cross-section of a transfer apparatus (1) tray containing 20 × SSC; (2) sheet of glass or plastic; (3) thick filter paper with wicks; (4) sheet of thick filter paper; (5) plastic strips surrounding gel; (6) gel; (7) sheet of nitrocellulose; (8) sheet of light filter paper; (9) stack of paper towels; (10) glass plate.

own manuals and these must be read before changing to a new type of membrane. Different membranes often cost substantially different amounts, both to purchase and to use.

2.10.1 Apparatus for capillary transfer

A suitable transfer apparatus is shown in *Figure 4*. The dimensions of the tray can be chosen to suit the size of gels used, but the authors have found that a deep plastic tray (27 × 46 cm external dimensions) will accommodate the largest gels normally used, or several smaller ones. If desired, only a small portion of the surface area of the transfer apparatus need be used at one time as long as the unused surface is covered to prevent drying, for example with a polythene sheet. In this apparatus a sheet of plate glass overlaps, and is supported by, the short ends of the tray. A sheet of thick filter paper (e.g. Whatman No. 17) overhangs the long sides of the glass so that the hanging parts, which will act as wicks, dip into the tray which is filled with 20 × SSC (1 × SSC is 0.15 M NaCl, 0.015 M sodium citrate, pH 7.0) or the appropriate buffer for the membrane in use. A second sheet of the same filter paper covers the horizontal surface of the first. This can be replaced easily from time to time if it becomes damaged or dirty.

2.10.2 Alternatives to capillary transfer

There are several alternatives to capillary transfer, outlined in this section; however, the original procedure is still hard to beat for producing a good result simply and inexpensively although it may take significantly longer. The other methods used for blotting are as follows:

i. Electrophoretic transfer (25)

An electric field is used to drive the DNA on to the membrane. Transfer must take place at low ionic strengths, or high currents will cause excessive heating. This is not possible with many membranes. Electrophoretic transfer is necessary for transfers out of acrylamide gels, because the gel matrix is too dense for capillary flow to move the DNA (except for very small fragments). Many commercial devices are available for electroblotting (for details see Section 5 of Chapter 8).

ii. *Vacuum blotting (27)*

A low vacuum sucks solution through the gel, accelerating the flow and speeding up the transfer of DNA fragments. The procedure requires a special apparatus and careful control of the vacuum. The size of the vacuum bed limits the area of transfer.

iii. *Centrifugal transfer (26)*

Moderate centrifugal force drives the buffer together with the DNA out of the gel and on to the membrane. The procedure is fast and simple, but requires a centrifuge with microtitre tray carriers. The gel size is limited to the size of a microtitre tray.

2.10.3 Preparation of the gel for transfer

For efficient transfer, the single strand size of the DNA must not be too high. Therefore, large molecules (greater than 20 kb) must first be nicked either by using a 254 nm UV lamp during gel photography or, if 302 nm lamps are used, the DNA can be partially depurinated by immersing the gel in 0.24 M HCl for up to 30 min prior to denaturation (28). The efficiency of the treatment in cleaving the DNA to the required size should be checked for different gel types and concentrations, and it may be necessary to check for excessive cleavage. For Zeta-Probe two 5 min treatments are recommended (Bio-Rad). Individual UV lamps may need to be calibrated.

Prior to transfer, denature the DNA. Normally this is done by immersing the gel in alkali, then neutralizing the gel. An alternative is to do the transfer in an alkaline solution if using a suitable Nylon membrane (N.B. nitrocellulose disintegrates in alkali). The standard procedure is described in *Protocol 13*.

Protocol 13. Denaturation of DNA in gels

1. Place the gel on a glass or plastic sheet and immerse it in 1.5 M NaCl, 0.5 M NaOH in a dish. Rock the dish every few minutes for 30 min.

2. Drain off the solution, rinse briefly with water and immerse the gel in 3 M NaCl, 1.0 M Tris-HCl (pH 8.0). Rock the dish every few minutes for 30 min.

3. Drain off the solution. The gel is now ready for the DNA transfer.

2.10.4 Capillary transfer of DNA from gels

Protocol 14. Capillary transfer of DNA

1. Pour 20 × SSC on to the surface of the transfer apparatus until it is just flooded.

2. Slide the gel on to the surface without trapping air bubbles.

3. Surround the gel with strips of plastic or glass, of about the same thickness as the gel and about 2 mm away from it.

Protocol 14. *continued*

4. Cut a sheet of nitrocellulose 1 cm wider and 1 cm longer than the gel; wet it by floating it on 2 × SSC.

5. Rinse the nitrocellulose sheet in 2 × SSC, drain it for 30 sec and lay it squarely on top of the gel and its surround without trapping bubbles of air beneath it. Note that if the surface of the apparatus is too wet, liquid bridges may form between the surround and the gel; excess liquid should be mopped up prior to laying the nitrocellulose sheet down. It is most important to keep the filter clean; handle it using gloves or forceps. Avoid changing the position of the filter once it has made contact with the gel.

6. Prepare a sheet of light filter paper (e.g. Whatman 3MM) as for the nitrocellulose and lay this on top of the nitrocellulose sheet.

7. Cover the unused surface of the apparatus with polythene sheet.

8. Stack paper towels on top of the filter paper, 5−10 cm high. Place a glass or plastic sheet on top, and weight this down lightly; a weight of about 100−500 g is sufficient, depending on the gel size. A 3-mm-thick glass sheet the same size as the gel is suitable. The towels soak up the 20 × SSC causing it to flow through the gel, transferring the fragments on to the nitrocellulose membrane. It is usually convenient to allow transfer to take place overnight, although small DNA fragments will be transferred more quickly. The completeness of transfer can be checked by restaining and photographing the gel.

2.10.5 Fixation of DNA to the filter after transfer

After transfer and prior to hybridization, the DNA must be fixed on the membrane; this can be done by baking or UV light.

i. Fixation of DNA by baking

Protocol 15. Baking DNA on to nitrocellulose filters

1. Remove the towels and filter paper.

2. Mark the gel position and orientation on the nitrocellulose with a ballpoint pen.

3. Carefully peel the nitrocellulose sheet back from the gel.

4. Rinse it for a few minutes in 2 × SSC until it is free of any pieces of agarose, etc.

5. Place the nitrocellulose sheet between pieces of thick filter paper, in a sandwich held together with paper clips.

6. Bake in a vacuum oven at 80°C for 2 h.

The sheet can be stored after baking. On prolonged storage the nitrocellulose will begin to oxidize and the sheet becomes brittle; this can be inhibited by sealing the filter in an air-tight plastic bag.

ii. Fixation of DNA using UV light

Ultraviolet fixation is an alternative to baking for attaching DNA to Nylon membranes such as Hybond-N. Two factors may affect the reliability of this process. First, the published protocol recommends wrapping the membrane in Saran Wrap (Dow Chemical Company), and it should be noted that other types of wrapping film may have different UV transmission characteristics. Second, transilluminators vary in the energy of their UV emission and in their spectral characteristics, so it is best to calibrate them using the following procedure (kindly provided by Amersham International).

Protocol 16. Calibration of a transilluminator for UV fixation of DNA to Nylon filters

1. Prepare 500 pg dots of denatured bacteriophage lambda DNA on Hybond-N in 6 × SSC. Use a strip of 6 dots for each chosen UV exposure time.

2. Allow the DNA dots to dry at room temperature.

3. Wrap the blots separately in Saran Wrap and place, DNA side down, on the transilluminator. Each blot should be exposed to UV light for one specific length of time to produce a time course. Time points of 30 sec, 1 min, 2 min, 5 min, 10 min, and 20 min are recommended.

4. Prepare a labelled lambda-DNA probe by standard procedures; for example, nick translation or random oligonucleotide primed labelling.

5. Hybridize the filters under standard conditions (see Amersham International manual).

6. After hybridization and washing, the dot blots may be exposed to film for 3 h or more as appropriate; optimum UV exposure time may be judged by eye. However, more accurate results will be obtained if the blots are dried and each dot individually counted in a scintillation counter.

7. A graph of exposure time versus mean counts per strip of dots should give a clear peak; optimum exposure time may be read off the x-axis.

A typical graph produced at Amersham International, using this procedure is shown in *Figure 5*.

2.10.6 Alkaline blotting (29)

As mentioned previously, some membranes can be blotted directly in alkali, eliminating all the steps between depurination and placing the gel on the transfer apparatus. In the case of Zeta-Probe the alkaline blotting solution is 0.4 M NaOH. Because of the derivatization of the Nylon membranes, the DNA links covalently to the membrane on contact so neither baking nor UV irradiation is necessary to fix the DNA after transfer. The filter should be rinsed in 2 × SSC containing 15 ml/litre neutralizing solution (see *Protocol 13*) before being allowed to air dry

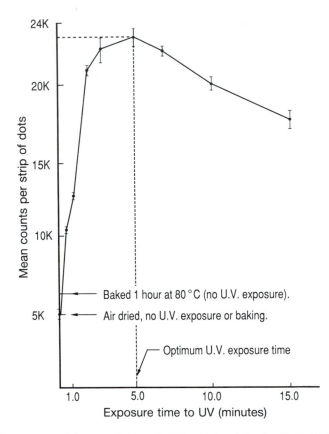

Figure 5. Time course of the cross linking of DNA to Hybond-N using UV light. (Figure kindly provided by Amersham International.)

between sheets of filter paper (e.g. Whatman 3MM). For a description of alkaline blotting using GeneScreen Plus (DuPont) see Section 4.1. of Chapter 3.

2.10.7 Filter hybridization

The hybridization process can be divided into three steps, namely: pre-hybridization, hybridization with a radioactive probe (the use of non-radioactive probes will not be considered here), and washing. In the pre-hybridization step the filter is incubated in a solution designed to block all the sites on it that would bind the probe non-specifically and cause a high level of background radioactivity. The original pre-hybridization mixture was 5 times Denhardt's solution (30) in 5 × SSC. Denhardt's solution contains 0.02% Ficoll (mol. wt 400 000), 0.02% polyvinylpyrollidone (mol. wt 360 000), 0.02% bovine serum albumin. Heterologous DNA can be included (10 – 50 μg/ml) to reduce the background even further. The DNA can be sonicated and/or denatured by heating in boiling water for 10 min followed by cooling rapidly in ice. Wet the filter by floating it on 2 × SSC for a few minutes and then incubate it in the pre-hybridization mixture for 2−20 h at 68°C; usually 0.1 ml/cm^2 of filter is sufficient. Longer pre-hybridization times reduce background but may lead to some loss of the filter-bound DNA. Alternatives to Denhardt's solution are discussed later.

It is convenient to carry out the pre-hybridization in the same vessel as will be used for the hybridization. The authors have used several methods.

i. Method 1

The filter is wrapped around the inside of a cylinder or tube, made of glass or plastic. This is loaded into a device that holds it horizontally and rotates (0.5−2 rpm) so that the liquid sweeps over the filter surface. Cut down measuring cylinders, sealed with a silicone bung or specially made caps, or commercially available tubes may be used. There are many types and sizes that are supplied complete with liquid-tight caps; a small hole should be made in the cap centre before use. The procedure for loading filters into these tubes is given in *Protocol 17*.

Protocol 17. Loading of filters into tubes for hybridization

1. Fill the tube/cylinder with 2 × SSC.
2. Dislodge any air bubbles sticking to the inside.
3. Wet the filter by floating it on 2 × SSC for a few minutes.
4. Roll the wetted filter and immerse it in the 2 × SSC in the tube/cylinder.
5. Allow the filter to unfold, and carefully drain out the liquid so that the filter adheres to the inside wall of the vessel without trapping any air bubbles.
6. Add the pre-hybridization mixture (0.1 ml/cm^2 of filter) and incubate for 2−20 h at 68°C.

---

ii. Method 2

The filter is put into a plastic bag, which is constantly agitated to keep the hybridization mix in motion. Suitable bags are Sears 'Seal-N-Save' Boilable Cooking Pouches, or bags can be made from relatively stiff plastic sheet such as that used to make bags for autoclaving biohazardous waste (Sterilin). The bags are sealed using a commercially available bag sealer. The bag may be placed inside a second bag in case of leakage. For incubation the bag can be clipped to a rotating plate inside the incubator, or immersed in a shaking water bath, or placed in a suitably sized plastic box containing a few centimetres depth of water inside a gently shaking orbital incubator. The last of these methods has the advantages that many independent hybridizations may be carried out in the same incubator, as well as pre-hybridization and washing. Any leakage of radioactivity is contained inside the box. The procedure used is as described in *Protocol 18*.

Protocol 18. Loading of filters into a plastic bag for hybridization

1. Wet the filter with 2 × SSC.

Protocol 18. *continued*

2. Place the wet filter in the plastic bag.

3. Heat seal the bag except for one corner.

4. Add the pre-hybridization mixture (0.1 ml/cm^2 of filter).

5. Gently squeeze out any air bubbles and heat-seal the corner.

6. Incubate the filter in the bag for 2−20 h at 68°C.

7. At the end of the pre-hybridization period drain off the liquid after cutting off the corner of the bag.

---

iii. Method 3

Small filters can be incubated easily by laying them in the bottom of shallow plastic boxes (e.g. 10 × 10 cm square Petri dishes). These can be placed, unsealed, inside larger, sealed plastic boxes whose atmosphere is kept saturated by also having a wet sheet of filter paper inside. Many small boxes can be placed inside each larger box, which may be placed on a rocking platform or in an orbital incubator. The major attraction of this method is the ease with which boxes can be opened and closed, allowing many filters to be processed simultaneously.

Whichever method is used, it is important that the filter is not allowed to dry out at any time if a high background is to be avoided.

For hybridization dissolve the denatured radioactive probe in 5 times concentrated Denhardt's solution, 5 × SSC, 0.1% SDS to give a volume of about 0.07 ml/cm^2 of filter and add it to the tube or bag, which is then resealed. Again, it may be advantageous to include heterologous DNA. The addition of dextran sulphate to 10% gives an accelerated rate of hybridization (28) if this is required. When adding the hybridization solution containing dextran sulphate, allowance should be made for the pre-hybridization mix retained by the wet filter; this is equivalent to 0.02 ml/cm^2 of filter. Hybridization is carried out at 68°C for 4−72 h (overnight is convenient and usually sufficient).

There are several alternatives to using both dextran sulphate and Denhardt's solution. The most important of these (31) replaces both with 7% SDS; the high detergent concentration blocks non-specific binding and micelle formation is thought to cause an acceleration of hybridization rates. A typical formulation for the hybridization solution is: 7% SDS, 1 mM EDTA, 0.5 M Na$_2$HPO$_4$ (pH 7.2). A substitute for Denhardt's solution is 0.5% BLOTTO (29), stocked as a solution of 10% non-fat powdered milk containing 0.2% sodium azide, stored at 4°C.

At the end of the hybridization period remove the probe solution and wash the filter. The first step is to dilute out unhybridized probe. Then the stringency of the main wash (Step 2 of *Protocol 19*) is chosen to be just lower than that required to maintain the hybrid DNA of interest and to remove those with lower melting temperatures. For a more extensive discussion of melting temperatures, rates of hybridization and stringency, refer to the volume *Nucleic Acid Hybridization* in this series (36). A typical washing procedure is as described in *Protocol 19*.

Protocol 19. Procedure for washing filters after hybridization

1. Rinse the filter twice in 200 ml of $0.1 \times$ SSC, 0.1% SDS at a temperature up to 68°C for 5 min.

2. Immerse the filter in 5 litres of $0.1 \times$ SSC, 0.1% SDS at 68°C for 60 min with occasional stirring.

3. Rinse in 1 litre of $2 \times$ SSC at room temperature.

4. Drain the filter. Do not allow it to dry if it is to be reprobed. Damp filters may be autoradiographed inside thin plastic bags.

5. Expose the filter to X-ray film (see Section 6.5 of Chapter 1).

Problems of high background may be due to one of the following factors:

- dirt, agarose, grease, etc., on the filter;
- air trapped on the filter, caused by poor wetting, or gas coming out of solution during hybridization;
- insufficient pre-hybridization;
- insufficient washing;
- contaminants in the probe (e.g. protein);
- contaminants in the gel or electrophoresis buffer (e.g. DNA).
- hybridization of probe to heterologous DNA used in the hybridization mixture.

2.10.8 Reprobing filters

There are two procedures that are commonly used to remove hybridized probe. The more efficient of these uses alkali to denature hybrids, but this is not suitable for nitrocellulose or Nylon membranes that bind DNA under these alkaline conditions. The procedure for Hybond-N is first to incubate at 45°C for 30 min in 0.4 M NaOH, followed by incubation at 45°C for 30 min in $0.1 \times$ SSC, 0.1% SDS, 0.2 M Tris-HCl (pH 7.5). The alternative is to incubate at high temperature, low stringency, and neutral pH, and in the presence of detergent. For example, washing twice for 30 min in 500 ml of 1 mM EDTA, 0.5% SDS, 10 mM Tris-HCl (pH 7.5) at 95°C.

It is advisable to check that the probe has been efficiently removed by autoradiographing the membrane overnight. Do not allow it to dry out!

3. Large-scale preparative gel electrophoresis of DNA

The early part of this chapter has dealt with some methods of recovering DNA from slices cut out of slab gels. Such methods are useful where only a few fractions (slices) are being dealt with and it is not necessary to recover more than a few micrograms per fraction. In order to overcome these limitations, several workers have designed

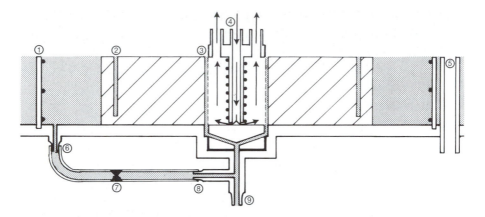

Figure 6. Cross-section of the annular electrophoresis apparatus. The terminals and the sides of the main vessel are not shown and dimensions are approximate. Buffer (stippled area) fills the apparatus to the height of the gel (hatched area). (1) outer electrode; (2) sample well; (3) fraction chamber; (4) bobbin top showing three ports allowing buffer circulation (arrows) between dialysis membrane (dashed line) and central electrode; (5) overflow tube; (6 – 9) fractions are collected via a port (9) with a magnetic valve (7) closing the tubing connecting ports (6) and (8). Between fractions the valve opens allowing buffer to flow from the outer compartments into the fraction chamber via ports (6) and (8).

apparatus where the DNA is eluted from the end of the gel and collected as a series of fractions over time (32,33). In this section two such devices are described. The principle of fractionation is the same in both cases. At the anode end of the gel, between the gel and the anode, is a dialysis membrane where the DNA gathers as it leaves the gel. At chosen time intervals the current polarity is briefly reversed, the DNA detaches from the dialysis membrane and is collected by removing the buffer in the space between the gel and the membrane. The buffer is then replaced, normal polarity is restored and the cycle is repeated. The first apparatus described here has an annular shape with a circular cathode outside the gel and a central anode. It can be used to fractionate up to 50 mg of DNA. The second is a linear version of this design and is useful for fractionating up to 10 mg quantities; it has four channels and can be used to carry out four fractionations simultaneously.

Very similar types of annular (34) and linear (17) apparatus have been described by one of us previously. In this chapter the annular apparatus is a variant of the published design and the linear apparatus described incorporates a few improvements on the original. The principle of operation is unchanged.

3.1 The annular preparative apparatus

3.1.1 Construction of the apparatus

The apparatus (*Figure 6*) consists of the following components:

- a main vessel which holds the gel, the two electrodes and the buffer;

Figure 7. The main vessel of the annular apparatus with the two electrodes and the outer moulding ring in position.

- a high-speed pump which circulates the buffer between the two electrode compartments from a reservoir;
- a low-speed pump which drains the sample from the collection slot to the fraction collector;
- a magnetic valve which closes off the tube connecting the collection slot to the outer electrode compartment during the period that the low speed pump is draining the slot;
- a fraction collector which operates on a time base, and which has an output that can be used to initiate a slave timer. The output can be an AC pulse or a circuit maker, such as is used for the event marker on a chart recorder.
- a slave timer which operates the collecting cycle described below. This can be either electromechanical or electronic;
- a set of moulding pieces for casting the gel;
- a DC supply of about 30 V and 1−2 A.

The main vessel

The main vessel (*Figure 7*) is square. It is about 5 mm wider than the outer electrode and 2 cm higher than the height of the gel. The base should be made of fairly thick Perspex (at least 6 mm) to provide strength and heat insulation. The base has two circular grooves to take the outer electrode and the outer moulding piece, a hole with a grommet to take the overflow tube, a port just inside the outer electrode for the tube that connects the outer buffer compartment to the sample collection slot, and a manifold fitted below a central hole. The central hole forms the seating for the inner moulding piece while the gel is cast, and for the centre electrode during electrophoresis. The manifold under the central hole has two ports; one to connect to the outer buffer compartment and one to connect to the sample collection pump (*Figure 6*). The inlet and outlet ports for the tube connecting the outer buffer compartment and the collection slot have fairly large holes (2 mm) to allow a good flow-rate with a small head of liquid.

The main vessel is supported by three legs. Their height is such that the magnetic pinch-cock clips on to the connecting tube while standing on the benchtop. Feet with levelling screws are attached to the bottom of the legs.

Outer electrode

This is a tube of PVC (24 cm i.d. × 5 cm) with a helical winding of platinum wire (0.25 mm diam.) around the inside. The windings should be close-pitched (about 3 mm) to give a uniform field at the periphery of the gel. The top end of the wire is soldered to a terminal. It is stuck to the surface with spots of glue placed every 4 cm. Six holes (1 cm diameter), equally spaced, in the lower half of the electrode, allow buffer to circulate into the space between the electrode and the outer part of the main vessel.

The central electrode

This is the most complex and fragile component of the apparatus. The bobbin which forms the core of the central electrode should be made of a tough plastic such as Nylon or Kel-F. The example shown (*Figure 8*) was made of Perspex but it is not really strong enough and shows signs of wear after several runs.

The platinum wire of the electrode is wound in a helical groove cut into the spindle of the bobbin. The pitch should be close enough (3 mm) to give a uniform field at the surface of the hole in the gel. The top end of the platinum wire passes through the head of the bobbin and is soldered to a terminal. The bottom end is attached to the spindle.

The head and base of the bobbin are grooved to take 'O' rings. These 'O' rings seal the dialysis tubing to the bobbin. Wet dialysis tubing is slipped over the base and fixed there by an 'O' ring outside the membrane, then fed up and over the top and fixed by a similar 'O' ring. It is important to get the dimensions of the head, the base and the collar right; the tubing is difficult to get on if it is too tight but it leaks if it is too loose. The apparatus uses 2.5-cm-wide dialysis tubing.

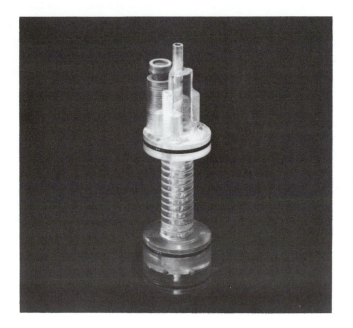

Figure 8. The central electrode for the annular apparatus.

During electrophoresis, a high flow rate of buffer should be circulated through the space between the central electrode and the dialysis tube and out to flow around the outside of the gel. The buffer is pumped through a port in the centre of the top, travels down the centre of the bobbin, out into the space via two ports at the bottom of the spindle and exits via two ports on either side of the entry port at the top.

The volume between gel and dialysis membrane is drained via eight radial channels in the bottom of the bobbin. These join centrally into a downward channel that connects with the ports in the main vessel manifold. The bottom of the bobbin seals into its socket with an 'O' ring.

Buffer flow system

The buffer is pumped from a 10−20 litre reservoir (plastic pail, bowl or bottle) through the bobbin electrode. It exits at the top of the electrode through two flexible tubes. The ends of these tubes are clipped to the inside of the outer electrode, diametrically opposite each other, so that they cause a swirling flow of buffer in the space between the gel and the outer electrode. The pump should have a capacity of at least 200 ml/min. Suitable types are centrifugal pumps with plastic heads for use with aqueous solutions (TEP model BM25/3, Totten Electrical Sales), or large capacity peristaltic pumps (Watson-Marlow, MRHK).

The overflow is a 9 mm i.d. Perspex tube pushed through an 'O' ring in the base of the main vessel, with a flexible tube attached to its lower end leading back to the buffer reservoir.

Connecting tube and magnetic valve

The connecting tube keeps the buffer in the collection slot at the same level as that outside the gel. Without this connection, electroendosmosis through the gel or the membrane could cause a change in the height of liquid in the collection slot. During electrophoresis, current passes through the buffer in the tube to ensure that any sample leaking from the collection slot into the tube is carried back to the collection slot. The connecting tube should be made of resilient tubing that reopens fully after being pinched off during sample collection. The inner bore should be large enough (2−3 mm) to allow rapid refilling of the sample collection slot when the valve is opened.

The magnetic valve must be quite powerful to stop the flow through the connecting tube. The LKB Ultro Rac valve, Flow Stopper is satisfactory but has the disadvantage that it is open when activated. This means that it becomes very hot and must be insulated from the vessel to prevent uneven heating of the gel. The authors have adapted the LKB valve to close when activated to avoid this problem. The magnetic valve is connected to the slave timer as described below.

The sample collection pump

This pumps drains the sample from the collection slot. It should be self-priming and have the capacity to pump at least 10 ml/min. It is connected by fine-bore plastic tubing to the port in the manifold and to the fraction collector. When it is not pumping it serves to seal off the drain hole of the collection slot. Suitable pumps are the Watson Marlow MRHK and the Biotec LP 600 (Biotec). A magnetic valve can be used instead of a pump. In this case the slot is drained by gravity. This system can work well but could fail if an air-lock forms in the drain tube, so a pump is preferable.

The fraction collector

The tubes for the fraction collector should be large enough to hold the contents of the collection slot, which may vary from 3−20 ml. The fraction collector is also used as a master-timer to set the interval between fractions. It should therefore operate on a time-base, and have an output that can be used to activate the slave timer. The Gilson Microcol TDC 80 (Gilson), can be used for fraction volumes up to 9 ml and the LKB Ultro Rac 7000 (Pharmacia-LKB AB) for volumes up to 20 ml. Both these models have an output that can be used to activate the slave timer. The Gilson event marker gives either a mains AC pulse or a short circuit at each change-over: the LKB 2000 has an outlet marked 'Live except during stepping' that can be used for this purpose.

The slave timer

At the end of a fraction period, the fraction collector changes tubes and sends a pulse to the slave timer, which then goes through a 2 min cycle during which it makes the following switches:

(a) the polarity of the DC current is reversed for 30 sec.

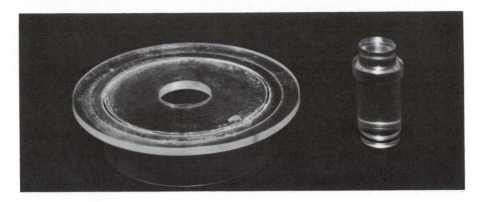

Figure 9. Moulding pieces for the annular apparatus. The items shown are the one-piece lid/slot former and the centre rod.

(b) the DC current is switched off, the magnetic valve closed, and the pump to the fraction collector started; 1 min is allowed for this step.

(c) the pump of the fraction collector is switched off, the magnetic valve is opened and 30 sec is allowed for the collection chamber to refill with buffer from the outer compartment.

(d) the slave timer returns to its rest position, switching the DC current on again with normal polarity.

The authors use an electromechanical timer, although electronic timers and the design described by Brownstone (32) could also be used with this apparatus. Suitable timers are commercially available.

The moulding pieces

The mould used to cast the gel is made from three pieces (*Figure 9*). The outer piece is plain tube 5 cm i.d. ×4.3 cm high that sits in the groove cut in the base of the main vessel. The central hole in the gel is formed by a rod about 3.5 cm diam. ×7.5 cm that fits tightly into the central hole in the base. The lid and slot moulds are in one piece. The lid is a disc cut from 6-mm-thick Perspex sheet, the slot mould is made from a cylinder 12.8 cm i.d., 3.5 cm long, 2 mm thick. The lid has a central hole that is a tight fit on the centre rod. There is a groove around the circumference that locates on the outer moulding ring. There are two holes in the lid, one just inside and one just outside the slot mould. These allow the addition of agarose or acrylamide solution and escape of steam and air during gel casting.

Dimensions and tolerances

The apparatus described here was designed with a particular purpose in mind. For other applications other dimensions may be more suitable. There is no limit to the diameter of gel that may be used, but gels above a certain height will not be self-supporting. Gels from 2 – 10 cm high have given satisfactory results but in order to use gels of differing height a more complex design of this apparatus is needed (34). For reasonable stability the wall of gel surrounding the loading slot should be about 1 cm thick. The gap between the gel and the membrane determines the fraction volume and the smaller this volume the better. However, with a gap smaller than 2 mm it is difficult to prevent the membrane from making contact with the gel, which prevents complete draining of the sample and leads to some intermingling of fractions.

It is important that the central electrode, the hole in the gel, the loading slot and the outer electrode are all accurately centred and stand square to the base. The design given here ensures accurate location around the centre, provided components are correctly machined (tolerance 0.1 mm).

3.1.2 Setting up the apparatus

The gel is prepared as described in *Protocol 20*.

Protocol 20. Casting an agarose gel for the annular apparatus

1. Dissolve agarose in buffer (e.g. 100 mM sodium acetate buffer, 10 mM EDTA, pH 7.5) and cool the mixture to 45 – 50°C; just above its setting point.

2. Place the outer, cylindrical moulding piece and the central moulding rod in position and use about 20 ml of agarose solution to seal them in place.

3. Tilt the apparatus by a few degrees and pour in the rest of the agarose slowly and evenly, avoiding bubbles.

4. Place the lid/slot former in position with the access holes on the upper side.

5. Leave the gel to set. The gel shrinks as it sets so this is compensated for by adding more agarose solution through the access holes in the lid/slot former. When the gel stops shrinking it is allowed to cool, preferably overnight, at room temperature.

6. Carefully remove the lid, rocking it backwards and forwards to allow air down the slot to avoid splitting the gel at the bottom of the slot. The use of a fine syringe filled with buffer to loosen the gel from the Perspex will also help.

7. Remove the inner and outer moulding pieces and clear away any loose pieces of agarose.

The apparatus has been used once with an agarose – acrylamide composite gel (C.Tyler-Smith, personal communication). A 5% polyacrylamide gel strengthened with 1.5% agarose was used to fractionate a mixture of 65 mg bovine serum albumin and 65 mg ovalbumin. In this case the procedure used is as shown in *Protocol 21*.

Protocol 21. Casting an agarose-polyacrylamide composite gel

1. Warm the apparatus to 65°C to stop the agarose setting.

Protocol 21. *continued*

  2. Pour the gel mix into the apparatus.

  3. After the acrylamide has polymerized, cool the apparatus to room temperature and allow the agarose to set.

  4. Carry out the electrophoresis at 4°C; other conditions are similar to those described for DNA fractionations using agarose gels.

Using this procedure, the separation and recovery were comparable to that reported by Brownstone (32) for a single run using his apparatus.

Protocol 22. Setting up the buffer flow for the annular apparatus

  1. Place the central electrode with the membrane in position in the hole in the gel.

  2. Attach the pump and the outlet tubes.

  3. Start the pump with the connecting tube pinched off, so that the collection slot does not fill up.

  4. While the collection slot is empty, and the dialysis tubing is distended, measure the conductivity between the two electrodes to check that the dialysis tubing is clear of the gel all the way up the collection well. If it touches the gel at any point or if it leaks, it will give a measurable conductivity. It may be possible to clear contact between gel and membrane by reducing the flow-rate of buffer. Otherwise it is necessary to refit the dialysis tubing.

  5. Connect the pump used to collect the sample and open the connecting tube so that the collection slot fills with buffer.

  6. Adjust the buffer level to the height of the gel.

3.1.3 Loading and running the sample

The loading procedure is very similar to that described for analytical gels. The DNA is dissolved in distilled water and one-tenth volume of Ficoll/marker/buffer (Section 2.5) is added; its final volume should be slightly less than the slot volume. The sample can then be layered into the middle of the buffer-filled slot, with a shallow layer of solution above it. This reduces 'edge effects' which can lower resolution. With the sample slot filled, the DC supply is switched on with normal polarity. When the front marker dye has almost reached the collection slot the fraction collector is switched on. A notch (5 mm) should be cut with a scalpel in the top of the gel between the sample slot and the outer compartment to allow buffer to flow into the slot when the buffer level is slightly less than the gel height. This will compensate for buffer loss from the slot caused by electroendosmosis.

3.2 The linear preparative apparatus

3.2.1 General description of the apparatus

The smaller apparatus (*Figures 10–13*) has four channels. Each channel holds a block of gel, cast *in situ*, with a sample well moulded into it. A dialysis membrane

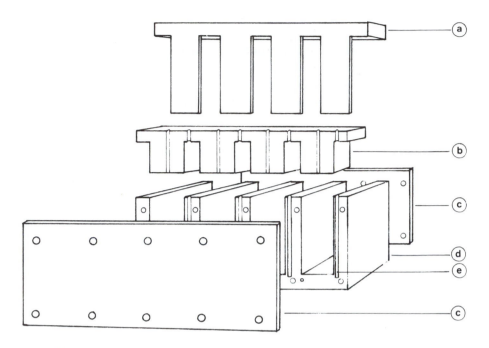

Figure 10. Moulding pieces for small preparative gel apparatus. Slot formers (a), are placed behind the cover (b) which can be placed at varying heights for gels of different sizes. This cover is left in position after the gel has set and during the run. (c) Front and back which are removed after the gel has set. (d) Main body; the grooves (e) down the front of the uprights form seatings for the tubing to the pump used to collect fractions.

is held 1.5 mm away from the end of the gel, leaving a space from which the samples are collected. This space is connected by a channel running underneath the gel to the buffer in the cathode compartment; thus the buffer in the space is maintained at a constant level except during collection. When the sample is pumped out of the space, the connection is closed off by a gate which is pushed over the ends of the channels at the cathode side (*Figure 12*). The gate is operated by a motor-driven cam and lever. The space between the gel and the membrane is also connected to a pump which removes the samples to a fraction collector. Most of the components are made from Perspex. The block of four channels locates in the centre of a tank containing the electrodes and is held in position at each side by a knurled screw. The pumps, fraction collector, slave timer and DC supply described in Section 3.1.1 are also suitable for this apparatus.

The cam drive mechanism. The cam is driven by a geared electric motor (Crouzet, type 82-344, 10 rpm). The power supply to this is activated via a relay triggered by a signal from the main control unit (see Section 3.1.1). As the cam rotates it activates a microswitch when the 'gate' is closed causing the rotation to stop while

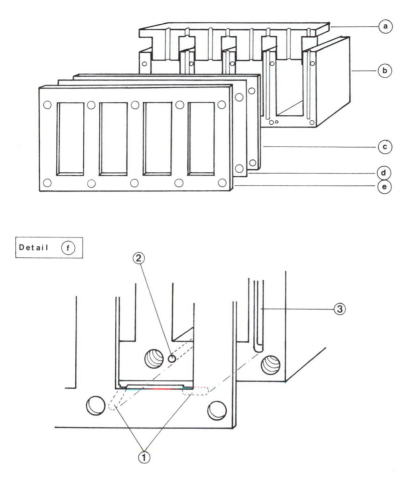

Figure 11. Front assembly of the small preparative gel apparatus. (a) Cover, shown here in the highest position. (b) Main body. (c) Spacer, made from PTFE sheeting. This piece holds the dialysis membrane (d) 1.5 – 2.0 mm away from the gel and forms the chambers from which fractions are collected. It is grooved (see detail) with three channels in each sector, two at the bottom to connect the buffer in the chamber to the collecting tube and to the channel running through to the back of the apparatus, one at the top to let air in when the sample is pumped out. (e) Clamping frame which holds the front assembly to the apparatus; Nylon screws are used to avoid corrosion. (f) Detail, showing grooves (1), in the face of the spacer. The 'windows' in the spacer should be about 1 mm narrower than the width of the gel to prevent the gel from sliding forward against the membrane. The grooves in the spacer are aligned with the channel (2) in the main body through which buffer flows to refill the fraction chamber, and the groove (3) through which fractions are removed.

the fraction chamber is emptied. On receiving a second signal from the timer, 30 sec later, the cam moves on, opening the 'gate' and then triggering a second microswitch which ends its revolution. The two microswitches are activated by the

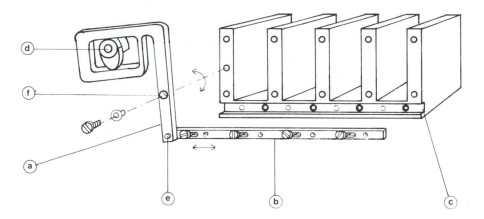

Figure 12. Rear assembly of the small preparative gel apparatus. (a) Lever which moves the gate (b) across at each fraction collection. The gate is held, with light pressure, against the 'O' rings by the Nylon screws which pass through slots in the gate to allow movement across the 'O' rings. At each fraction collection, the motor driven cam (d) raises the lever (a) thus moving the gate to the left. The channels are then sealed off while the fraction is pumped out of the collection chambers. When these are completely emptied, the cam is driven back to its original position, placing the holes in register with the channels and allowing buffer from the cathode reservoir to flow into the collection chambers. (e) Fulcrum of the lever. (f) Hinge holding the lever to the gate.

bolthead in the cam shaft which fixes the cam to the motor drive. The whole mechanism is mounted in a Perspex box to one side of the main chamber (*Figure 13*).

3.2.2 Setting up the apparatus

Protocol 23. Casting an agarose gel for the linear apparatus

1. Screw a flat plate onto each end of the gel container, and lower the cover to give the desired height to the gel(s) (*Figure 11*).

2. Dissolve the agarose in electrophoresis buffer (e.g. 100 mM Tris-acetate, pH 7.8, 10 mM EDTA) and cool the mixture to just above the setting point.

3. Pour the agarose into as many channels as are needed for the separation and remove any bubbles trapped beneath the cover by tilting the apparatus.

4. Lower the slot former into the gel behind the cover and leave the gel to set.

5. Remove the slot former and the end plates and clear away any loose pieces of agarose.

Once you have cast the gel then proceed directly to assembling the apparatus ready for use as described in *Protocol 24.*

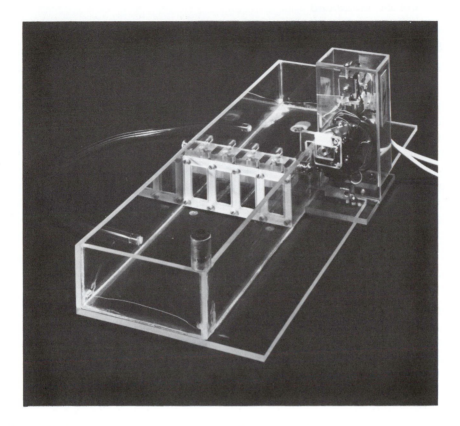

Figure 13. The linear electrophoresis apparatus. The apparatus is shown completely assembled but without buffer, gel, or dialysis membrane.

Protocol 24. Setting up the linear apparatus

1. Sandwich a moist dialysis membrane between the spacer and the clamping frames (*Figure 11*). Pierce holes in the membrane at the position of the holes in the frame.

2. Screw this assembly against the front of the gel container. The membrane is stretched as the screws are tightened, starting from the centre and working outwards.

3. Trim off excess membrane, attach the gate to the back, and place the whole assembly in position in the main vessel (*Figure 13*).

4. Push the tubes to the sample-collecting pump into their grooves, down the front of the gel assembly.

Protocol 24. *continued*

   5. Fill the vessel with buffer to the height of the gel.

   6. Load the sample and begin electrophoresis.

   When the sample has run into the gel the buffer level can be raised above the top of the gel to prevent it drying out. Saran Wrap or polythene sheet is floated on the surface of the buffer to stop evaporation. Buffer is circulated from a reservoir into one of the electrode compartments and flows, via tubing connecting the ports in the side of the main vessel, into the other electrode compartment. In the apparatus shown, the buffer is returned to the sump via a syphon but it is better to raise the apparatus and fit an overflow tube as in the annular design. Loading of the sample is carried out in the same way as for the annular apparatus; the sample well dimensions are approximately $0.3 \times 1.5 \times 3.0$ cm giving a maximum sample volume of about 1 ml with the cover in its highest position.

3.3 Practical aspects of preparative separations

In general, the observations made regarding analytical separations of double-stranded DNA apply to the use of both types of preparative apparatus and the following points should be taken in conjunction with them. Many of the points below are demonstrated in the two fractionations described later. The authors have not used denaturing buffers in either apparatus; for polyacrylamide, see earlier observations.

3.3.1 Gel capacity

At typical operating voltages (20–30 V) the annular apparatus has a low current density (1.5 mA/cm^2) at the sample well and a more usual current density of 5 mA/cm^2 at the fraction chamber (34). This, combined with the high ratio of well area to gel volume, gives the design a higher capacity than might be expected from experience with analytical gels. The capacity of the linear apparatus may be deduced directly from the value for analytical gels under similar running conditions. For a complex DNA digested with a restriction enzyme recognizing a four base sequence, the capacity of one channel on the linear apparatus may be as high as 5 mg, and of the annular-apparatus, 50 mg. For the resolution of a few large fragments of a simple DNA, the upper limits are 0.1 mg and 1 mg, respectively.

3.3.2 Choosing the gel concentration

In an analytical gel, all the DNA is electrophoresed for the same time and different sized fragments cover different distances; with the preparative devices the fractions all travel the same distance and are eluted at different times. One consequence of this is that a complete fractionation takes much longer than an analytical separation under equivalent conditions, but there is a corresponding increase in the potential resolution of fragments, especially those of higher molecular weight. The resolution of all fragments also increases with gel concentration, but so does the duration

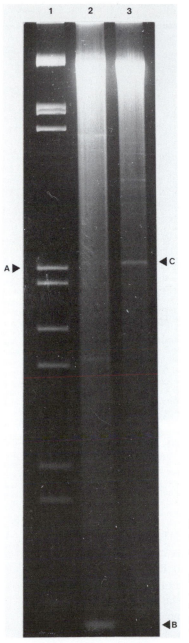

Figure 14. Analysis of *Bam*HI restriction fragments of DNA from mouse/human hybrid cell line HORL 9X. Samples were analysed on a 2%, 20-cm-long neutral agarose gel. Lane 1, *Eco*RI/*Hind*III digested λcI *ts* 857. Band A is 2.02 kb in size. Lane 2, *Bam*HI digested HORL 9X DNA. Band B can also be seen in fraction 19 of *Figure 15*. Lane 3, *Bam*HI digested TIL 1 DNA. Band C, 2.04 kb in size indicates the position of the sequence sought in the preparative fractionation of HORL 9X DNA.

of the fractionation. Our practice has been to arrange for the fractions of greatest interest to be eluted after about 24−48 h in the course of a separation covering 3−5 days.

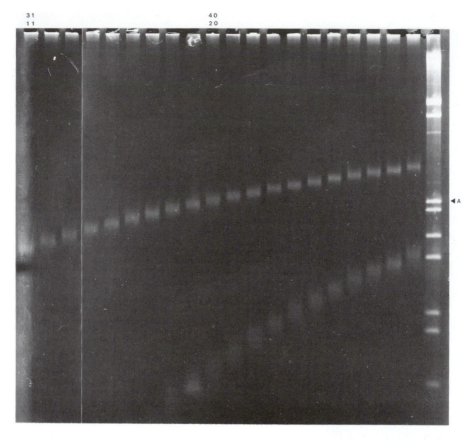

Figure 15. Analysis of the eluate from the annular apparatus. Fractions 11 – 50 of *Bam*HI digested HORL 9X DNA, recovered from the annular apparatus and analysed on a 2%, 20 cm neutral agarose gel. Two fractions were loaded into each well. Band A is as in *Figure 14*.

3.3.3 Fraction interval

Fraction intervals of 10 min to 24 h have been used. The time taken for a DNA fragment to pass through the gel is proportional to its molecular weight. Thus, a programme of increasing intervals is generally used for a preparative run, for example, 20 min fractions for the first 12 h, then 60 min fractions for the next 24 h, then 2 h fractions for 48 h. A simpler programme can be used for isolating a single fraction of known mobility since the elution time on a preparative gel can be accurately predicted from its mobility on analytical gels.

3.3.4 Fraction volume

This is about 10 ml for the annular apparatus and 1.5 ml for each channel of the linear apparatus, as described here.

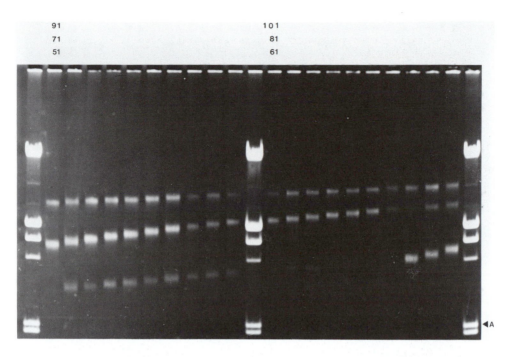

Figure 16. Analysis of the eluate from the annular apparatus. Fractions 51 – 110 of *Bam*HI digested HORL 9X DNA recovered from the annular apparatus were analysed on a 1%, 20-cm-long neutral agarose gel. Three fractions were loaded into each well. Band A is as in *Figure 14*.

3.3.5 Fraction storage

To avoid bacterial growth, the fractions should be removed to a cold room as soon as possible after collection, and a drop of chloroform added. Long-term storage is best at −20°C after the addition of two volumes of 95% ethanol.

3.3.6 DNA recovery

The DNA can be recovered from fractions by ethanol precipitation or high speed centrifugation.

3.4 Examples of fractionations

3.4.1 Using the annular apparatus

DNA (3 mg) from the mouse/human hybrid cell line HORL 9X was digested to completion with the restriction endonuclease *Bam*HI, generating the profile seen in *Figure 14*. This was loaded on a 2% annular agarose gel in a volume of 15 ml with the objective of preparing a fraction centred on 2.02 kb in size and corresponding to a band observed in a *Bam*HI digest of the cell line TIL 1 (a human cell line with four 'X' chromosomes.). The buffer was 10 litres of 100 mM Tris-acetate, 1 mM EDTA (pH 7.6) at 20°C; pH was monitored and the buffer changed every 48 h.

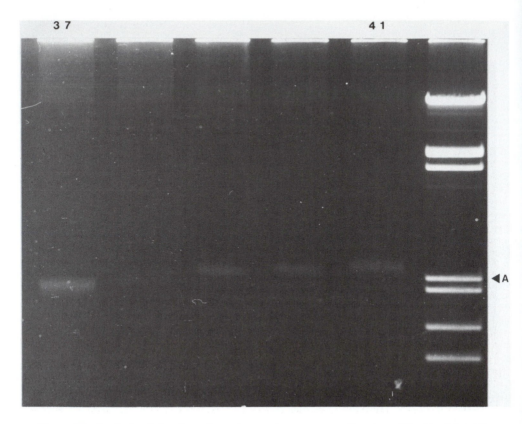

Figure 17. Analysis of the eluate from the annular apparatus. Fractions 37 – 41 of *Bam*HI digested HORL 9X DNA recovered from the annular apparatus were centrifuged for 16 h at 100 000 *g*. The pelleted DNA was resuspended in distilled water and analysed on a 2%. 10-cm-long neutral agarose gel. Band A is as in *Figure 14*.

The voltage used was 28 V. The first 67 fractions were collected at 20 min intervals and subsequent fractions at intervals up to 4 h. Fractions were examined using a series of analytical gels (*Figures 15 – 17*). Aliquots of the fractions (about 20 μl from 10 ml) were loaded in series with two or three fractions per analytical slot. The effect of increasing the fraction interval is clearly seen at fractions 67/68 (*Figure 16*). The fractions corresponding to about 2 kb in size were recovered by centrifugation at 100 000 *g* for 16 h and re-analysed (*Figure 17*). Three of these fractions (38 – 41) were pooled and could be successfully nick translated and ligated. Microdensitometry of this gel indicates that fractions 37 and 41 are separated by about 100 bp with less than 10% cross-contamination. One of these fractions, containing a few micrograms of DNA, represents an enrichment for the desired sequence of at least 100-fold.

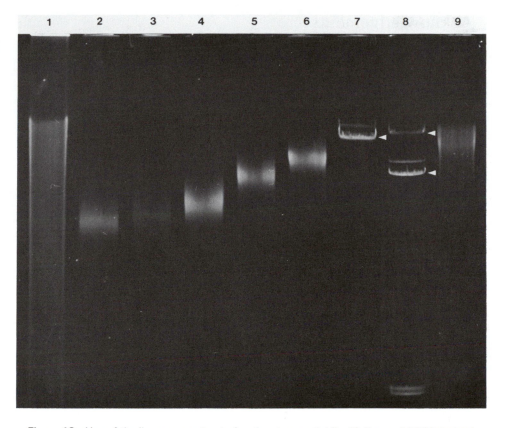

Figure 18. Use of the linear apparatus to fractionate a partial *Eco*RI digest of PG19 derived cell DNA. A 0.8% neutral agarose gel 20 cm long was used. Lane 1, unfractionated digest; lanes 2−6 aliquots of fractions collected at 18, 22, 26, 30, and 34 h, lane 7, λ WES 5 DNA 39 kb in size; lane 8, λcI *ts* 857 plus *Eco*RI/*Hind*III digested λcI *ts* 857 DNA. The indicated bands are 48.54 kb and 21.24 kb in size. Lane 9, pooled fractions from 38−50 h.

3.4.2 Using the linear apparatus

A partial digest of DNA from mouse-derived cells (*Figure 18*) was fractionated on a 0.8% agarose gel; 750 μg in a volume of 0.45 ml was loaded. The voltage used was 38 V and the buffer was 100 mM sodium acetate (pH 7.6), 1 mM EDTA. The objective was to prepare a fraction of 25−50 kb in size for cloning in a cosmid vector. The orange G was eluted at about 4 h; fractions were collected at 2 h intervals up to 18 h and subsequently at 4 h intervals. Aliquots of fractions were analysed on gels as before (*Figure 18*). The DNA from fractions eluted at intervals from 38−50 h was pooled, recovered by alcohol precipitation in the presence of carrier tRNA and subsequently ligated. (Data provided by A.T.H.Burns.)

Acknowledgements

We would like to thank Duncan Fletcher for constructing the devices described, Val Jones (Amersham International) for her contribution of the Quick dot blot protocol, Anne Deane for typing the manuscript, Barbara Smith for help with the diagrams and Norman Davidson for photographs of the preparative devices.

References

1. Fangman, W. L. (1978). *Nucleic Acid Res.*, **5**, 653.
2. Jovin, T. M. (1971). In *Methods in Enzymology* (ed. L. Grossman and K. Moldave). Vol. 21, p. 179, Academic Press, London and New York.
3. Fischer, S. G. and Lerman, L. S. (1979). In *Methods in Enzymology* (ed. R. Wu), Vol. 68, p. 183. Academic Press, London and New York.
4. McDonell, M. W., Simon, M. N., and Studier, F. W. (1977). *J. Mol. Biol.*, **110**, 119.
5. Bailey, J. M. and Davidson, N. (1976). *Anal. Biochem.*, **70**, 75.
6. Gould, H. J. and Hamlyn, P. H. (1973). *FEBS Lett.*, **30**, 301.
7. Maniatis, T., Jeffrey, A., and van de Sande, H. V. (1975). *Biochemistry,* **14**, 3787.
8. Cummins, J. E. and Nesbitt, B. E. (1978). *Nature (London)*, **273**, 96.
9. Mocharla, R., Mocharla, H., and Hodes, M. E. (1987). *Nucleic Acids Res.*, **15**, 10589.
10. Sanger, F., Coulson, A. R., Hong, G. F., Hill, D. F., and Peterson, G. B. (1982). *J. Mol. Biol.*, **162**, 729.
11. Sutcliffe, J. G. (1978). *Nucleic Acids Res.*, **5**, 2721.
12. Streeck, R. E. and Zachau, H. G. (1978). *Eur. J. Biochem.*, **89**, 267.
13. Bersaude, O. (1988). *Trends in Genetics,* **4**, 89.
14. Brunk, C. F. and Simpson, L. (1977). *Anal. Biochem.*, **82**, 455.
15. Southern, E. M. (1979). *Anal. Biochem.*, **100**, 319.
16. Schaffer, H. E. and Sederoff, R. R. (1981). *Anal. Biochem.*, **115**, 113.
17. Southern, E. M. (1979). In *Methods in Enzymology* (ed. R. Wu). Vol. 68, p.152. Academic Press, London and New York.
18. Elder, J. K. and Southern, E. M. (1983). *Anal. Biochem.*, **128**, 227.
19. Horz, W., Oefele, K. V., and Schwab, H. (1981). *Anal. Biochem.*, **117**, 266.
20. Feinberg, A. and Vogelstein, B. (1983). *Anal. Biochem.*, **132**, 6.
21. Feinberg, A. P. and Vogelstein, B. (1984). *Anal. Biochem.*, **137**, 266.
22. Wieslander, L. (1979). *Anal. Biochem.*, **98**, 305.
23. Michaels, F., Burmeister, M., and Lehrach, H. (1987). *Science,* **236**, 1305.
24. Yang, R. C.-A., Lis, J., and Wu, R. (1979). In *Methods in Enzymology* (ed. R.Wu). Vol. 68, p. 176. Academic Press, London and New York.
25. Arnheim, N. and Southern, E. M. (1977). *Cell,* **11**, 363.
26. Wilkins, R. J. and Snell, R. G. (1987). *Nucleic Acids Res.*, **15**, 7200.
27. Olszewska, E. and Jones, K. (1988). *Trends in Genetics.*, **4**, 92.
28. Wahl, G. M., Stern, M., and Stark, G. R. (1979). *Proc. Natl. Acad. Sci. (USA)*, **76**, 3683.
29. Reed, K. C. and Mann, D. A. (1985). *Nucleic Acids Res.*, **13**, 7207.

30. Denhardt, D. T. (1966). *Biochem. Biophys. Commun.*, **23**, 641.
31. Church, G. M. and Gilbert, W. (1984). *Proc. Natl. Acad. Sci. (USA)*, **81**, 1991.
32. Brownstone, A. D. (1969). *Anal. Biochem.*, **27**, 25.
33. Polsky, F., Edgell, M. H., Seidman, J. G., and Leder, P. (1978). *Anal. Biochem.*, **87**, 397.
34. Southern, E. M. (1979). *Anal. Biochem.*, **100**, 304.
35. Thurston, S. J. and Saffer, J. D. (1989). *Anal. Biochem.*, **178**, 41.
36. Hames, B. D. and Higgins, S. J. (1985). *Nucleic Acid Hybridization: A Practical Approach,* IRL Press, Oxford and Washington DC.

3

Pulsed field gel electrophoresis

RAKESH ANAND and ED M. SOUTHERN

1. Introduction

The practical usefulness of conventional agarose gel electrophoresis of DNA as described in the preceding chapter is limited to the separation of DNA up to 50 kb in size. A different type of electrophoresis called pulsed field gradient gel electrophoresis, capable of separating DNA molecules up to 2000 kb (2 Mb) long, was first described by Schwartz and Cantor in 1984 (1). In this method, DNA travels through a concentrated agarose gel under the influence of two electric fields. The angle between the two fields is almost perpendicular, the fields are non-uniform in strength and are alternatively switched or pulsed. It is generally assumed that in high gel concentrations and at high voltage gradients, DNA molecules must be stretched along the direction of the field in order to penetrate the pores of the gel and make a net forward movement. Schwartz and Cantor explain the separation they achieved by alternating field directions as resulting from the need for DNA molecules to reorient themselves in the gel in order to travel in a direction approximately at right angles to the axis along which they are stretched. The longer the DNA molecule, the longer the time taken to find the new orientation, and the more the molecule is held back in the gel.

Several modifications of the original method of Schwartz and Cantor have followed rapidly in an effort to improve the resolution as well as the running characteristics (2−4). These have led to several different names being used to refer to this technique depending on the type of apparatus and the arrangements of the electrodes. For historical reasons, as well as to avoid confusion, the authors will use the term PFGE (pulsed field gel electrophoresis) except when referring to a particular type of apparatus, in which case its 'given' name will be used. The size of fragment that can be separated has also been extended and DNA molecules larger than 6 Mb in size have been clearly resolved (5). The mechanism of separation also seems to be better understood (6).

The practical applications of this technique are many. In some lower eukaryotes such as yeast and trypanosomes, the chromosomes are small enough to be resolved as individual bands. In many of these species, chromosomes never condense to become visible by microscopic methods, so this technique makes karyotyping possible for the first time. In addition, any cloned DNA sequence can be assigned to chromosomes very easily. The technique is also useful for the purification of DNA from individual chromosomes in order to make chromosome-specific libraries.

In higher eukaryotes such as mammals, chromosomes have not yet been separated as individual bands. The smallest human chromosome is estimated to be about 45 Mb. However, there are commercially-available restriction nucleases, for example *Sfi*I and *Not*I, which have an 8 bp recognition sequence and therefore sites for these are relatively rare in DNA. Another useful class of enzymes are those which are methylation sensitive and generate large fragments from vertebrate, bacterial and plant DNA (see Section 2.6.2). Analysis of the large DNA fragments by blotting (7) and hybridization can provide maps of large stretches of DNA which can help in establishing physical linkage between genetic loci, complementing the methods of classical genetics. This approach has been very useful in the study of large genes like the gene associated with Duchenne muscular dystrophy and Becker muscular dystrophy which spans more than 2 Mb of the Xp2.1 region of the human X-chromosome (for a review see ref. 8). Once an area has been mapped, the position of all chromosomal translocations can be localized and the extent of deletions accurately established by PFGE. Another important application is the identification of large DNA fragments, which when purified from a gel, yield an enriched source of DNA sequences for cloning (9,10). There are many other examples of the usefulness of PFGE, but in this chapter the authors will restrict discussion mainly to the experimental aspect of this technique.

1.1 Types of apparatus

Since the 1984 publication of Schwartz and Cantor (1), modifications of the original arrangement of electrodes and experiments aimed at understanding the mechanisms of separation showed that the angle between the two fields had to be obtuse, and that the inhomogeneity in the field was not necessary but was responsible for the distortions seen in the early separations. Some of the types of apparatus that have been developed are described briefly in the following paragraphs. For more detailed descriptions of the electrophoretic fields, DNA resolution and possible construction of the different types of apparatus the reader is referred to the original articles.

1.1.1 Single inhomogeneous field

Schwartz and Cantor originally used one homogeneous and one inhomogeneous field on the same gel (1). The inhomogeneous field was achieved using an array of electrodes as cathode and a single point electrode as anode (*Figure 1a*). They have since modified their original gel box and now use a small gel box with an array of diode-isolated electrodes. Calculations to achieve a constant field angle throughout the gel have yielded a geometry that results in almost straight tracks (11). Another modification of this field geometry together with an altered switch pattern has also been described (12) and this too results in a reduced distortion of the DNA tracks.

1.1.2 Double inhomogeneous field

The first modification of the original arrangement of electrodes was by Carle and Olson (2) who used a double inhomogeneous field. The cathodes were long continuous

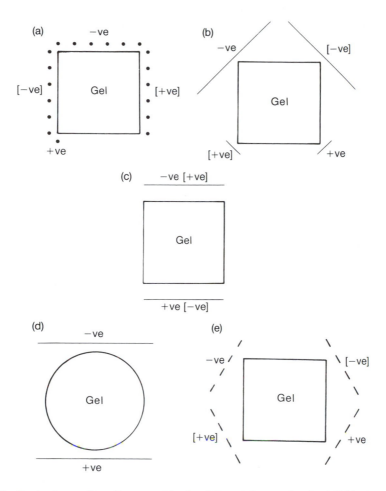

Figure 1. Electrode configurations used in the different types of pulsed field apparatus. Changes in polarities after the duration of a pulse are shown in brackets. (a) Diode isolated electrodes (●) providing one homogeneous and one inhomogeneous field (1) was the first electrode configuration to be used. (b) Double inhomogeneous field using long cathodes and short anodes (2) was the first modification of the original field. (c) Field inversion gel electrophoresis (14,15) using a homogeneous electrical field. (d) Rotating gel system (3,6) also uses a homogeneous field. In this system, the polarity remains the same but the gel is turned at each pulse interval. (e) Contour clamped homogeneous electric field system (4) uses a hexagonal electrophoresis chamber with electrodes on four sides of the hexagon. The field homogeneity is produced electronically by fixing voltages at the various electrodes which are connected by a series of resistors.

platinum wires and the anodes were short, but not point, electrodes (*Figure 1b*). The DNA in the field produced by this electrode geometry runs symmetrically around the centre line though the outer tracks are curved. This distortion can be eliminated by standing the gel vertically with the two fields running through the thickness of the gel (13). In this way, the apparatus resolves DNA into straight tracks, as in homogeneous field electrophoresis. However, some small molecules are lost from the gel during long pulses.

1.1.3 Field inversion gel electrophoresis (FIGE)

The use of field inversion to improve electrophoretic resolution led Tombs (14) to its use for separating large DNA molecules. Extensive studies on the separation characteristics of this system have been done by Carle and Olson (15). They used a conventional horizontal gel apparatus and the polarity of the field was switched at each cycle (*Figure 1c*). Net forward movement of DNA is achieved by using a greater part of the switching cycle for forward than for reverse migration or a higher voltage in the forward than in the reverse direction. In this system it is found that, under a given set of switching conditions, a range of DNA sizes are fractionated but, surprisingly, some larger sizes show greater mobility than shorter fragments. To overcome this problem a 'ramp' is used where the time of the switching cycle is increased throughout the run according to a preprogrammed sequence (see Section 3). This system has the advantage that tracks are straight. However, the relationship between size and mobility is not understood, making accurate sizing difficult, especially for fragments which migrate close to the region of minimum mobility.

1.1.4 Homogeneous crossed field electrophoresis

The main advantage of this form of electrophoresis is that the DNA migrates in straight tracks which makes it easier to compare bands between tracks. The first apparatus to achieve this was a circular gel in a square horizontal gel box (3,6). The gel is turned at each switching cycle so that the DNA experiences a change in field angle of greater than 90° (*Figure 1d*). It was initially designed to test a model of the mechanism of DNA separation (6), and all examples of DNA separation presented in this chapter have been achieved using this apparatus. A similar apparatus has also been described by Serwer (16). Another form of homogeneous field apparatus is the one described by Chu *et al.* (4). This is a hexagonal box with platinum electrodes connected by a series of resistors (*Figure 1e*). The electric fields are at 120° to one another and since the electrode potentials are clamped to provide a near homogeneous field, the technique has been named contour-clamped homogeneous electric field (CHEF) gel electrophoresis.

1.2 Choice of apparatus

The choice of apparatus depends on the requirements of the investigator. A version of the double inhomogeneous field apparatus is available from Pharmacia-LKB (Pulsaphor), who also supply a kit to convert this system to a CHEF type of apparatus. The field inversion system apparatus (14,15) is available from several suppliers including Hoefer and Acronym Pvt. Ltd. (Boronia 3155, Victoria, Australia). The vertical apparatus (13) is available from Beckman Instruments and CHEF from Bio-

Rad. The rotating gel apparatus is currently available from Tribotics Ltd (Oxford) and Hoefer. There are some major advantages in obtaining straight tracks across the whole width of the gel since one can run size markers on the two end tracks and all the remaining tracks on a gel can be used for sample analysis. This simplifies the accurate sizing of fragments and a further advantage is that the gels can be used routinely for preparative runs. Finally, a point which might influence the choice of apparatus is the capability of resolving large size DNA. As far as the authors are aware, the double homogeneous field (5), CHEF (17), the vertical apparatus and the rotating gel apparatus (*Figure 5*) have all been used successfully to resolve molecules up to at least 6 Mb in size. There is no obvious reason to assume that the other systems will not be able to resolve large DNA molecules but, as yet, they have presumably not been used for this successfully. As far as band sharpness and general resolution is concerned, the reader is referred to the articles describing these separations. However, it is clear that in the double inhomogeneous field apparatus of Schwartz and Cantor, bands are sharpened as they run into a weak field, whereas FIGE gives broad bands in the high molecular weight region.

2. Preparation and processing of samples

A prerequisite for this type of electrophoretic analysis is the preparation of very large size DNA as well as adequate size markers. Extra care is essential to avoid any nuclease contamination and therefore, where possible, all solutions should be autoclaved. Solutions which cannot be autoclaved should be made up in glass double-distilled, autoclaved water and then filtered through a Millipore filter. Although some of these methods have been described in earlier reviews (3,18,19), detailed descriptions are given in the following sections.

2.1 Chemicals and solutions

2.1.1 Solutions for sample preparations and gel electrophoresis

Low gelling temperature (LGT) agarose

The authors use Sea Plaque LGT agarose (FMC corporation) since this has proved to be quite a reliable source. However, sometimes there may be batch to batch variations in the purity of LGT agarose, especially from other sources. Therefore, it may be desirable to test a few batches to find one which yields the most intact DNA in agarose plugs, and has no restriction enzyme inhibitors present. This can be done by embedding bacteriophage lambda DNA in agarose plugs and then testing for degradation and digestion. Alternatively, agarose solution can be purified by treatment with AG-1-X2 (200−400 mesh) or DEAE-cellulose ion-exchange resin. This can be done using 100−200 ml of 1−1.5% agarose which can then be stored at 4°C for subsequent use. FMC corporation also produce a batch-tested, guaranteed nuclease-free grade of LGT agarose (Insert agarose).

NDS (1% lauryl sarcosine 0.5 M EDTA, 10 mM Tris, pH 9.5)

This solution is used extensively in the preparation and storage of DNA embedded in agarose. To make 500 ml of this solution, add 93 g of EDTA (mol. wt 372) to approximately 350 ml of distilled water. Add 0.605 g of Tris base and then add solid NaOH pellets to bring the pH above pH 8.0 in order to dissolve the EDTA (10 − 12 g of NaOH pellets are needed for this). Dissolve 5.0 g of lauryl sarcosine in 50 ml of distilled water and add this to the first solution. Adjust the pH to pH 9.5 using concentrated NaOH and make the volume up to 500 ml. Filter through a 0.2 μm Millipore filter and store at 4°C.

Proteinase K (Boehringer Mannheim)
Dissolve in NDS at a concentration of 20 mg/ml just before use.

Pronase (Boehringer Mannheim)
Dissolve in NDS at a concentration of 20 mg/ml just before use. The authors find that this pronase works just as well as proteinase K and is much cheaper. However, it may be advisable to inactivate contaminants that may be present by autodigesting this pronase solution at 37°C for 30 min before use.

Zymolase-20T (Seikagaku Kogyo Co. Ltd)
Dissolve in 1.2 M sorbitol at a concentration of 20 mg/ml just before use.

Lyticase (Sigma Chemical Co)
Dissolve in 50% glycerol to a concentration of 20 units/μl and store at −20°C.

Gelatin (Oxoid Ltd)
Dissolve in water at a concentration of 1 mg/ml. Dispense 500 μl aliquots into microcentrifuge tubes, pierce a hole in their tops, autoclave and store at −20°C.

Spermidine and dithiothreitol (Sigma Chemical Co)
Make 0.1 M solutions of each in sterile distilled water. Filter through a 0.2 μm Millipore filter and store frozen at −20°C.

Phenylmethylsulphonyl fluoride (PMSF) (Sigma Chemical Co)
This is extremely toxic and should be handled with care. Using a microcentrifuge tube, make a 1 M solution in DMSO (dimethyl sulphoxide, spectrosol grade, BDH). This needs to be warmed to about 25°C to get a clear solution which can then be diluted to the working concentration of 0.1 mM.

2.1.2 Solutions for hybridization
20% dextran sulphate
Dissolve 20 g in sterile distilled water, filter through a 0.45 μm Millipore filter and store frozen at −20°C.

Disodium hydrogen phosphate
Prepare as a 2 M solution, filter through a 0.2 μm Millipore filter and store at 4°C. This solution crystallizes and needs to be warmed up to dissolve all the crystals before use.

Sodium dihydrogen phosphate
Prepare as a 2 M solution, filter using 0.2 μm Millipore filter and store at 4°C.

Salmon sperm DNA (Sigma Chemical Co)
Dissolve in water to a concentration of 10 mg/ml. Sonicate using 30-sec bursts until there is a considerable drop in viscosity. Generally, 5 – 10 bursts are sufficient. This DNA should have an average size of about 500 bp. Prepare 1 ml aliquots in microcentrifuge tubes and store frozen at −20°C.

40 × Denhardt's solution (0.8% bovine serum albumin, BSA, 0.8% Ficoll and 0.8% polyvinyl pyrrolidone)
Dissolve polyvinyl pyrrolidone to a concentration of 1.6% in water and mix it with an equal volume of a solution containing 1.6% BSA and 1.6% Ficoll. Filter through a 0.45 μm Millipore filter and store at 4°C.

20 × SET (3 M NaCl, 20 mM EDTA and 0.4 M Tris-HCl, pH 7.8)
Filter through a 0.2 μm Millipore filter and store at 4°C.

SDS (BDH Chemicals, product No. 30175)
Make a 10% solution, filter through a 0.2 μm Millipore filter and store at room temperature.

Sodium pyrophosphate (BDH Chemicals, 'Analar' BDH product No. 10261)
Make a 10% solution, filter through a 0.2 μm Millipore filter and store at room temperature.

2.2 Equipment for loading DNA on to gels

There are several ways of preparing high molecular weight DNA embedded in blocks of agarose. One of the procedures is to pour the molten agarose and cell mixture on to a Petri dish, allowing it to solidify and then cut to the appropriate size of plug. The authors find that using a plug mould has several advantages over this method. A plug mould produces very regular-shaped blocks which are reproducible and easy to handle. Moreover, a sample well-former (comb) which matches the size of the blocks, enables easy loading of samples on to the gel. The authors use a Perspex plug mould which gives 100 plugs, each ~6 mm × 1.7 mm × 10 mm in size (*Figure 2*). The size is not critical as long as the resultant plug matches the size of the sample well of the gel. A gel comb for sample wells can be made out of Perspex, PTFE, or similar material. For ease of loading, the size of wells should be slightly larger than the size of the plug after it has been cut and digested (*Figure 2*).

2.3 Yeast chromosome markers

Yeast chromosomes are the most frequently used DNA size markers for pulsed field gel electrophoresis. *Saccharomyces cerevisiae* strains are used in the range of 100 kb

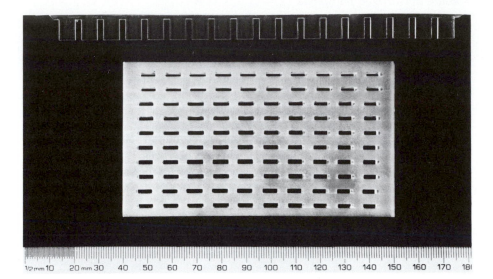

Figure 2. A Perspex plug mould used for preparing DNA embedded in agarose and a gel comb for making sample loading slots in the gel to accommodate the DNA agarose plugs.

to 2 Mb and *Schizosaccharomyces pombe* strains are used in gels for resolving DNA fragments of 3 Mb or larger. The following protocols should work for most strains of yeast used as size markers in pulsed field gels. The strains used by the authors are given in brackets.

Protocol 1. Preparation of chromosomal DNA markers from *Saccharomyces cerevisiae* [X2180-1B and YNN318]

1. Streak cells on to a YPD-agar plate (2% agar in YPD) and grow at 30°C.

2. Pick one colony and grow in 50 ml of YPD medium (2% glucose, 2% bactopeptone and 1% yeast extract) overnight at 30°C.

3. Inoculate two 2-litre flasks (each containing 500 ml of YPD) with $5-10$ ml of the overnight culture and grow the cells at 30°C to an O.D. (optical density) of 0.6 at 600 nm (O.D. of 0.3 at 600 nm is equivalent to 3.3×10^6 cells/ml). This should yield sufficient cells to make between $150-200$ marker sample plugs.

4. Harvest the cells by centrifuging at approximately 500 *g* for 10 min. Resuspend the cells in a solution containing 1.2 M sorbitol, 20 mM EDTA and 14 mM 2-mercaptoethanol and do a cell count. Centrifuge to pellet the cells and resuspend to give a concentration of approximately $2-3 \times 10^9$ cells/ml.

5. Heat to dissolve LGT agarose to a concentration of 1% agarose in 1.2 M

108

Protocol 1. *continued*

sorbitol, 20 mM EDTA. Cool to 37°C and keep at this temperature; add 2-mercaptoethanol to 14 mM.

6. Warm the yeast cell suspension to 37°C and add Zymolyase 20T to 1 mg/ml or lyticase to 20 units/ml.

7. Mix equal volumes of agarose solution and yeast cell suspension and fill the plug mould (see Section 2.5). Allow to set on ice for approximately 10 min.

8. Remove the agarose plugs into a 50 ml Falcon tube containing 20 ml of sorbitol, EDTA, 2-mercaptoethanol solution containing 1 mg/ml Zymolyase-20T or 20 units/ml of lyticase. Incubate at 37°C for 2 h.

9. Replace the solution with 20 ml of yeast lysis solution (1% lithium dodecyl sulphate, 0.1 M EDTA and 10 mM Tris-HCl (pH 8.0), filtered through a 0.2 μm filter). A convenient method of changing solutions is to cover the mouth of the Falcon tube with a Nylon mesh tea strainer to drain off the solution and retain the plugs. Incubate the plugs in the lysis solution at 37°C for 30−60 min.

10. Pour off the solution and add another 20−30 ml of yeast lysis solution. Incubate overnight at 40−45°C.

11. Pour off the solution. The plugs can now be stored in 10 ml of yeast lysis solution at room temperature. This solution will precipitate when cooled and therefore should not be stored in the refrigerator. To store these plugs cold the authors use 20% NDS. Plugs should be washed twice for 2 h in 20% NDS, using approximately 20 ml of solution per wash, to remove most of the lithium dodecyl sulphate. Plugs are stable for several years.

The procedure for *Schizosaccharomyces pombe* [972] is similar to that described previously in *Protocol 1* except for the following differences:

Protocol 2. Preparation of chromosomal DNA markers from *Schizosaccharomyces pombe* [972]

1. Grow the yeast in YPD medium containing 0.5% yeast extract, 0.5% bacto-peptone and 0.5% glucose at 30°C.

2. After harvesting, wash cells twice in 50 ml of 1.2 M sorbitol, 20 mM EDTA, 14 mM 2-mercaptoethanol and 20 mM citrate−phosphate buffer (pH 5.6). Resuspend cells in this buffer to approximately $2-3 \times 10^9$ cells/ml.

3. Add Novozyme to a final concentration of 0.5 mg/ml and incubate at 37°C for 60 min. At this stage it is advisable to check for the formation of spherical protoplasts and if this has been achieved in approximately 50% of cells, proceed to the next step, otherwise continue incubation for another 30−60 min. Apparently, Zymolyase can also be used instead of Novozyme to form *pombe* spherical protoplasts (5).

Protocol 2. *continued*

4. Gently mix the protoplast suspension with an equal volume of 1% LGT agarose solution made up in the yeast suspension buffer and cooled to 37°C. Gently fill the plug mould (see Section 2.5) and allow to set on ice for 10 min.

5. Remove the plugs into yeast lysis solution and continue from Step 9 of *Protocol 1*.

2.4 Bacteriophage lambda DNA oligomers as markers

In the authors' experience, commercially-available bacteriophage (phage) lambda DNA is not suitable for making oligomers since the cohesive ends are not sufficiently intact to form large multimers. It is preferable to make phage lambda DNA and use autoclaved solutions for oligomerization in order to avoid any nuclease contamination. Phage lambda cI857Sam7 DNA (48.5 kb) should be made using the phage maxi-prep method as follows:

Protocol 3. Preparation of phage lambda DNA oligomers

1. Use the induction method described in ref. 20 which gives a high yield and is the method of choice.

2. After CsCl banding of the phage lambda virions, recover the virions and dialyse against 10 mM Tris-HCl (pH 8.0), 10 mM NaCl and 5 mM $MgCl_2$ to remove most of the CsCl.

3. To this solution containing the virions add EDTA to 20 mM, SDS to 0.5% and pronase to 0.5 mg/ml followed by incubation for 2 h at 37°C.

4. Extract the phage lambda DNA using standard phenol/chloroform extractions followed by dialysis overnight against about 2L of TE buffer (10 mM Tris-HCl, pH 7.5, 1 mM EDTA) at 4°C.

5. Concentrate the DNA solution using butanol extractions to give a final concentration of about 0.25 mg/ml DNA. Take extra care when preparing this DNA and use gentle mixing and pipetting to minimize damage by shearing.

6. Carry out oligomerization by incubating this phage lambda DNA at a concentration of 200 μg/ml in 2 × SSC (1 × SSC is 150 mM NaCl, 15 mM sodium citrate, pH 7.0), containing 3% Ficoll and a little orange G as dye, at 37°C for 30 min followed by incubation at room temperature overnight. These oligomers can be stored at 4°C for several months; only 5 μl of this solution is needed to see oligomers up to 1 Mb. Since only a small volume is needed, this marker can be loaded in thin wells (0.5 mm thick) to improve band sharpness.

Oligomers can also be formed by embedding phage lambda virions in agarose plugs. Sufficient virions should be used to give a final DNA concentration of approximately

30 μg DNA/ml. These plugs should be treated like embedded cells (see Section 2.5). Then equilibrate the plugs in 2 × SSC by washing them 3−4 times in a tenfold excess of 2 × SSC and allow oligomerization to proceed at room temperature for 1−2 weeks. If poor oligomerization or slight DNA degradation is observed, include 10 mM EDTA and 10 mM Tris-HCl (pH 7.5) in the 2 × SSC. it is advisable to use autoclaved solutions for the oligomerization reactions in order to avoid any nuclease contamination.

2.5 Preparation of DNA plugs from cultured cells

The procedure described here is for tissue culture cells but it can also be used for other cell types (e.g. lymphocytes).

Protocol 4. Preparation of DNA plugs from cultured cells

1. Harvest the cells by centrifugation at 500 g for 10 min, wash them in Dulbecco's isotonic saline and suspend the cells in 1.0 ml of the same solution. Try to avoid any cell clumping and, if it occurs, disrupt as many of the clumps as possible using dispensing tips or plastic transfer pipettes. Take an aliquot and count in a haemocytometer. Dilute the cell suspension to 5−6 × 10^7 cell/ml; this initial cell concentration can be reduced to 3 × 10^7 cells/ml and this gives sharper bands and a generally improved gel resolution. The authors strongly recommend this, although there is a consequential increase in the autoradiographic time required for the detection of single copy sequences. When working with repeated DNA sequences, the cell concentration can be further reduced to 1 × 10^7 cells/ml for a highly improved gel resolution.

2. Make 1% LGT agarose in Dulbecco's saline. Cool to 37°C and keep at this temperature.

3. Stick tape (the authors use plastic electrical insulation tape) on to one surface of the clean sample plug mould (the mould should have been cleaned by boiling in 0.25 M HCl followed by several washes in distilled water to remove all traces of acid).

4. Warm the cell suspension to 37°C, mix equal volumes of cell suspension and 1% agarose solution and dispense into the plug mould. This mixture may be kept at 37°C while transferring it to the plug mould. It is advisable to place the plug mould on ice while transferring the cell suspension. This enables the agarose to set soon after it is dispensed and so reduces the risk of cells settling during gelling.

5. Leave the mould on ice for 5−10 min to set the agarose plugs completely. Remove the tape and gently push the plugs out, using a bent glass Pasteur pipette, into a Falcon tube containing NDS and 1 mg/ml pronase (proteinase K can also be used but the authors find that pronase from Boehringer works equally well at the same concentration and is much cheaper).

6. Leave the tube at room temperature for 20−30 min and then incubate at 50°C

111

Protocol 4. *continued*

overnight. Replace the NDS containing pronase with fresh solution and continue
incubation at 50°C for another 24 h.

7. Rinse the plugs in NDS, twice for 2 h, to remove most of the pronase. Store
in NDS at 4°C.

8. The plugs are stable for several years in NDS. Each plug has sufficient DNA
for at least 3 separate restriction digests.

2.6 Restriction enzyme digests of DNA in plugs

DNA embedded in agarose plugs can be digested using the following procedure.

Protocol 5. Restriction enzyme digestion of agarose embedded DNA

1. Cut a complete plug into three parts such that the dimensions still match the
sample loading well of the gel. Each part has sufficient DNA for one restriction
digest.

2. Pool all the plugs from each cell line in a siliconized, autoclaved glass test-
tube or sterile Falcon tube. Wash the plugs at least three times for 30 min in
25−30 ml of cold TE (10 mM Tris-HCl, pH 7.5, 1 mM EDTA) at 0°C. One
can add 0.1 mM PMSF to the first two washes to inactivate any residual pronase
or proteinase K. If the plugs are from different cell lines, they should be washed
individually using 5 × 1.5 ml of TE in separate microcentrifuge tubes. Ideally,
the first TE wash should be overnight to remove most of the NDS.

3. Separate the plugs into individual microcentrifuge tubes and wash once for
30 min at 0°C in 500 μl of the appropriate restriction buffer (follow the enzyme
manufacturer's recommendation) without gelatin (or BSA), DTT (or
2-mercaptoethanol) or spermidine.

4. Replace the restriction buffer with 60 μl of buffer containing 100 μg/ml gelatin,
1 mM DTT and spermidine (if necessary; see Section 2.6.1). Incubate at room
temperature (approx. 20°C) for 10 min.

5. Add the restriction enzyme. Mix and incubate at the appropriate temperature
for 2 h. The amount of enzyme needed varies from one enzyme to another,
but generally 5−20 units are sufficient. The enzyme can be added in two
aliquots; the second aliquot being added after an hour of incubation.

6. After 2 h, drain off the restriction buffer and add about 500 μl of STOP buffer
(0.5 × TAE, 10 mM EDTA and 0.1 mg/ml Orange G as marker dye; the
composition of TAE is given in Section 3). Store at 0°C (for a maximum of
2 h) ready for loading on to the gel. If digested plugs need to be stored for
longer, they should be placed in NDS at 4°C (low molecular weight DNA will
diffuse out and be lost during prolonged storage). These plugs should be re-
equilibrated in STOP buffer before loading on a gel.

2.6.1 Precautions for restriction digests of DNA in plugs

- All restriction buffers should be made up as $10 \times$ stock solutions without spermidine, gelatin and DTT. Aliquots of 1 ml can be autoclaved in microcentrifuge tubes (puncture the top before autoclaving) and stored sealed and frozen. TE buffer and distilled water should also be autoclaved. Once opened, these solutions should only be used for one complete experiment.

- Since the DNA in plugs is very large, it is very sensitive to degradation. A control incubation without restriction enzyme is essential to monitor the extent of degradation. If the amount of degradation is minimal, the incubation time used for restriction digests can be increased for expensive and difficult enzymes.

- Spermidine may sometimes help restriction digests, especially when a batch of DNA plugs is giving problems with digestion. If used, its final concentration should not exceed 2 mM in buffers with $50-100$ mM salt and 5 mM spermidine in buffers containing concentrations of 100 mM or more of salt.

- All microcentrifuge tubes and dispensing tips should be autoclaved. Disposable plastic gloves should be worn during all manipulations involving plugs and enzymes in order to avoid nuclease contamination from the skin and consequential DNA degradation.

2.6.2 Useful enzymes and mapping strategies

Since the technique of pulsed field electrophoresis is used mainly for the separation of DNA molecules larger than 50 kb in size, the most frequently used enzymes are those which have an 8 bp recognition-sequence (*Table 1*). However, methylation of the genomic DNA renders some 6 bp recognition-sequence enzymes very useful also. A good example is the G:C base-pair methylation of vertebrate genomic DNA (21). The CpG dinucleotide occurs in vertebrate DNA at approximately 25% of the frequency expected from base composition and, in addition, it is often methylated at cytosine. This methylation inhibits cleavage by many enzymes which have one or more G:C base-pairs in their recognition sequence (*Table 1*). Similarly, methylation of adenine in bacterial DNA, and cytosine in plant DNA in the sequence ^{m}CNG (where N is any nucleotide), can also inhibit some restriction enzymes. A systematic search for rare cutting enzymes in bacterial genomes has been published (22). Other frequently cutting enzymes can be useful sometimes especially in the study of simple sequences (e.g. satellite DNA) which can be devoid of sites for these enzymes over long stretches of DNA (23).

The length of DNA that can be successfully mapped by PFGE depends on the number of probes available and their physical distribution in the region of interest. When starting from a single random probe, partial and complete digests as well as double digests provide only limited information. The direction of mapping can only be established if characterized genetic markers, such as deletions and translocations, exist within the largest hybridizing fragment resolved by PFGE. To extend the map further, new probes need to be generated either by random cloning from a DNA

Enzyme	Recognition sequence	Enzyme	Recognition sequence
*Not*I	GCGGCCGC CGCCGGCG	*Aat*II*	GACGTC CTGCAG
*Sfi*I	GGCCNNNNNGGCC CCGGNNNNNCCGG	*Cla*I*	ATCGAT TAGCTA
*Rsr*II	CGG(A/T)CCG GCC(T/A)GGC	*Eco*47 III*	AGCGCT TCGCGA
*Bss*HII	GCGCGC CGCGCG	*Nae*I*	GCCGGC CGGCCG
*Mlu*I	ACGCGT TGCGCA	*Nar*I*	GGCGCC CCGCGG
*Nru*I	TCGCGA AGCGCT	*Sal*I*	GTCGAC CAGCTG
*Pvu*I	CGATCG GCTAGC	*Sma*I*	CCCGGG GGGCCC
*Sac*II/*Sst*II	CCGCGG GGCGCC	*Sna*BI*	TACGTA ATGCAT
Sp/I	CGTACG GCATGC	*Xho*I*	CTCGAG GAGCTC

source enriched in the region of interest or by cloning ends of deletions or trans-locations in the region. Such an approach has been used successfully in mapping about 4.5 Mb of the Xp2.1 region of human X-chromosome containing the gene associated with Duchenne and Becker muscular dystrophies (8). If, however, one starts from a probe at or near the end of a chromosome, complete and partial digests can be probed to obtain a directional map starting from the telomere and extending towards the centromere. This approach has been used successfully to map the pseudoautosomal region of the Y-chromosome (24). A similar approach can also be used to map from some translocation junctions; for example, in one of the X:21 translocations associated with Duchenne muscular dystrophy, the derived-X and derived-21 chromosomes both have ribosomal DNA repeats on one side of the junction (25). Although ribosomal DNA has many *Not*I sites, *Not*I does not cleave the X-chromosome DNA within about 1 Mb from the translocation point. If *Not*I complete digests are followed by partial and complete digests with other rare cutting restriction enzymes and probed with an X-chromosome probe at, or close to the two junctions, sequential restriction sites can be mapped.

2.6.3 Tests for the completeness of a restriction digest

Although in some well-resolved pulsed field gels the DNA bands can be visualized by ethidium bromide staining, it is generally difficult to distinguish between a complete and a partial digest. Standard techniques like increasing the enzyme concentration and incubation time can be used to ensure a complete digest. However, a more rigorous test is to use a second enzyme which can test the complete cleavage of a specific site: an enzyme that cuts frequently, with sites flanking the infrequent site being tested, is used to digest some uncut plug DNA as well as some of the plug DNA already digested with the infrequent cutter enzyme. These DNA samples are then run on a conventional gel, blotted and probed (26). This test is only feasible if one has a probe for the DNA between the two 'frequent-cutter' sites. An alternative is to use an internal control probe which is known to hybridize to a DNA fragment of pre-determined size on a pulsed field gel blot. It should be noted, however, that complete cleavage at one site for an enzyme does not guarantee complete cleavage at all sites since it is known that some sites are cleaved faster than others.

2.6.4 Partial digests

Although partial digests are often a hindrance, they can sometimes be extremely useful. This is especially true when mapping consecutive restriction sites or establishing a physical linkage between probes known to be close to one another by genetic evidence. When the DNA is in solution, reducing the enzyme concentration or incubation time can usually achieve a reasonable partial digest, but as digests for pulsed field gels are performed on DNA embedded in agarose, a combination of the two conditions is preferable. The final conditions will vary depending on the enzyme used. However, the most convenient way of arriving at the appropriate conditions is to preincubate the DNA plugs for 30−60 min on ice with three different enzyme concentrations (2 plugs for each concentration). Once the enzyme has equilibrated through the plugs, incubate at the required temperature for two different time points. It is always advisable to do a limit digest at the same time as the partial digest.

3. Electrophoresis conditions

The rate of migration of DNA through a gel depends on the gel concentration, voltage gradient, and temperature. Generally, electrophoresis is in 0.5 × TAE (5 × TAE contains 24.2 g Tris base, 5.71 ml glacial acetic acid and 20 ml of 0.5 M EDTA, pH 8 per litre) or 0.5 × TBE (5 × TBE contains 54 g Tris base, 27.5 g boric acid and 20 ml of 0.5 M EDTA, pH 8 per litre) at a constant temperature and with a fixed voltage gradient across the electrodes. The authors prefer TAE buffer because TBE buffer can lead to subsequent problems with transferring the DNA if blotting on to nitrocellulose filters. The voltage gradients for optimum separations depend on the size range to be resolved and the gel concentration (see *Figures 3−5*). The appropriate voltage gradient for a particular apparatus can be found in the various

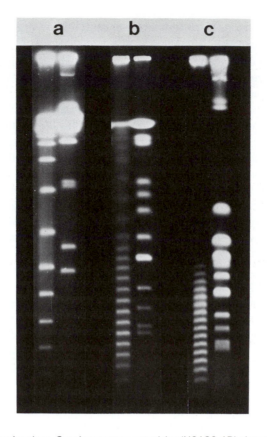

Figure 3. Effect of pulse time. *Saccharomyces cerevisiae* (X2180-1B) chromosomes and phage lambda (c1857Sam7) oligomers run on a 1.5% agarose gel in 0.5 × TAE at 6 V/cm and 20°C. The pulse times used were (a) 20 sec, (b) 60 sec and (c) 100 sec. The high resolution showing polymorphism (possibly due to mitotic recombination) in chromosome 3 is clearly visible in the 20 sec pulse. In the 100 sec pulse, chromosome 12 polymorphism resolves as a doublet near the top of the gel. The general band broadening seen with this pulse time can be reduced by lowering the voltage gradient.

published articles. Temperature has similar effects in all systems, that is increasing temperature increases the mobility of DNA. Therefore, to enable results from different gel runs to be comparable, it is essential to maintain a constant temperature. A detailed analysis of temperature and its effects on the Pulsaphor PFGE system has been published (27). The effect of gel concentration and pulse time is discussed in subsequent sections, and is based on the authors' experience with the rotating gel system. Although the effect of pulse time is comparable between the various pulsed field systems, this is not so for the field inversion system since part of the switching cycle is used for reverse migration of DNA. Increasing the duration of the switching cycle (forward and reverse DNA migration) increases the upper limit

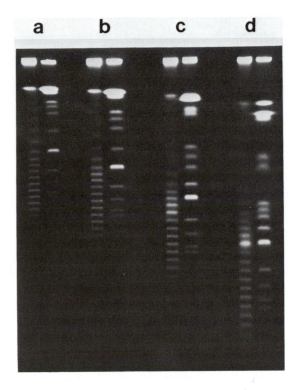

Figure 4. Effect of agarose gel concentration. A composite gel with agarose concentrations of (a) 1.5%, (b) 1.25%, (c) 1.0% and (d) 0.75% was run in 0.5 × TAE at 20°C. The voltage gradient was 4 V/cm with a total run time of 35 h and a pulse duration of 90 sec. The samples were phage lambda (c1857Sam7) oligomers and *Saccharomyces cerevisiae* (X2180-1B).

of the DNA size that can be resolved; that is, in a 1% gel run at 7.5 V/cm with a forward to reverse ratio of 2:1, a switching cycle of 3 sec resolves molecules in the range of 50−100 kb, whereas a switching cycle of 12 sec resolves molecules in the 100−300 kb range. By building a ramp between these two intervals (3−12 sec) linear separations in the size range of 50−300 kb can be achieved (28). For further details the reader is referred to ref. 15 or to the recommended conditions provided with the commercially-available apparatus.

3.1 Effect of pulse time

In general increasing the pulse time results in an increase in the size range of DNA molecules separated. There is, however, a consequential loss of resolution, that is the separation between molecules reduces with increasing pulse time. As can be seen from *Figure 3*, at 6 V/cm DNA molecules up to 400 kb are separated with a 20 sec pulse and DNA ≥ 1000 kb with a 60 sec pulse. However, increasing the pulse time to 100 sec results in a slight smearing of the larger size DNA molecules. In order

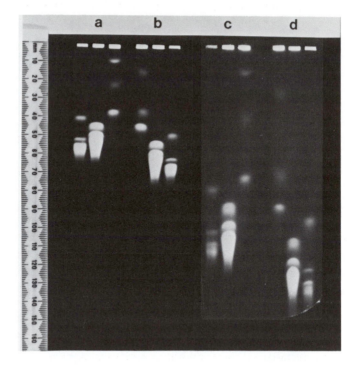

Figure 5. Separation of very large size DNA. *Saccharomyces cerevisiae* YNN318, X2180-1B and *Schizosaccharomyces pombe* 972 chromosomes run on 0.7% agarose (a and c) and 0.6% agarose gels (b and d). Electrophoresis was in 0.5 × TAE at 5°C and a voltage gradient of 1.2 V/cm was used with 60-min pulses. The total run time was 7 days (a and b) and 14 days (c and d) Chromosome 12 of YNN318 is longer than that of X2180-1B and is separating as such. However, the rest of the chromosomes of these two strains are of almost the same size but the X2180-1B DNA is considerably retarded clearly demonstrating the drag effects caused by excessive DNA loading in these gels.

to achieve good resolution of these larger DNA molecules, the voltage gradient should be reduced. This results in an increased run time and has to be accompanied by a proportional increase in pulse time as well. For example, if a 1.5% agarose gel is run in 0.5 × TAE at 6 V/cm, the total run time on a rotating gel system with a 22 cm diameter gel is 34 h at 20°C. The same gel at 5 V/cm would require a total electrophoresis time of 49 h and therefore a pulse time of 100 sec at 6 V/cm will need to be increased to 144 sec at 5 V/cm to achieve a comparable separation but sharper resolution. A detailed analysis of the effect of pulse time using the Pulsaphor system has also been published (29).

3.2 Effect of gel concentration

At a fixed voltage gradient, decreasing the gel concentration results in an increase in the rate of DNA migration. The effect is similar, though not identical to the effect of increasing the pulse time. Mathew *et al.* (27) have shown a small but significant improvement in resolution by increasing the gel concentration from 0.9% to 1.2%. The authors normally use 1.5% agarose gels for all separations up to 1.5 Mb purely because of the ease of handling these relatively rigid gels. Lower gel concentrations can be used (*Figure 4*) but the pulse time and voltage gradients should be adjusted to achieve the required resolution.

3.3 Separation of very large DNA

As mentioned in Section 3.1, at a fixed voltage gradient, increasing the pulse time resolves larger DNA molecules. This relationship holds true up to a point beyond which the larger DNA molecules begin to smear. This problem is particularly severe when attempting to resolve molecules larger than 1 Mb. For these separations, low voltage gradients have to be used in combination with low concentration agarose gels. The authors routinely use 0.6% agarose gels at 1.2 V/cm voltage gradient to resolve molecules in the size range of 1 Mb to ≥ 6 Mb. These runs are carried out at 5°C using a pulse duration of between 30 − 60 min. The total duration of the run is between 1 − 2 weeks depending on the resolution required (*Figure 5*).

4. Gel processing, DNA transfer, and hybridization

4.1 Gel blotting procedures

As PFGE gels contain large DNA molecules, it is essential to break up the DNA before transfer. The authors use acid depurination as opposed to UV irradiation. The DNA can be transferred either to Nylon or nitrocellulose membranes. Since the feasibility of multiple hybridization on the same blot is an obvious advantage, the authors recommend DNA blots on to Nylon membranes using SSC or alkaline solutions.

Protocol 6. Blotting on to Nylon membranes

1. Stain the gel with ethidium bromide (0.5 μg/ml) for 1 − 2 h (prepare the staining solution from a 10 mg/ml stock solution).

2. Destain for 1 − 2 h and then photograph using a UV transilluminator. The authors use a Photodyne transilluminator (model 3-3002) and photograph gels on to a Kodak Professional Technical pan film through a Wratten red and blue-green filter combination or through an interference filter with peak transmission at 590 nm and a half band width of 10 nm.

3. Cut an appropriate size of GeneScreen Nylon membrane. Carefully layer on 2 × SSC to wet the membrane completely.

4. Depurinate the DNA by immersing the gel twice, each time for 15 min, in 500 ml of 0.25 M HCl.

5. Denature the DNA by immersing the gel twice, each time for 15 min, in 500 ml of 0.5 M NaOH, 1.5 M NaCl.

Protocol 6. *continued*

6. Now transfer the DNA on to the GeneScreen Nylon membrane by blotting over-night using 0.4 M NaOH or 10 × SSC as the transfer solution.

7. Carefully remove the membrane and wash it in 2 × SSC containing 0.2 M Tris-HCl (pH 7.5) to neutralize the membrane (after alkaline blotting) and to remove any gel fragments that may have stuck to it.

8. Wash the membrane in 2 × SSC and place it, with the DNA side up, on a filter paper. Allow to dry at room temperature. Drying can be speeded up by placing the filter in a warm room or an incubator.

9. Bake the membrane in a vacuum oven at 80°C for 2 h or expose to UV light (see Section 2.10.6 of Chapter 2) to fix the DNA.

To transfer the DNA on to nitrocellulose, gels must be neutralized after denatura-tion (see *Protocol 6*) by immersing the gel twice, each time for 15 min, in 500 ml of 1.5 M NaCl, 0.5 M Tris-HCl (pH 7.0). The DNA can now be transferred on to nitrocellulose using 10 × SSC (1.5 M NaCl, 0.15 M sodium citrate, pH 7.0) as the transfer solution, over 24−48 h. After transfer, rinse the filter in 2 × SSC, air dry, and bake it in a vacuum oven at 80°C for 2 h.

4.2 Probing blots by hybridization

There are many different hybridization protocols. Most of them should work for pulsed field gels. The ones used successfully by the authors are described below.

4.2.1 GeneScreen

For GeneScreen mix concentrated stock solutions to the following final concentrations for pre-hybridization as well as hybridizations at 65−68°C.

- 5 × Denhardt's solution (0.1% Ficoll, 0.1% polyvinyl pyrrolidone, 0.1% BSA)
- 0.5 M phosphate buffer (pH 7.5)
- 1% SDS
- 10% dextran sulphate
- 100 μg/ml denatured sonicated salmon sperm DNA (optional).

Pre-hybridization is for 3−6 h followed by fresh hybridization solution and hybridization for 18−24 h. The Nylon membrane filters are then washed twice, each time for 5 min, in 2 × SSC at room temperature then twice for 30 min in 2 × SSC containing 1% SDS at 65°C followed by two final stringency washes at 65°C for 30 min in 0.1 or 0.2 × SSC. If the filter is to be re-used for subsequent hybridizations, it must be sealed damp for autoradiography. If the filter is allowed to dry, some of the probe will bind in an irreversible manner and it will not be possible to strip off all the probe from the filter. To remove the signal (stripping) for subsequent hybridizations it is best to use the protocols described in the GeneScreen handbook. See also Section 2.10.9 of Chapter 2.

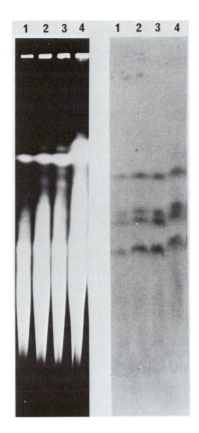

Figure 6. Effect of DNA loading. On the left is an ethidium bromide stained gel photograph of four tracks of *Sfi*I digested human female (X:21 translocation) DNA. On the right is the corresponding autoradiograph of these same tracks probes with a human X-chromosome probe which hybridizes to four fragments of 565 kb, 690 kb, 715 kb and 840 kb. The DNA loadings in tracks 1 to 4 were 1.5 μg, 3.0 μg, 3.0 μg and 6.0 μg.

4.2.2 Nitrocellulose

For pre-hybridization use a solution containing:

- 5 × SET
- 5 × Denhardt's solution
- 0.1% SDS
- 0.1% sodium pyrophosphate

The hybridization mix has all the above constituents plus 10% dextran sulphate. Pre-hybridization is at 65°C for 1−6 h followed by hybridization at 65°C for 18−24 h. Nitrocellulose filters should be washed twice, each time for 30 min, at

65°C in 2 × SSC containing 0.1% SDS and 0.1% sodium pyrophosphate followed by the final stringency washes, washing twice at 65°C for 30 min in 0.1 or 0.2 × SSC containing 0.1% SDS and 0.1% sodium pyrophosphate.

5. General precautions

The method of pulsed field gel electrophoresis has extremely wide applications. However, one must be aware of the limitations and pitfalls which arise mainly from the experimental procedures involved in its use. Some of the problems associated with partial and limit restriction enzyme digests have already been discussed. It is also found that methylation patterns vary substantially between different cell lines and therefore one must be careful in interpreting and comparing results obtained using enzymes which are sensitive to methylation. Another point relates to the sizing of fragments. DNA migration in pulsed field gels is sensitive to the loading concentration: overloading results in retardation of DNA fragments causing an overestimation of their size (*Figure 6*). Although preliminary work can be done quicker by using high DNA concentrations and thus achieving fast autoradiographic results, final, accurate sizing must always be done on gels where DNA concentration drag effects are minimal. Finally, the authors have encountered a problem where DNA migration in all the pulsed field gels was retarded. The effect was most noticeable in the marker tracks and especially affected the small members of the phage lambda DNA oligomeric series. The problem was finally tracked down to a new batch of agarose that had been used for making the PFGE gels. Therefore it is recommended that at least one test gel with various size markers should be run with each new batch of agarose used.

Acknowledgements

We gratefully acknowledge the contributions of Chris Tyler-Smith, William Brown, Paul Whittaker, and other members of the research group who contributed to these methods. We should also like to thank Martin Johnson for building the rotating gel apparatus, Jill Ogden and Ron Davis for the yeast strains, Mary Hine for typing the manuscript and Ken Johnson for the photographic work. R.A. acknowledges support by the Muscular Dystrophy Group of Great Britain and Northern Ireland whilst he was at Oxford University. R.A. is now at the Department of Biotechnology, ICI Pharmaceuticals, Alderley Park, Macclesfield, Cheshire, U.K.

References

1. Schwartz, D. C. and Cantor, C. R. (1984). *Cell,* **37**, 67.
2. Carle, G. F. and Olson, M. V. (1984). *Nucleic Acids Res.,* **12**, 5647.
3. Anand, R. (1986). *Trends in Genetics, 2*, 278.
4. Chu, G., Vollrath, D., and Davis, R. W. (1986). *Science,* **234**, 1582.
5. Smith, C. L., Matsumoto, T., Niwa, O., Klco, S., Fan, J.-B., Yanagida, M., and Cantor, C. R. (1987). *Nucleic Acids Res.,* **15**, 4481.

6. Southern, E. M., Anand, R., Brown, W. R. A., and Fletcher, D. S. (1987). *Nucleic Acids Res.,* **15**, 5925.
7. Southern, E. M. (1975). *J. Mol. Biol.,* **98**, 503.
8. Monaco, A. P. and Kunkel, L. M. (1987). *Trends in Genetics,* **3**, 33.
9. Michiels, F., Burmeister, M., and Lehrach, H. (1987). *Science,* **236**, 1305.
10. Anand, R., Honeycombe, J., Whittaker, P. A., Elder, J. K., and Southern, E. M. (1988). *Genomics,* **3**, 177.
11. Cantor, C. R., Warburton, P., Smith, C. L., and Gaal, A. (1986). *Electrophoresis 1986: Proceedings of the 5th meeting of the International Electrophoresis Society* (ed. M. J. Dunn), p. 161. VCH Publishers, Federal Republic of Germany.
12. McPeek, F. D. Jr., Coyle-Morris, J. F., and Gemmill, R. M. (1986). *Anal. Biochem.,* **156**, 274.
13. Gardiner, K., Lass, W., and Patterson, D. (1986). *Somatic Cell and Mol. Genetics,* **12**, 185.
14. Tombs, M. P. International patent application WO-87-00635.
15. Carle, G. F., Frank, M., and Olson, M. V. (1986). *Science,* **232**, 65.
16. Serwer, P. (1987). *Electrophoresis,* **8**, 301.
17. Vollrath, D. and Davis, R. W. (1987). *Nucleic Acids Res.,* **15**, 7865.
18. Van Ommen, G. J. B. and Verkerk, J. M. H. (1986). *Human Genetic Diseases: A Practical Approach* (ed. K. E. Davies), p. 113. IRL Press, Oxford.
19. Smith, C. L., Warburton, P. E., Gaal, A., and Cantor, C. R. (1986). *Genetic Engineering* (ed. J. K. Setlow and K. Hollaender), Vol. 8, p. 45. Plenum, New York.
20. Maniatis, T., Fritsch, E. F., and Sambrook, J. (1982). *Molecular Cloning: A Laboratory Manual,* Cold Spring Harbor Laboratory Press, NY.
21. Bird, A. P. (1986). *Nature,* **321**, 209.
22. McClelland, M., Jones, R., Patel, Y., and Nelson, M. (1987). *Nucleic Acids Res.,* **15**, 5985.
23. Tyler-Smith, C. (1987). *Development,* **101 suppl.,** 93.
24. Brown, W. R. A. (1988). *EMBO J.,* **7**, 2377.
25. Bodrug, S. E., Ray, P. N., Gonzalez, I. L., Schmickel, R. D., Sylvester, J. E., and Worton, R. G. (1987). *Science,* **237**, 1620.
26. Brown, W. R. A. and Bird, A. P. (1986). *Nature,* **322**, 477.
27. Mathew, M. K., Smith, C. L., and Cantor, C. R. (1988). *Biochemistry,* **27**, 9204.
28. Bostock, C. J. (1988). *Nucleic Acids Res.,* **16**, 4239.
29. Mathew, M. K., Smith, C. L., and Cantor, C. R. (1988). *Biochemistry,* **27**, 9210.

4

The electrophoresis of synthetic oligonucleotides

J. WILLIAM EFCAVITCH

1. Introduction

The introduction of simple synthetic methodologies and automated synthesizers has led to an unprecedented availability of synthetic oligodeoxynucleotides and the concomitant need to isolate a discrete fragment from a mixture of similarly sized oligodeoxynucleotides. Electrophoresis is the method of choice for the analysis and isolation of microgram quantities of synthetic fragments since it provides a greater dynamic range in size than most chromatographic methods and allows the simultaneous processing of several samples. Because of the infinite variety of sequences encountered, the observed mobilities of synthetic fragments are sometimes quite anomalous. This chapter focuses on the special considerations that are needed in the handling of both routine and anomalous oligodeoxynucleotides, and describes some observations unique to chemically-synthesized DNA. This chapter also discusses the isolation of fluorescently-labelled oligodeoxynucleotides used as primers for non-radioactive Sanger sequencing methods.

Although there is a growing need for and availability of synthetic oligoribonucleotides, this chapter does not specifically concern itself with techniques for their electrophoresis. Aside from the differences required by the synthetic methodologies and the increased susceptibility to degradation, the methods used for the electrophoresis of these molecules are essentially the same as those for synthetic oligodeoxynucleotides.

2. Basic techniques

The electrophoresis of synthetic oligodeoxynucleotides is really a subset of the techniques used to sequence DNA. Thus the basic techniques, equipment and buffer systems are the same (1). Denaturing polyacrylamide gels provide separations based on molecular weight assuming a constant ratio of charge to mass. Synthetic oligodeoxynucleotides can range in size from several nucleotides (nts) to 200 nts in length. The purity required is dependent on the application. Thus, for use as primers and probes the degree of purity required is less stringent. For other applications

125

the purity required may need to be greater than 98%. Purity of a synthetic oligo-deoxynucleotide is defined by two criteria: first, size homogeneity and second, chemical authenticity. While a good electrophoretic technique can guarantee size homogeneity to within one nucleotide, chemical authenticity is more a function of the synthetic chemistry. Thus, while a gross chemical modification which alters the molecular weight or the charge can be resolved, it is beyond the capability of polyacrylamide gel electrophoresis to ensure chemical homogeneity. Fortunately, the current synthetic methods afford a reasonably high level of chemical homogeneity.

The salient problem in analysing or purifying a synthetic oligodeoxynucleotide *n* nucleotides long from a crude synthesis mixture is the separation of similar molecular weight contaminants. Predominantly these consist of (*n*-1)mer and (*n*-2)mer oligodeoxynucleotides although sometimes (*n*+1)mers may be present also. Resolution of the other contaminants is usually trivial even though their cumulative mass may represent a significant fraction of the total contaminant mass. Thus electro-phoretic conditions for the purification of synthetic oligodeoxynucleotides should optimize the separation between the desired product and its (*n*-1)mer while ignoring the separation between the other components.

Electrophoresis is used in several different ways to analyse and isolate synthetic oligodeoxynucleotides:

- crude syntheses and purified products may be analysed and visualized by UV light (see Section 2.8.2.). Detection by UV light is a fast and easy detection method but requires a considerably larger amount of sample than is usually required for autoradiography.
- crude syntheses and purified products may be analysed by enzymatically labelling the 5'-OH or 3'-OH termini with ^{32}P or ^{35}S and detected by autoradiography.
- crude syntheses may be purified by preparative electrophoresis and analysed by UV.

2.1 Equipment

2.1.1 Slab gel apparatus

The most desirable format for the electrophoretic analysis and preparative isolation of synthetic oligodeoxynucleotides is the slab gel. There are many commercially-available formats which essentially consist of a sandwich of two glass plates held apart by two side spacers with a sealed bottom and a multi-welled sample well-former (comb) at the top. Details of this type of apparatus have been given in Chapters 1 and 2. The lengths and widths of the gel are not critical as long as the minimum length is 20 cm, with 40 cm being the best choice. A practical consideration in choosing a width for the plates is the number of samples usually run at one time which therefore dictates the volume of acrylamide solution used and hence the cost of each gel. Some types of commercially-available apparatus use a clamping gel-casting stand to seal the bottom of the gel; most use a bottom spacer or better still are taped along the lower edge.

2.1.2 Spacers

In general, the thickness of the most useful spacers are between 0.4 mm and 1.6 mm. The thinnest are used for analytical gels of ^{32}P-labelled oligodeoxynucleotides while the latter are best for preparative isolations. Since the amount of sample which can be loaded is a function of the surface area of the well, the larger thickness should be chosen for preparative separations. There are two formats for spacers, one for glass plates without notches and the other for notched plates. The latter are somewhat simpler to use but require a slightly more expensive set of glass plates which are more prone to breakage. The former sometimes require the application of water-insoluble grease to prevent leakage which makes them harder to clean for subsequent use.

2.1.3 Sample well former

The most critical choice to be made is in the width of the wells of the sample well-former (comb). Maximum resolution of preparative samples is obtained by having the sample band as thin as possible. This is achieved by having the maximum surface area for sample loading. Well widths should be greater than 1 cm with a width of 2.5 cm providing a comfortable margin. Widths of less than 1 cm are only useful for analytical runs or for sample loads of less than 1 A_{260} units. Combs containing eight to ten 2.5 cm wells are commercially available and these are usually suitable because it is rare to process more than ten preparative samples at a time.

2.1.4 Power supply

Since the resolution of synthetic oligodeoxynucleotides is not adversely affected by the field strength, it is desirable to perform the electrophoresis at the highest possible voltage to obtain the shortest run time. Additionally, higher field strengths result in heating of the gel which provides denaturing conditions that disrupt any secondary structure. Power supplies capable of providing 3000 V and 300 mA are best. Since it is desirable to control the heating of the gel, a constant power option for the power pack is also recommended.

2.1.5 Temperature control

The optimum temperature for the electrophoresis of synthetic oligodeoxynucleotides is 50−70°C. This is sufficiently hot enough to crack glass plates unless there is some way to dissipate some of the heat. The easiest way to accomplish this is by clamping an aluminium plate to one of the glass surfaces. For glass plates with the dimensions 33 × 40 cm, a 30 × 35 cm aluminium plate 1 cm thick can be used.

2.2 Stock solutions

2.2.1 Acrylamide stock solutions

A stock solution of 38% acrylamide, 2% bisacrylamide should be prepared in advance. High quality acrylamide and bisacrylamide are available from several

Table 1. Approximate sizes of oligodeoxynucleotides co-migrating with tracking dyes in denaturing polyacrylamide gels.

Percentage of polyacrylamide	Size of oligodeoxynucleotides (nts)	
	Size co-migrating with bromophenol blue	Size co-migrating with xylene cyanol
5	35	130
6	26	106
8	19	70 – 80
10	12	55
20	8	28

Table 2. Velocities of tracking dyes.

% Polyacrylamide	Bromphenol blue	Xylene cyanol
20%	8.3 cm/h	4.3 cm/h
12%	9.5 cm/h	4.8 cm/h
8%	10.6 cm/h	5.7 cm/h

Measurements were made on a 33 × 40 cm gel at a constant power of 60 W.

commercial sources and can be used without further purification. This solution is prepared in the following way. Weigh out 380 g of acrylamide and 20 g of bisacrylamide, dissolve and make up to 1 litre with double-distilled water. *Acrylamide in the monomeric form is a neurotoxin so be sure to wear gloves and a dust mask when weighing it out.* After dissolving the two monomers, filter the solution through Whatman 3MM paper. This solution should be stored at 4°C and is stable for at least a month.

2.2.2 Electrophoresis buffer (10 × TBE)

This solution is used in the gel itself and with dilution as the electrode buffer. 10 × TBE is 0.89 M Tris base, 0.89 M borate and 25 mM EDTA. To prepare this buffer dissolve 108 g Tris base, 55 g boric acid and 9.3 g disodium EDTA dihydrate and make up to 1 litre with distilled water.

It is not necessary to adjust the pH of this solution; it should be about pH 8.3. The solution should be stored at 4°C. This stock is stable for many months but is susceptible to bacterial growth and should be inspected visually from time to time.

2.2.3 1.6% Ammonium persulphate

This solution is prepared as a 1.6% solution in distilled water. Since it is the initiator of polymerization it is advisable to make a fresh stock solution every week, although it can be used successfully for several weeks. A volume of 100 ml is a good compromise for a laboratory running six to ten gels a week. Store this solution at 4°C.

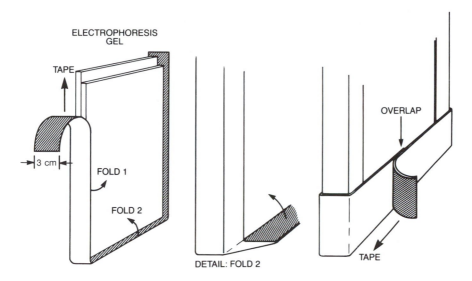

Figure 1. Procedure for sealing the slab gel assembly with tape. After aligning the spacers and clamping the glass plates at the top the edges are sealed with one continuous strip of tape 3 cm wide. The excess tape is folded in on to the plates with lapped corners. A second strip of tape is applied around the bottom from side to side to provide additional support.

2.3 Choice of gel concentration

The choice of the acrylamide percentage is a compromise between speed and resolution. It is recommended that one uses the highest possible percentage gel for a given size of oligodeoxynucleotides. Since diffusion does not appear to be a significant factor in the loss of resolution, the longer electrophoresis time required for a higher percentage gel is not detrimental. Best resolution is achieved by allowing the sample to migrate as far as possible. *Table 1* shows the mobility of various sizes of oligodeoxynucleotides relative to the percentage of acrylamide. *Table 2* is a listing of the velocities of the bromophenol blue and xylene cyanol dyes in various gel percentages.

2.4 Preparation of the gel plates

The following assembly procedure is a generic one for a system which does not utilize a bottom spacer or a gel assembly stand.

Protocol 1. Assembly of the gel mould

1. Clean the glass plates well with hot water and detergents.
2. Thoroughly wipe the plates with acetone or ethanol to dry them and remove any organic contaminants. (Siliconizing the glass is not generally necessary

Protocol 1. *continued*

for thicker gels; with 0.4 mm gels it may be desirable to siliconize one of the two plates.)

3. Support the larger of the two plates on a raised platform; a 12.5 cm diameter recrystallization dish is useful.

4. Lay the spacers along each side, position the top plate and clamp the plate at the top with a bulldog clamp.

5. Using one continuous strip, apply 3-cm-wide sealing tape (Permacel, New Brunswick, NJ 08903) along the right, left and bottom thicknesses of the gel plate assembly; position the tape so that an equal width is above and below the spacer (*Figure 1*).

6. Press the tape firmly along the perimeter to ensure a good seal.

7. Starting at the top right-hand corner carefully fold the tape over the edge, moving down towards the bottom right-hand corner.

8. Now fold over the left-hand edge, starting at the top left-hand corner moving towards the bottom. Fold the bottom edge over, forming neat square corners at each side.

9. Now turn the whole assembly over and repeat the process on the other side.

10. Make a secondary seal binding the lower sides together and covering the corners. Using one continuous strip of tape overlap the lower seal on one side, the right-hand corner, and the lower seal on the other side and the left-hand corner, finishing by overlapping the ends of the tape strip.

11. Apply clamps positioning the jaws over the spacers along each side. It is necessary to only cover the lower half of the sides.

Table 3. Recipes for polyacrylamide gels[a].

Gel thickness	Total volume	% Poly-acrylamide	40% Acrylamide	Urea	10 × TBE	Ammonium persulphate	TEMED
0.4 mm	100 ml	8%	20 ml	42 g	10 ml	4.4 ml	60 μl
		12%	30 ml	42 g	10 ml	4.4 ml	60 μl
		20%	50 ml	42 g	10 ml	4.4 ml	60 μl
0.8 mm	150 ml	8%	30 ml	63 g	15 m	6.6 ml	100 μl
		12%	45 ml	63 g	15 m	6.6 ml	100 μl
		20%	75 ml	63 g	15 m	6.6 ml	100 μl
1.6 mm	250 ml	8%	50 ml	105 g	25 ml	11 ml	150 μl
		12%	75 ml	105 g	25 ml	11 ml	150 μl
		20%	125 ml	105 g	25 ml	11 ml	150 μl

[a] Recipes are for 33 cm × 40 cm format gels.

2.5 Preparing the gel

Before assembling the gel plates it is advisable to begin preparing the gel solution since urea is slow to dissolve. Always prepare a greater volume of gel solution than is needed in case of an accidental leakage. *Table 3* gives the volumes of ingredients needed for various thicknesses and percentages of gel for a 33 cm × 40 cm gel mould. Add together the components as described in *Protocol 2*. It is not necessary to degas the acrylamide solution as this leads to very rapid polymerization which can interfere with the pouring process. A well-sealed gel mould should not leak while the gel is polymerizing.

Protocol 2. Preparation of the polyacrylamide slab gel

1. Weigh out the amount of urea specified in *Table 3* into a conical flask.

2. Add the acrylamide and 10 × buffer. Add water to bring to the correct volume.

3. Stir with mild warming until the urea is dissolved, do not overheat! Stop heating as soon as the urea has dissolved. While stirring the mixture, assemble the gel plates.

4. When ready, add the TEMED then the ammonium persulphate with stirring. Immediately proceed to next step.

5. Holding the gel mould assembly at an angle of 45°, quickly pour the gel solution down one side until the level is just below the top of the lower plate.

6. Bring the mould to a vertical position to dislodge any trapped air bubbles. Tapping the surface gently with a wooden object helps to free stubborn bubbles.

7. Quickly return the gel mould to a near horizontal position (approx. 5°) and insert the comb, taking care not to trap any bubbles along the edges of the wells.

8. Clamp the top of the plates with one bulldog clamp placed in the centre of the plates.

The gel should polymerize within 10−15 min. Gels can be used after an hour but it is convenient to allow them to polymerize overnight as this gives better reproducibility. Once poured, gels are stable at 4°C for at least one week if the top is wetted with a small amount of water and wrapped with Saran Wrap to prevent dehydration.

2.6 Preparation of the sample

2.6.1 Processing the crude synthetic oligodeoxynucleotide mixture

It is very important that the concentration of oligodeoxynucleotide in the total synthesis is accurately determined to prevent overloading of the gel. After synthesis the crude synthetic oligodeoxynucleotide in concentrated NH_4OH must be incubated at 55°C to remove the exocyclic amine protecting groups. After heating, allow the solution to cool to room temperature. Loosen the cap to prevent vacuum formation and chill on solid CO_2 for $10-15$ min. Carefully pipette out 1% of the volume into a microcentrifuge tube and evaporate to dryness. Dissolve the residue in 1.0 ml of water; further dilution may be necessary to allow quantitation by UV absorption at 260 nm. The rest of the NH_4OH solution can be evaporated to dryness or a portion can be removed and the remainder stored at $0°C$.

For preparative purposes the first step is to determine how much purified oligo-deoxynucleotide is required for the given experiment(s). The quantity of material to be loaded on to the gel is determined by the gel thickness, the well surface area, the quality of the synthesis and the length of the product. Remember that the yield of product from a chemical synthesis is related exponentially to the step yield and the number of steps. Thus for an octadecamer synthesized at a 98.5% step yield the theoretical yield of product available before isolation is $0.985^{(18-1)}$ or 77.3% of the crude synthesis mixture. Thus for every absorbance unit loaded, only 0.77 is available as product. The sensitivity of the theoretical yield to this function is demonstrated by the fact that for a 36-nucleotide compound synthesized at the same 98.5% step yield the theoretical yield is 58.9%.

The maximum mass to well surface area ratio seems to be $10-12$ absorbance units/cm^2. It is strongly advised that two loadings be made of each sample; one at $10-12$ absorbance units/cm^2 and one at half that amount. This range guarantees that at least one loading will be optimal. For syntheses up to $40-50$ nts long, 1 absorbance unit of crude mixture may be loaded on a surface area of 0.08 cm^2 (0.8 mm $\times$ 1 cm) in a volume of 5 μl.

The mass to surface area ratio is also a function of the amount of material to be found in any single size species. Thus, a longer synthetic oligodeoxynucleotide will contain less of the desired product than a shorter one; this means that a larger amount of crude mixture can be safely loaded on to the same surface area without a loss of resolution. This also means that a synthesis with a lower step yield which contains less of the desired product will tolerate a larger loading.

The next stage of the preparation of the sample is determined by the state of the 5′ terminus. If the 5′ terminus is blocked with a dimethoxytrityl group it must be removed. There are situations when the trityl group is left on during electrophoretic purification but these are not commonly encountered. If purification is done exclusively by gel electrophoresis, the 5′-O-dimethoxytrityl should be removed on-line by the automated DNA synthesizer. However, the procedure for manual deblocking requires little effort and is described in *Protocol 3*.

Protocol 3. Removal of 5′-O-dimethoxytrityl blocking groups

1. Evaporate the sample to dryness.
2. Add $5-100$ μl of 80% acetic acid.
3. Vortex.
4. Leave the solution for 15 min at room temperature.

Protocol 3. *continued*

 5. Evaporate to dryness.

 6. Add 1−2 drops of ethanol:water (1:1) and vortex.

 7. Evaporate to dryness.

 8. Repeat Steps 6 and 7 once more.

If the sample is to be detected by UV (Section 2.8.2) it requires no further preparation other than evaporation to dryness. However, if the sample is to be labelled with ^{32}P-phosphate it is necessary to calculate the number of picomoles needed. It is possible to estimate the volume of the 1 ml−1% dilution to be used by the following:

$$\text{No. of } \mu\text{l of 1\% aliquot} = \frac{\text{No. of nucleotides} \times \text{No. of pmols} \times 10^{-2}}{A_{260}}$$

This formula uses an extinction coefficient based on the number of nucleotides but ignores hypochromicity effects. It also uses an average extinction coefficient of 10^4 A_{260}/mole. This assumes a balanced purine to pyrimidine content. Also the calculation leads to an underestimation of the number of ends because it ignores the size heterogeneity present in a crude synthesis. Ammonium salts inhibit the activity of T4 bacteriophage kinase. Evaporate the sample to dryness and add several drops of ethanol−water (1:1). Re-evaporate and repeat once more to ensure their removal.

2.6.2 5′-end-labelling of oligodeoxynucleotides

Oligodeoxynucleotides, both mixtures and purified samples, can be analysed by 5′-end-labelling and subsequent electrophoresis. This method of analysis provides the highest resolution. Detection by autoradiography also has a very large dynamic range which allows the visualization of trace contaminants. For analytical purposes it is sufficient to use 1−10 picomoles of crude oligodeoxynucleotides. The procedure for end-labelling used by the author is given in *Protocol 4*.

Protocol 4. Procedure for the 5′-end-labelling of oligodeoxynucleotides

 1. Prepare stock solutions:
    - 10 × buffer (100mM MgCl$_2$, 1 mM KCl, 50 mM DTT, 100mM Tris-HCl, pH 7.6)
    - 660 μM ATP
    - 10 mM spermidine
    - Stop solution (98% deionized formamide containing 0.1% bromophenol blue, 0.1% xylene cyanol, 10 mM EDTA).

 2. Add the following reagents to the dried sample:
    - 1 μl 10 × buffer
    - 1 μl 660 μM ATP

Protocol 4. *continued*

- 1 μl 10 mM spermidine
- 1 μl [γ-^{32}P]-ATP, specific activity ~111 GBq/μmol (3 Ci/μmol)
- 5 μl autoclaved distilled water.

3. Vortex vigorously for 5$-$10 sec.

4. Centrifuge briefly (11 000 *g* for 10 sec) to bring all of the droplets to the bottom of the tube.

5. Add 1 unit of T4 kinase (1 U/μl).

6. Vortex vigorously for 5$-$10 sec.

7. Centrifuge briefly as before to bring all of the droplets to the bottom of the tube.

8. Incubate at 37°C for 30 min.

9. Add 25 μl of sample loading buffer (Section 2.6.3) to the reaction buffer and vortex to mix.

10. Load 5$-$20 μl of solution on to the gel.

It is not necessary to remove the unincorporated radioactive ATP; it will, however, migrate into the lower buffer chamber of the electrophoresis apparatus requiring proper disposal of the buffer and decontamination of the apparatus.

2.6.3 Sample loading buffer

Samples to be run on a gel are taken up and heated in a denaturing medium which is also dense enough to underlay the electrophoresis buffer. For small oligodeoxy-nucleotides (<30 nucleotides) 98% formamide$-$water is recommended. Longer oligodeoxynucleotides have limited solubility in formamide$-$water. Short oligodeoxy-nucleotides with a high G or G+C content also have limited solubility in this solution even with boiling. In these cases 7 M urea is a better loading buffer.

For analytical samples which have been 5'-end-labelled include bromphenol blue and xylene cyanol tracking dyes in the sample buffer. However for visualization by UV illumination it is unwise to include these dyes since they may co-migrate with the desired species and obscure the band. It is usually sufficient to load one lane with dyes for reference purposes. Prepare the samples as described in *Protocol 5*.

Protocol 5. Sample preparation

1. Add the appropriate volume of sample buffer to the dried residue. The concentration should be 1$-$2 absorbance units/μl; a minimum volume of 5 μl is practical.

2. Vortex vigorously.

3. Place in boiling water for 2 min.

4. Vortex again.

5. Cool on ice before loading.

2.7 Electrophoresis conditions

Protocol 6. Electrophoresis procedure

1. Fix the gel plate assembly to the gel apparatus.
2. Fill the upper buffer tank and check for leaks. If there are none, fill the lower buffer tank.
3. Rinse out any pieces of polyacrylamide in the wells, using a pipette.
4. Clamp an aluminium plate to the surface of the plates making sure that there is good contact with the plate.
5. Apply a constant power of 60 W and allow the gel to run for at least 30 min.
6. Turn off the power. Rinse out any urea which has diffused from the gel into the wells.
7. Carefully add the sample to the well by allowing it to stream from a syringe or pipette tip on to the lower surface of the well. Be careful to prevent any bubbles from disturbing the sample application.
8. Quickly turn on the power and run at 60 W constant power until the tracking dyes have reached the desired position.

The exact run time depends on the size of the oligodeoxynucleotide. The time should be sufficient to allow the desired species (the slowest moving) to migrate at least two-thirds the length of the gel. After the separation has been developed sufficiently, turn off the power and remove the gel mould from the apparatus.

2.8 Detecting the sample

2.8.1 Autoradiography

Samples which have been 5'-end-labelled and run on a 0.4-mm-thick analytical gel are visualized after electrophoresis by autoradiography as follows:

Protocol 7. Autoradiography of gels

1. Remove one of the glass plates, being careful not to pull the gel from the other plate.
2. Place one thickness of Saran Wrap over the gel surface and pull it snugly around the edges of the plate.
3. Place the plate inside a suitable light-excluding film cassette containing an intensifying (enhancer) screen.
4. In a darkroom, lay a sheet of Kodak X-OMAT AR film evenly on the surface of the gel.

Protocol 7. *continued*

5. Close the cassette lid being sure that the film is not buckled and pulled away from the surface of the gel.

6. Clamp evenly around the perimeter or press evenly with suitable weights.

7. Place the film cassette into a freezer maintained at −70°C for the appropriate time.

The use of an intensifying screen greatly reduces the time required to obtain a satisfactory exposure. Remember that the total radioactivity incorporated is distributed amongst several to many species depending upon the length of the synthetic oligodeoxynucleotide. This 'dilution' of the radioactivity means that a longer exposure time is needed to obtain a dark enough image. The large dynamic range in incorporation in a mixture also requires long exposure times to visualize minor components of the mixture. Develop the film according to the manufacturer's instructions.

2.8.2 UV analysis of gels

For a preparative gel, the oligodeoxynucleotides can be visualized directly by UV shadowing. In this technique the gel is laid on a fluorescent background and visualized by incident illumination with a short wavelength (254 nm) UV source. The nucleic acid is seen as a dark shadow against the fluorescing background and can be subsequently photographed if desired. Although this method is most useful for visualizing preparative amounts of oligodeoxynucleotides, it is quite sensitive with a single band detection limit of about 0.5 absorbance units. A number of fluorescent surfaces can be employed as long as the surface area is sufficiently large and the mechanical strength is good enough to support the gel. An ideal surface is provided by an intensifying screen such as the DuPont Kronex. The screen is available in several different sizes, with the largest able to support 33−40 cm format gels. A low-cost alternative is a 20 × 20 cm silica gel TLC plate containing a fluorescent indicator, covered tightly with Saran Wrap.

It is strongly recommended that the preparative separations are recorded routinely by photographing the UV shadowed image. This record provides valuable information on the quality of the synthesis which may be useful in analysing problems arising from a fragment that is used in subsequent biological procedures. Quite reasonable images are obtained using Polaroid 555 3.5″ × 4.5″ instant film with ½ second exposure times with a green filter (58) in front of the camera lens.

2.9 Recovering the sample from the gel

After visualizing a preparative separation, the next step is the extraction of the desired species from the gel matrix. Although there are a number of methods described for the extraction of nucleic acids from gels, the following two are the most appropriate for the recovery of synthetic oligodeoxynucleotides from polyacrylamide gels.

The band containing the required species should be carefully excised from the gels using a clean razor blade. High percentage acrylamide gels cut cleanly with a drawing action of the blade with little ripping of the surrounding gel. Low percentage gels need to be 'chopped' since a drawing motion of a blade will pull the surrounding gel. Make sure that you are looking as vertically as possible down at the band in question; parallax can result in the excision of an adjacent band and contamination of the final product. Start by making vertical cuts in between the individual lanes. Finish with the horizontal cuts above and below the desired band. Be careful to overlap each corner completely or a small fragment of the gel slice will tear off.

As in any preparative technique, the degree of fractionation determines the yield and purity. A wide gel slice will provide higher yield with the possibility of lower purity. Since the most critical contaminating (n-1)mer, migrates faster (lower molecular weight) than the product, it is best to take an asymmetric slice containing more of the trailing edge of the product band. However, there is at least one report which describes a heterogeneous product with a slightly higher molecular weight due to a chemical modification during synthesis (3). Careful electrophoresis allowed this higher molecular weight contaminant to be slightly separated away from the product.

2.9.1 The 'crush and soak' method

This technique essentially involves passive diffusion of the nucleic acid into the liquid phase. The original technique involved homogenizing the gel slice (see Section 2.8 of Chapter 5) but this is not necessary with low molecular weight synthetic oligo-deoxynucleotides. Avoiding homogenization also minimizes extraction of polymeric acrylamide. The procedure is as follows:

Protocol 8. Extraction of samples from gels using the 'crush and soak' method

1. Transfer the gel slice to a small vial and cover with at least 3 ml of extraction liquid. A number of formulations have been described: 0.5 M NaCl, 2 M triethylammonium acetate, 50 mM triethylammonium acetate and plain water. There seems to be little difference in the efficiency of extraction by any of the different formulations. In the first instance, try using a solution of 0.5 M NaCl, 0.1 M Tris-HCl (pH 7.0) containing 1 mM EDTA.

2. Incubate the slice at room temperature overnight.

3. Decant off and save the solution from the gel slice.

4. After the sample has been eluted from the gel slice, remove dissolved urea, salts and extracted acrylamide by the use of a reverse-phase mini-column or by gel exclusion chromatography using Sephadex.

Procedures for the use of a SEP-PAK (Waters Associates) have been described (2). *Protocol 9* is a procedure for the use of an alternative cartridge called an oligonucleotide purification cartridge (OPC) (Applied Biosystems) (9).

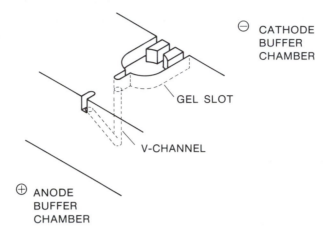

Figure 2. Diagram of an electroelution apparatus. A slice of gel material containing the desired nucleic acid is placed in the horseshoe-shaped slot surrounded by 0.5 × TBE buffer. 75 μl of a very high ionic strength solution is placed in the V-shaped channel (see Section 2.9.2.). After electroelution the sample is concentrated at the interface of the high salt buffer and is removed by pipetting.

Protocol 9. Procedure for desalting samples using an oligonucleotide purification cartridge (OPC)

1. Attach the OPC to the end of a 20 ml syringe using the adapter provided.
2. Rinse the cartridge with 5 ml of acetonitrile, taking 1−2 min to pass the liquid through the cartridge.
3. Equilibrate with 5 ml of 2 M triethylammonium acetate.
4. Load the sample on to the cartridge.
5. Wash with 15 ml of 0.1 M triethylammonium acetate.
6. Elute the sample into a microcentrifuge test-tube with 1−2 ml of 20% acetonitrile in water.

2.9.2 Electroelution

Electroelution uses an externally applied field to hasten the migration of the nucleic acid from the gel into a collecting solution. There are a number of methods described in the literature. The author has had success with a device available from International Biotechnologies Inc., called the Unidirectional Electroeluter, Model UEA schematically shown in *Figure 2*. This apparatus allows the simultaneous electroelution of six samples. The oligodeoxynucleotide is eluted into free solution and concentrated against a very high salt solution (e.g. 10 M ammonium acetate) which effectively halts the migration of the nucleic acid. The salt solution is removed and the product is recovered by ethanol precipitation. The procedure for the use of the device is described in *Protocol 10*. More detailed instructions and information are to be found in the product literature available from the manufacturer.

Protocol 10. Extraction of samples from gels using an electroeluter

1. Prepare high salt buffer (10 M ammonium acetate solution) as follows:
   - 5.8 g ammonium acetate
   - 1.0 mg bromophenol blue
   - water to 10 ml.

2. Fill the apparatus with approximately 500 ml of low salt buffer, 0.5 × TBE buffer (Section 2.2.2).

3. Place one gel slice per chamber. Position this as close as possible to the V-channel.

4. Fill each chamber with just enough low salt buffer to cover the gel slice. Flush any air bubbles out of each V-channel.

5. Load the high salt buffer into all V-channels. Carefully expel 100 μl of high salt solution into the bottom of the V-channel. Use a syringe which will reach to the bottom of the V-channel. Mixing should be avoided and a sharp interface should be formed.

6. Connect the electrodes and begin electroelution at 100 V. Continue for about 30 min. Monitor the exact time by eluting a gel slice containing the marker dye (bromophenol blue or xylene cyanol) closest in migration to the sample as a control.

7. Turn off the power. Carefully remove a $50-100$ μl aliquot of the interface between the low salt buffer and high salt buffer. The eluted oligodeoxynucleotide should be located just at or below the interface.

8. Add 1 ml of cold 95% ethanol and mix by inverting several times.

9. Store at -7°C for 30 min.

10. Centrifuge at 11 000 g for 15 min.

11. Resuspend the material in 50 μl of autoclaved distilled water and add one-tenth volume of 7 M ammonium acetate.

12. Repeat Steps 8 and 9 once more.

---

3. Applications

3.1 Resolution

The goal of electrophoresis in the analysis and purification of synthetic oligodeoxy-nucleotides is the maximization of resolution. Resolution of the product from its

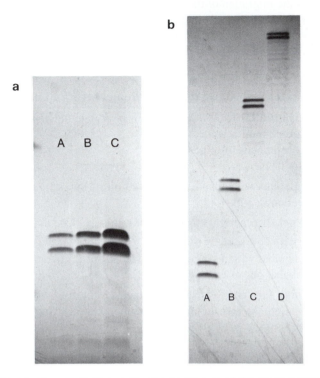

Figure 3a. Mass load – resolution study of a co-mixture of the crude syntheses of an octadecamer and a heptadecamer. Samples were analysed on a 12% polyacrylamide gel 0.8 mm thick, with 1-cm-wide wells and visualized by UV shadowing. Lane A, a total of 1.1 absorbance units of mixed heptadecamer and octadecamer. Lane B, 2.2 absorbance units. Lane C, 4.4 absorbance units.

Figure 3b. Run time – resolution study of a co-mixture of the crude syntheses of an octadecamer and a heptadecamer. The samples were analysed on a denaturing 12% polyacrylamide gel, 0.8 mm thick, and visualized by UV shadowing. A total of 1.1 absorbance units of the mixed octadecamer and heptadecamer were loaded in each well at 1 h intervals. Lane A, 4 h run time. Lane B, 3 h run time. Lane C, 2 h run time. Lane D, 1 h run time.

contaminants is relatively easy, since for the most part the chemical synthesis of oligodeoxynucleotides produces species with discrete molecular weights. Thus optimization of factors such as mass load and run time should provide adequate results routinely.

3.1.1 Mass load

The most common cause of poor resolution in the electrophoresis of synthetic oligodeoxynucleotides is due to overloading of the gel. Most often this is due to the lack of or improper quantitation of the crude reaction mixture.

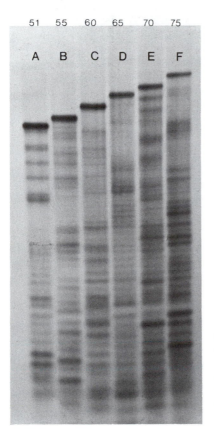

Figure 4. Analysis of enzymatic ^{32}P-labelled crude mixtures of synthetic oligodeoxynucleotides. The samples were subjected to electrophoresis on a denaturing 12% polyacrylamide gel, 0.4 mm thick. The excess $[\gamma\text{-}^{32}\text{P}]$-ATP was not removed after the enzymatic labelling. The sizes of the oligodeoxynucleotides are shown above the lanes.

Figure 3a demonstrates the effect of increased mass loading on the separation between two oligodeoxynucleotides differing by one nucleotide in length. Too much mass load per well surface area can result in the smearing of the bands and the subsequent loss of resolution. If the wells are too narrow or too thin for a given mass load then overloading can occur. Best results are obtained by using wide spacers (2−3 cm) with thick gels (1.5−3.0 mm) and adhering to a mass to surface area ratio of 10−12 absorbance units/cm^2. Two loadings per sample at two different mass to surface area ratios will provide a safety margin for optimum results.

3.1.2 Run time

The second most common cause of poor resolution is insufficient migration as a result of too short a run time. Too often, attention is not paid to the mobility of the product band relative to the commonly used marker dyes, resulting in an inadequate separation of closely sized species. *Figure 3b* shows the relationship of the separation of two species with mobility. The smallest separation between the co-mixed heptadecamer and octadecamer is 0.3 cm (migrating distances of 8.1 cm

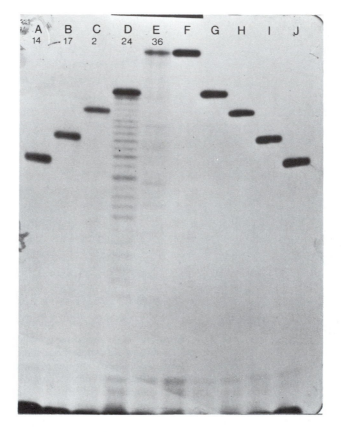

Figure 5. Autoradiogram of crude mixtures and gel purified synthetic oligodeoxynucleotides. The 5'-*O*-enzymatic ^{32}P-labelled samples were analysed by a denaturing 20% polyacrylamide gel, 0.4 mm thick. The excess [γ-^{32}P]-ATP was not removed prior to electrophoresis. Lanes A – E are aliquots of crude synthesis mixtures. Lanes F – J are aliquots of the products from UV-shadowed preparative gel electrophoresis. The sizes are shown above the lanes.

and 8.4 cm) which is achieved after electrophoresis for an hour; the maximum separation is 0.9 cm (migration distances of 26.0 cm and 26.9 cm) which is achieved after electrophoresis for 4 h. Run time should be estimated by the time it takes for a given species to migrate at least two-thirds the length of the gel and not judged by the arbitrary separation of the marker dyes. Optimum separation often requires that one of the marker dyes migrates off the bottom of the gel. This results in the loss of some fast moving species but for preparative purposes this is not important. A consequence of the need for maximum separation is that too wide a range of sizes should not be run on the same gel unless multiple loadings (progressively longer run times) are used.

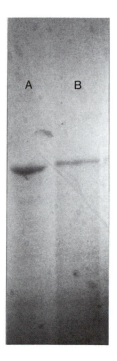

Figure 6. Preparative separations of 79 and 80 residue synthetic oligodeoxynucleotides. The samples were applied to a denaturing 12% polyacrylamide gel, 1.6 mm thick and visualized by UV shadowing. The separation was run for 6 h; the xylene cyanol marker migrated 32.75 cm, while the products migrated about 23 cm. Lane A, 79 residues; 2.7 absorbance units loaded. Lane B, 80 residues; 5.7 absorbance units loaded.

3.2 Typical results

Figure 4 is an autoradiogram of the crude reaction mixtures of several typical synthetic oligodeoxynucleotides. The gel demonstrates the type of resolution achieved by analytical electrophoresis of 5'-labelled oligodeoxynucleotides. The run time was sufficient to provide adequate separation between the various sizes and the range of sizes is relatively narrow. *Figure 5* is an autoradiogram demonstrating the analysis of several syntheses and the isolated products. This gel illustrates the level of purity which can be obtained by preparative gel electrophoresis and extraction of the product. *Figure 6* is a UV shadow photograph of an 8% polyacrylamide gel showing the preparative separation of synthetic oligodeoxynucleotides 79 (lane A) and 80 (lane B) residues long. The gel was run for 6.5 h which results in most of the smaller contaminants running off the end of the gel.

3.3 Anomalous results

3.3.1 Chemical effects

Quite apart from the effects related to the mobility of a charged species in a gel, there are anomalies due to the synthetic origin of oligodeoxynucleotides. These include incomplete deprotection and unwanted chemical modifications. *Figure 7* is a UV shadow photograph of a gel comparing the migration of a synthetic octadecamer

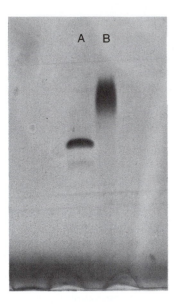

Figure 7. Comparison of deprotected and partially deprotected synthetic oligodeoxynucleotides. The samples were analysed on a denaturing 20% polyacrylamide gel, 0.8 mm thick with visualization by UV shadowing. Lane A, d(TpCpCpCpApGpTpCpApCpGpApCpGpTpTpGpT) after heating in concentrated NH_4OH at 55°C for 12 h. Lane B, an aliquot of the same octadecamer after being treated with concentrated NH_4OH at room temperature for 15 min.

which has not had the exocyclic amine protecting groups removed with the migration of the same oligodeoxynucleotide sequence completely deprotected. Since there are multiple sites for protected and deprotected nucleotides the product and contaminants exist as a heterogeneous population of closely related species resulting in a broad smear of a higher average molecular weight than the desired product. This gel serves to illustrate the powerful diagnostic insight provided by electrophoresis which may be used to check if there are any problems with the chemical synthesis process.

3.3.2 Base composition effects

One of the more common anomalies which affects the mobility of synthetic oligodeoxynucleotides is the effect of composition. It has been demonstrated that the mobility of short to medium sized oligodeoxynucleotides is a function of the base composition with each of the four nucleotides having a differential mobility. As a result, not all oligodeoxynucleotides of the same size will migrate with the same mobility. However, the mobility differences are very predictable and can be calculated in both a relative and absolute sense. The order of mobility is $C>A>T>G$ (4) with the mobility relationship between the homologous series being:

$$C_{10.9} = A_{10} = T_{8.5} = G_{6.9}$$

144

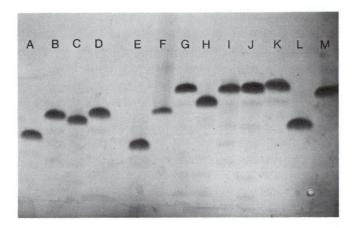

Figure 8. Analysis of base composition and sequence effects on the mobility of synthetic oligodeoxynucleotides. The samples were separated by a denaturing 20% polyacrylamide gel, 0.8 mm thick, with visualization by UV shadowing. The sequences in each lane are listed in *Table 4*.

Table 4. Oligodeoxynucleotide sequence.

Gel lane	Sequence
A.	$d(C)_{16}$
B.	$d(ApGpTpC)_4$
C.	$d(ApT)_8$
D.	$d(T)_{16}$
E.	$d(TpGpTpGpApGpCpTpApGpCpTpCpApCpA)$
F.	$d(G)_{15}$
G.	$d(TpCpCpCpApGpTpCpApCpGpGpApCpGpTpTpGpT)$
H.	$d(CpCpCpApGpTpCpApCpGpGpApCpGpTpTpGpT)$
I.	$d(ApGpGpGpTpCpApGpTpGpCpTpGpCpApApCpA)$
J.	$d(ApGpTpC)_4ApT$
K.	$d(T)_{18}$
L.	$d(TpGpTpGpApGpCpTpTpTpTpGpCpTpCpApCpA)$
M.	$d(TpGpTpGpApGpCpTpTpTpTpCpGpApTpCpApC)$

Figure 8 contains examples of differential mobilities for compounds for the same sizes. Lanes A, B, C, D, and F contain $d(C)_{16}$, $d(ApGpTpC)_4$, $d(ApT)_4$, $d(T)_{16}$, and $d(G)_{15}$ respectively (see *Table 4*). The slower mobility of $d(G)_{15}$ relative to either $d(C)_{16}$ or $d(T)_{16}$ demonstrates the severe retarding effect that the presence of dG has on the mobility of an oligodeoxynucleotide. Comparison of lanes B and D shows that the mobility of $d(T)_{16}$ deviates to the extent of about half a nucleotide from the mobility of a completely 'balanced' sequence. From an operative standpoint this means that sequences which are skewed in base composition from each other should be expected to show different mobilities.

3.3.3 Base sequence effects

In addition to the effect of base composition on the mobility of oligodeoxynucleotides there can be also a very strong influence arising from the sequence of the oligodeoxynucleotide. This mobility shift is due to the formation of secondary structure which is not totally destroyed by the combination of thermal and chemical denaturation. In *Figure 8* lane E contains a palindromic hexadecamer. A palindrome is a double-stranded sequence which contains an axis of symmetry. As a result of this symmetry the single-stranded oligodeoxynucleotide can exist in solution in three forms:

I.   TGTGAGCTAGCTCACA

II.   TGTGAGCTAGCTCACA
ACACTCGATCGAGTGT

III.   TGTG^{AG}C$_\text{T}$
ACAC$_\text{TC}$G^A

According to Maniatis and co-workers (5) form II should have a lower mobility than the single-stranded form I while the hairpin form III should have a higher mobility. In fact, on polyacrylamide gels containing 7 M urea form II is never seen and usually form III Is the predominant species. The presence of form III is dependent on the number of G:C base-pairs and the length of the stem; but even short oligodeoxynucleotides will show this form since it is very stable thermodynamically. The hairpin structure is very compact and migrates as a smaller species with a 2−3 nucleotide shift relative to a 'normal' hexadecamer (Lane B). These type of sequences are encountered quite frequently since most linkers used to modify or add restriction sites to the end of a double-stranded fragment are palindromic.

Another type of sequence which will migrate anomalously but is not a palindrome is a splinker (6). These oligodeoxynucleotides are self-complementary sequences which contain an inverted repeat sequence which forms a hairpin structure to give a known restriction site. In *Figure 8* lanes L, M show the drastic mobility shift introduced by a hairpin structure with a oligodeoxynucleotide of the same composition but without the inverted repeat.

3.4 Nucleotide analogues

3.4.1 5′ Fluorescent-labelled oligodeoxynucleotides

Figure 9 is a photograph of a UV-shadowed gel containing several 5′-modified oligodeoxynucleotides. The compounds shown represent a class of molecules which have been first modified by the addition of a terminal aminoethyl or aminohexyl spacer and then subsequently reacted with a fluorescent N-hydroxysuccinimide (NHS) ester dye (7). This process allows the labelling of oligodeoxynucleotides for the purpose of non-radioactive detection. The introduction of an amino group allows the incorporation of a wide variety of amino-reactive molecules.

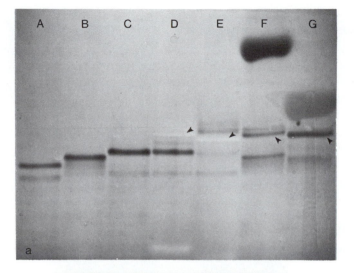

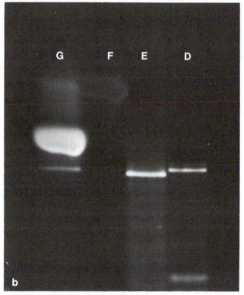

Figure 9. Preparative separation of 5′ dye-labelled octadecamers. The dye-labelling reaction mixtures were resolved on a denaturing 20% polyacrylamide gel, 1.5 mm thick with 2.5-cm-wide wells. Aliquots of the reaction mixtures containing 1 absorbance unit of 5′-alkylamino-oligodeoxynucleotide and a 20 – 40-fold excess of *N*-hydroxy-succinimide ester of four different fluorescent dyes were subjected to electrophoresis to separate the product from the reaction components. Panel (a): visualization by UV shadowing. Lane A, un-derivatized 18-mer. Lane B, 5′-aminoethyl octadecamer. Lane C, 5′-aminohexyl octadecamer. Lanes D – G, dye-labelling reaction mixtures; arrow heads show position of fluorescently-labelled oligonucleotides. Panel (b): intrinsic fluorescence by UV irradiation at 365 nm. Lanes D – G, dye-labelling reaction mixtures.

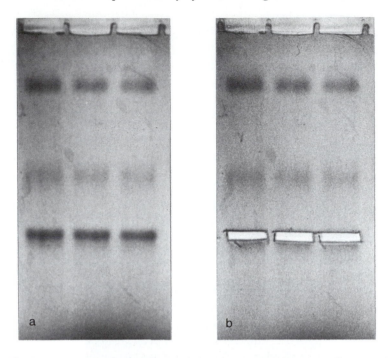

Figure 10. Preparative separation of methyl phosphonate octadecamers. Compounds were separated on a denaturing 20% polyacrylamide gel, 1.5 mm thick with visualization by UV shadowing. The electrophoresis was conducted for 6.5 h. Panel (a): all three lanes loaded with 8 absorbance units each of d(ApApTp*Cp*Gp*Gp*Gp*Cp*Ap*Tp*Gp*Gp*Ap*Tp*T), where p* represents *O*-methyl phosphonate. Panel (b): after excision of products.

Lanes D−G of *Figure 9a* shows preparative separations of 1 absorbance unit of crude reaction mixtures containing alkylamino-labelled octadecamers which have been reacted with a 20−40-fold excess of four fluorescent dyes. The large smears present in lanes F and G are the unreacted NHS ester dyes. The unreacted dyes in lanes D and E have a much higher mobility and have migrated off the end of the gel. This separation demonstrates the ability to isolate a non-nucleoside derivatized oligodeoxynucleotide from a complex reaction mixture.

Figure 9b is a photograph of the same gel under 365 nm UV irradiation showing detection of the bands by the fluorescent chromophore only. The weak appearance of the species in lane F is due to the fact that the photograph was taken through a green filter.

3.4.2 O-methyl-phosphonate oligodeoxynucleotides

Miller and co-workers (8) have described a class of modified synthetic oligodeoxy-nucleotides which contains a neutral phosphonate linkage instead of the normal anionic phosphodiester linkage found in DNA. For solubility purposes it is usually necessary to include at least one phosphodiester linkage, typically at the 5' end. The inclusion of a single charge confers sufficient electrophoretic mobility to allow them to be separated and analysed by acrylamide gel electrophoresis.

Figure 10a is a UV shadow photograph of a 12% polyacrylamide gel containing an oligodeoxynucleotide of the sequence:

d(ApApTp*Cp*Gp*Gp*Gp*Cp*Ap*Tp*Gp*Gp*Ap*Tp*T)

where p* represents a neutral O-methyl phosphonate linkage. Because there are only two charges in the molecule the mobility is drastically reduced relative to normal oligodeoxynucleotides. After loading, the gel was run at 60 W for 6.5 h. This means that the bromophenol blue ran off the end of the gel while the xylene cyanol migrated about 30 cm.

Figure 10b is a photograph of the same gel showing the location of the excised products which were subsequently eluted from the gel and worked up by standard techniques.

Acknowledgements

I would like to thank Jerry Zon and Scott Davidson for providing some of the synthetic oligodeoxynucleotides, Cheryl Heiner for the autoradiograms, and Jeff Springer for the dye-labelling reaction mixtures, Larry Kerila for the drawing of the electroeluter, and all of the research staff who volunteered suggestions.

References

1. Maniatis, T., Fritsch, E. F., and Sambrook, J. (1982). *Molecular Cloning: A Laboratory Manual*. Cold Spring Harbor Laboratory Press, Cold Spring Harbor, NY.
2. Atkinson, T. and Smith, M. (1984). In *Oligonucleotide Synthesis, A Practical Approach* (ed. M. J. Gait), p. 35, IRL Press, Oxford.
3. Eadie, J. S. and Davidson, D. S. (1987). *Nucleic Acids Res.*, **15**, 8333.
4. Frank, R. and Koester, H. (1979). *Nucleic Acids Res.*, **6**, 2069.
5. Maniatis, T. (1975). *Biochemistry*, **14**, 3787.
6. Kalisch, B. W., Krawetz, S. A., Schoenwaelder, K.-H., and Van de Sande, J. H. (1986). *Gene*, **44**, 263.
7. Connell, C., Fung, S., Heiner, C., Bridgham, J., Chakerian, V., Heron, E., Jones, B., Menchen, S., Mordan, W., Raff, M., Recknor, M., Smith, L., Springer, J., Woo, S., and Hunkapiller, M. (1987). *Biotechniques*, **5**, 342.
8. Miller, P. S., McParland, K. B., Jayaraman, K., and Ts'o, P. O. P. (1981). Biochemistry, **20**, 1874.
9. McBride, L. J., McCollum, C., Davidson, S., Efcavitch, J. W., Andrus, A., and Lombardi, S. J. (1988). *Biotechniques*, **6**, 362.

5

Two-dimensional gel electrophoresis of nucleic acids

RUPERT DE WACHTER, JACK MANILOFF, and WALTER FIERS

1. Introduction to two-dimensional gel electrophoresis of RNA

1.1 Factors governing the electrophoretic mobility of RNA on polyacrylamide gels

Quite a large number of two-dimensional systems have been described for the separation of RNA mixtures too complex for fractionation in a single dimension. A brief consideration of the factors determining electrophoretic mobility of RNA on gels may be helpful in selecting the combination optimally suited for a particular application.

The mobility of an RNA molecule on a polyacrylamide gel subjected to a voltage gradient is determined by the chain length, net charge, and conformation of the RNA on the one hand, and by the gel composition (mainly gel concentration, but also the degree of crosslinking) on the other. In *Figure 1* the contribution of each of the four bases to the net charge of an RNA molecule is plotted as a function of pH. Over the range pH 5.0 to pH 8.5 each residue bears a charge of -1, and the net negative charge of an RNA molecule is nearly equal to its chain length. In the acidic pH range, however, each of the four residues contributes a different charge and the net charge of the RNA will depend both on chain length and base composition. As to the secondary and tertiary structure of the RNA, this is gradually lost when the temperature is raised or when increasing amounts of denaturing agents such as urea or formamide are added. At lower pH values, classical base-pairing is lost but some other interactions may occur.

For gel electrophoresis in the neutral pH range, an approximate description of the mobility as a function of chain length is given by the equation:

$$M = a - b \log N$$

where M is the mobility expressed in cm^2/V sec, N is the chain length, and a and b are constants depending on gel composition and temperature. On gels containing no denaturing agent, however, conformational disparities between RNA molecules of the same chain length; for example, branch points may result in appreciable

151

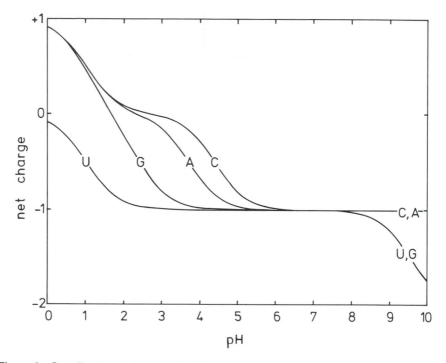

Figure 1. Contribution made by each of the four nucleotides to the net charge of an RNA molecule. This figure shows the net charge of uridylic — (U), guanylic — (G), adenylic — (A), and cytidylic — (C) residues as a function of pH.

differences in mobility. Moreover, two RNA fragments may remain bound by base pairing between complementary sequences and move as expected for a much larger molecule.

If one now considers gels run in the acid pH range, then the data in *Figure 1* show that the net charge of an RNA fragment is no longer determined only by its chain length, but also by its base composition. Under these conditions, mobility is no longer described by the above equation, and RNA fragments of the same length but unequal base composition will show different mobilities, even in the presence of urea. Incidentally, the fact of working at low pH itself has a denaturing effect, since the presence of positive charges on cytosine, adenine, and guanine disrupts base-pairing.

Finally, it is necessary to consider the effect of increasing gel concentration, which results in decreasing RNA mobility. Conceivably, the effect of gel concentration on the mobility of an RNA becomes critical near the point where the dimensions of the RNA molecule and the gel pore size are of the same magnitude. This point can be reached at a different gel concentration for two molecules with similar chain length but possessing different conformations. As a result, these molecules may have the same mobility on one gel but quite different mobilities on a more concentrated gel.

Having distinguished the main factors influencing RNA mobility on polyacrylamide gels it is now possible to classify the multitude of two-dimensional combinations described in the literature into a few main categories and, on this basis, the method best suited for a particular fractionation problem can be selected.

1.2 The main types of two-dimensional gel systems

In an optimal two-dimensional system the relative order of migration of the components of the mixture should be very different in the two dimensions of separation. If the migration order is only marginally different, the distribution of spots approaches a diagonal pattern and little is gained in comparison with a one-dimensional separation. The main factors that can be altered to obtain a different migration order are gel concentration, pH, and the presence or absence of a denaturing agent. Which parameters are best altered depends on the size of the RNAs in the mixture. *Table 1* summarizes the three types of combinations that are most frequently used.

In the 'urea shift' method, the first- and second-dimensional gels have the same concentration and are both run in the neutral pH range. Either the first- or the second-dimensional gel contains a high urea concentration, the other one is run in non-denaturing conditions. This combination has been used for the fractionation of tRNA mixtures, with a gel concentration of 16% polyacrylamide, and for the isolation of mRNAs, on 6% polyacrylamide gels. In both cases the separation is based on a differential effect of the conformational change induced by urea on the migration order of the components. The urea shift method was first described in a study of interactions between RNA fragments ranging in size from 13 to 80 nucleotides resulting from a partial T_1 RNase digest of 5S RNA (1). In this case, the first dimension was run in the absence of urea and so fragments containing hidden breaks migrate as base-paired complexes. The second dimension contained urea which results in dissociation of the complexes to give the constituents which run considerably faster. Conformation seems to play only a minor role in determining the mobility of such small molecules, since fragments containing no hidden breaks remained more or less on the diagonal.

In the 'concentration shift' method the only difference between the first- and second-dimensional gels resides in a change in polyacrylamide concentration, usually in a ratio of 1:2. In this case, nothing is done to change the RNA conformation between the first and second dimension, but the differential effect of a tighter gel matrix on RNA molecules with different conformations is exploited to change their order of mobility. One favourable side-effect of increasing the gel concentration after the first-dimensional separation is that, due to the decrease in mobility, spots are sharpened when migrating from the dilute to the concentrated gel. This tends to offset the effect of diffusion. The concentration shift method has been used for the fractionation of RNAs in the size range of 80−400 nucleotides.

The third type of two-dimensional gel separation combines a pH shift with a urea and a concentration shift. This method (7,8) was originally designed for separating complex mixtures of RNA fragments obtained when large RNA molecules are

Table 1. Two-dimensional gel systems for RNA separation

Main difference first/second dimension	First dimension			Second dimension			RNA mixture separated	Reference
	Gel concn.	pH range	Urea concn.	Gel concn.	pH range	Urea concn.		
Urea shift	12.5%	neutral	0 M	12.5%	neutral	8 M	5S RNA fragments	1
	15–16%	neutral	6–7 M	16%	neutral	0 M	tRNAs	2,3
	6%	neutral	0 M	6%	neutral	5 M	mRNAs	4
Concentration shift	10%	neutral	0 M	20%	neutral	0 M	small RNAs	5
	10.4%	neutral	4 M	20.8%	neutral	4 M	tRNAs and precursors	6
pH shift / Urea shift	10.3%	acid	6 M	20.6%	neutral	0 M	viral RNA fragments	8
Concentration shift	10.3%	acid	6 M	20.6%	neutral	0 M	viral RNA oligonucleotides	9

Gel concentration is expressed as % weight/volume including the crosslinker. The 'neutral pH range' is between pH 4.5 and pH 8.5; acid range is below pH 4.5. Most neutral gels are run at pH 8.0 or 8.3, and the acid ones at pH 3.3. The term 'RNA fragment' is used for products of partial RNase hydrolysis. Most oligonucleotide mixtures are products of complete T_1 RNase hydrolysis. The references are limited to those introducing a new procedure or applying it to a new type of RNA mixture.

partially digested with RNases with the aim of sequence analysis. However, its main area of application is now the fingerprinting of T_1 RNase digests, mainly of large viral RNAs. The first dimension is run at pH 3.3 in the presence of 7 M urea. Electrophoresis at this pH results in mobilities depending more on the base composition than on the chain length of the oligonucleotides, since the net charge per monomer residue is different for each base (*Figure 1*). Complementary base-pairing of fragments is precluded due to both the low pH and the presence of urea. The second dimension is run at pH 8.0 or pH 8.3 and at a higher gel concentration, in the absence of urea. Under these conditions mobility is determined mainly by chain length. Oligonucleotides obtained by T_1 RNase digestion are arranged in a characteristic pattern. *Figure 2* shows a graticule that allows one to derive the base composition of the shorter components from their map position. The separation at pH 3.3 combined with one at neutral pH is especially effective with oligonucleotides because these show extremes of base composition. With longer RNA fragments the base composition of the components seldom differs enough for the mobility to be influenced profoundly. In this case, the pattern approaches a diagonal and little is gained with the pH shift alone. Complexes, however, are dissociated into their constituent components under the denaturing conditions of the first-dimensional separation, and the latter can then be separated in the second dimension on the basis of a difference in size and/or secondary structure which forms under the neutral pH conditions.

2. Apparatus and experimental procedure for two-dimensional separations of RNA

This section describes the basic design of an apparatus and one of the most commonly used techniques for two-dimensional gel electrophoresis. It is an almost impossible task to give a complete survey of the large number of modifications that have been described, but the main variants are mentioned in Section 3 together with applications for which they have been used. Some authors have used commercially-available apparatus, but most work has been carried out with home-made apparatus. The apparatus described below can be easily constructed in a workshop. When constructing apparatus care must be taken to ensure its electrical safety.

2.1 Separation in the first dimension

2.1.1 Preparation of the apparatus

The gel mould consists of two thoroughly cleaned, 0.4-cm-thick glass plates, which form the front and back walls, kept at a suitable distance by two acrylic resin spacer strips (e.g. Perspex, Plexiglas, Lucite) which form the side walls, as shown in *Figure 3*. The size of the plates is frequently 20 × 40 cm, but obviously this is a matter of personal preference and the size can be adapted to the type of separation and electrophoresis conditions (see Section 3). The thickness of the spacers is usually between 1 mm and 4 mm. Thin gels seem to give superior resolution and are preferable for analytical purposes; thicker gels allow one to load a larger sample

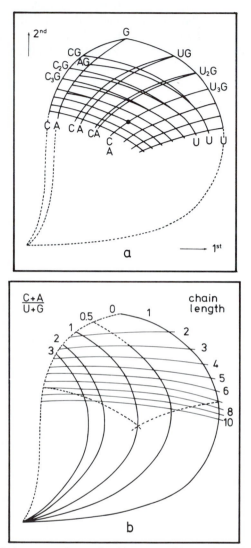

Figure 2. Relationship between oligonucleotide composition and position on a pH-shift two-dimensional gel. The gel system is the one described in Section 3.1 and *Table 4*. First dimension: 10% acrylamide, 6 M urea (pH 3.3). Second dimension: 20% acrylamide (pH 8.0). (a) Position of oligonucleotides obtained by RNase T_1 hydrolysis of RNA. From the position of Gp, three lines depart marked C, A and U. These lines connect the positions of the oligonucleotide series C_nG, A_nG and U_nG, respectively. Starting from these positions and counting the number of intersections in the directions U, C or A, one can find the position of any oligonucleotide. As an example, the black dot shows the position of oligonucleotides of composition $(C_x,A_{5-x},U_3)G$. (b) Oligonucleotide position as a function of base composition and chain length. The curves marked 0, 0.5, 1, 2, 3, connect all the positions corresponding to these C + A/U + G ratios. The curves marked 2 – 10 connect the positions corresponding to these chain lengths.

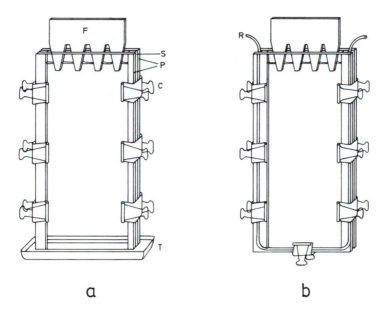

a   b

Figure 3. Gel mould during preparation of the gel. P, glass plates; S, acrylic strips serving as spacers; F, acrylic slot former or 'comb'; C, steel clips. During the casting of the gel the mould has to be closed at the bottom. This can be done in one of the following ways: (a) the assembled mould is pushed firmly into a trough (T) filled with plasticine; (b) before the clips are put in place the mould is surrounded on three sides by a rubber tube or thread (R) slightly thicker than the spacers. In this case the spacers do not have to be greased. After polymerization, the rubber tube is withdrawn by taking off and replacing the clips one by one. The bottom clip is not replaced.

for preparative separations. Gels thinner than 1 mm have been advocated, but it is somewhat tricky both to pour the gel without trapping air bubbles and to load it. The acrylic strips (1.5 cm wide) should be sparingly greased and put between the glass plates; the assembly is held together with a row of steel clips. Before casting the gel, the gel mould must be closed at the bottom. This can be done in several ways, two of which are illustrated in *Figure 3*. An acrylic sample well-former or 'comb' is inserted between the glass plates at the top of the gel mould to a depth of 3−4 cm. The thickness of the comb is the same as that of the spacers; the teeth are 5 mm wide at the extremities and there is at least 1.5 cm space between them. Some workers prefer to use Teflon or silicone rubber for the comb and spacers. Silicone rubber spacers have the advantage over rigid plastic spacers that they do not have to be greased, but they cannot be used in conjunction with a plasticine sealing strip at the bottom.

2.1.2 Preparation of the sample

The volume of the samples to be loaded should preferably be chosen such that the sample solutions in the wells are not deeper than 1 mm. This means approximately

Table 2. Stock solutions for use in gel electrophoresis

Type of solution	Components	Concentration	Comments
Gelling agent	acrylamide	400 g/litre	Use recrystallized chemicals. These are available commercially. Some authors recommend treatment with charcoal followed by filtration through a membrane filter before use. The ratio of bisacrylamide to acrylamide varies for some applications. Keep solution frozen.
	bisacrylamide	13 g/litre	
Denaturing agent	urea	9 M (540 g/litre)	Keep frozen.
Buffer[a,b]	citric acid	1 M	Keep frozen.
	Tris-citrate (pH 8.0)	1 M	Keep frozen.
	Tris-acetate (pH 8.4)	1 M	Keep frozen.
	Tris	0.89 M	Keep frozen.
	boric acid	0.89 M	
	EDTA	0.025 M	
Catalyst	$FeSO_4 \cdot 7H_2O$	25 g/litre	Subject to oxidation: prepare fresh solutions as required.
	ascorbic acid	100 g/litre	
	H_2O_2	300 g/litre	Keep in polythene bottle to avoid decomposition.
	ammonium persulphate $(NH_4)_2S_2O_8$	100 g/litre	Decomposes slowly upon storage; prepare fresh solution as required
Loading solution	sucrose	500 g/litre	Add 0.1–0.3 volumes to the sample before loading. Urea should only be present if the sample itself contains urea. Only one or two dyes are used in most procedures (see Section 3). Keep solution frozen.
	urea	300 g/litre	
	xylene cyanol FF	2 g/litre	
	bromophenol blue	2 g/litre	
	trypan red	5 g/litre	
	fluorescein	10 g/litre	

[a] This list of buffer stock solutions is not exhaustive. See *Tables 4–7* for complete gel compositions for all the procedures described in this chapter.
[b] Unless, otherwise stated, the molarity of Tris-containing buffers refers to the concentration of the base.

10 μl of sample volume on a gel 2 mm thick. The electrolyte concentration in the sample should be as low as possible because this results in a steep local voltage gradient and band-sharpening as the RNA penetrates into the gel. The density of the sample must be greater than that of the buffer. Finally, carrier RNA should be added to a final concentration of 10 μg/μl, especially for samples to be separated on acid pH gels. The following example of a loading procedure for an acid pH gel, containing 6 M urea, satisfies all these criteria. Mix the solution of radioactive RNA sample to be analysed with an appropriate amount of carrier RNA (e.g. tRNA) to make a total of approximately 100 μg of RNA; the volume of the sample should not exceed 500 μl at this stage. Allow the RNA to precipitate at −15°C after addition of a one-tenth volume of 5 M sodium acetate buffer (pH 5) and two volumes of ethanol. Recover the RNA by centrifugation (13 000 g for 5 min) and dissolve the pellet in 8 μl of 6 M urea. To this solution add 2 μl of a mixture containing 300 mg urea, 500 mg sucrose, 2 mg bromophenol blue, 2 mg xylene cyanol FF and 5 mg trypan red per millilitre.

The sample is now ready to be loaded. If the sample contains oligonucleotides too short to be precipitated with ethanol (a chain length of 5 nucleotides or less), the volume must nevertheless be kept below 10 μl. In this case it will be necessary to lyophilize the sample instead of precipitating it with ethanol. If the first-dimensional gel contains no urea, then the residue may be dissolved in water and urea may be omitted from the sucrose plus dye solution.

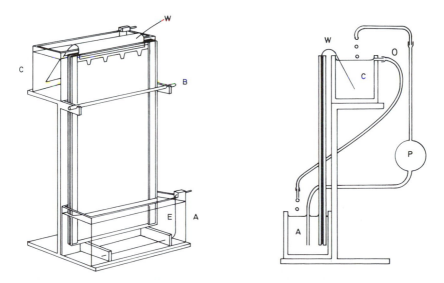

Figure 4. Electrophoresis stand made from acrylic plastic (Perspex), C, cathodic buffer compartment; A, anodic buffer compartment; W, paper wick; B, bar retaining the gel mould; E, platinum electrode; O, overflow; P, peristaltic pump for recirculating buffer. The steel clips (not shown) remain on the gel mould during the run.

2.1.3 Preparation and electrophoresis of gels

When the gel mould has been assembled, prepare a suitable gel mixture from stock solutions of acrylamide, bisacrylamide, buffer, denaturing agent, and catalyst for polymerization. *Table 2* lists suitable concentrations for stocks of the most frequently used solutions.

Protocol 1. Preparation of the first-dimensional gel

1. Mix appropriate volumes of stock solutions and dilute with water to obtain the desired gel composition for a particular application. Gel compositions are given in *Tables 3 – 7* for RNA separations (Section 3).

2. As soon as the catalyst has been added, pour the gel mixture into the gel mould using a funnel until the teeth of the comb are immersed to a depth of approximately 1.5 cm. After approximately 15 min, depending on the catalyst concentration chosen, the mixture begins to polymerise into a gel. This is accompanied by the evolution of heat and by the appearance of an interface of changing refractive index at the edges of the gel slab.

3. After another 15 min, pipette a layer of approximately 1 cm of electrophoresis buffer on top of the gel and carefully withdraw the comb so as not to damage the sample well surfaces.

Some authors recommend 'ageing' the gel at this stage by allowing it to stand for a few hours, or overnight, before starting electrophoresis, and nearly all workers recommend an electrophoretic 'pre-run' of a few hours before loading the samples.

Protocol 2. Electrophoresis of the gel

1. Transfer the gel mould for electrophoresis to the set-up illustrated in *Figure 4*. Fill the upper and lower reservoirs with buffer and remove any air bubbles trapped under the gel slab using a Pasteur pipette with a bent tip.

2. Connect the upper reservoir with the buffer layer on top of the gel by a paper wick consisting of one or more layers of Whatman 3MM chromatography paper soaked in buffer. When a urea-containing buffer is used, cover the wick with a thin plastic sheet to prevent evaporation and crystallization of the urea (a variant design which does not involve the use of paper wicks is described in Section 2.3).

3. Switch on the current and circulate buffer using a peristaltic pump from the lower to the upper reservoir, from where it drains back to the lower one. This minimises pH shifts in the reservoirs caused by electrolysis of the buffer.

4. At the end of the pre-run switch off the voltage and remove the wick from the top of the gel to allow the loading operation.

Protocol 2. *continued*

5. Take up the previously prepared samples (Section 2.1.2) into a capillary with a drawn-out tip, insert the tip into the buffer in the sample well to within a few millimetres of the gel surface, and slowly expel the contents. The dense, coloured sample should fall through the buffer layer and form a clearly defined layer in the well.

6. As soon as all the samples have been loaded, put the wick in place again and start electrophoresis. The conditions of electrophoresis depend on the type of separation carried out; examples are given in Section 3.

7. After completion of the run, remove the wick, suck up the buffer layer from the top of the gel with a Pasteur pipette, and lay the gel mould flat.

8. Take off the clips and loosen the top glass plate from the gel slab by inserting a knife blade or a spatula between the spacer and the plate.

9. Remove the spacers and rub the grease off the sides of the bottom glass plate.

At this stage an autoradiograph may be taken as described in Section 2.5. In a routine separation the position of the separated material will be known relative to that of the dye markers, or it may be roughly located with the aid of a radiation monitor if the radioactivity is sufficient. In this case, it is possible to proceed immediately with the second dimension.

2.2 Separation in the second dimension

Protocol 3. Gel electrophoresis in the second dimension

1. Cut out a rectangular strip of gel from the slab, as illustrated in *Figure 5a*. The width of the strip should be 1 cm and the length is determined by the distance migrated by the slowest and fastest moving material. It is important that the gel strip has smooth, straight edges, otherwise air bubbles can become trapped in the second-dimensional gel, which results in band distortion upon transfer. A smooth-edged strip is obtained by cutting the gel slowly with a scalpel along a ruler.

2. Transfer the excised gel strip onto a clean glass plate that will form the back wall of the second-dimensional electrophoresis gel mould.

3. Align the gel strip 4 cm from the bottom edge, as illustrated in *Figure 5b*. The second-dimensional gel should have the same thickness as the first one, the height and width depending on the length of the strips cut out of the first gel. For strips not exceeding 20 cm in length, a suitable size is 30 × 25 cm, so that an approximately square area remains above the gel strip for the second-dimensional separation.

4. Place spacers on the back glass plate on which the gel strip is lying, place a front glass plate on top and clip the assembly together.

Protocol 3. *continued*

5. Seal the assembly at the bottom as shown in *Figure 3* and place it upright.

6. Pour the gel mixture for the second-dimensional gel between the plates to within 3 cm of the upper edge.

7. After polymerization, transfer the gel mould to the electrophoresis apparatus (*Figure 4*), pipette a layer of buffer on top of the gel, and insert a wick. The direction of electrophoresis is from the bottom to the top if the first-dimensional strip has been cast in the second-dimensional slab near the bottom, as described here.

Working in this manner has an advantage in that, if the first-dimensional gel contains urea and the second does not, while the strip is immersed in the gel mixture prior to polymerization, some urea diffuses out of it and the denser solution sinks downwards, causing somewhat irregular polymerization below the strip. Since the RNA migrates upward during the second-dimensional run, the pattern cannot be disturbed by this phenomenon. If the second gel and not the first one contains urea, it is better to embed the gel strip near the top of the slab and to run the second dimension downwards.

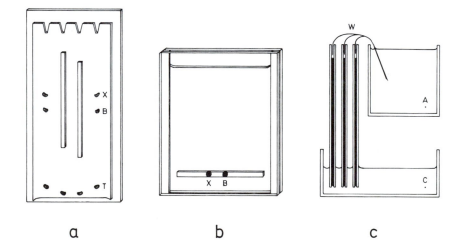

a   b   c

Figure 5. Separation in the second dimension. (a) A 1-cm-wide strip containing the separated material is cut out from the first-dimensional gel. X, B, and T are the dye markers xylene cyanol FF, bromophenol blue, and trypan red, respectively, used in the method described in ref. 8. (b) The strip is transferred on to a glass plate which is assembled into a new gel mould. A gel is cast in this mould as shown in *Figure 3*, enclosing the strip excised from the first-dimensional gel slab. (c) Up to three second-dimensional gels can be run simultaneously in an apparatus slightly modified from the design in *Figure 4*. W, paper wicks; A, anode; and C, cathode, are in this orientation only when the run is from bottom to top.

If the two gels have different acrylamide concentrations, the more dilute one should be run first. In this way, the mobility of the RNA decreases when migrating from the first-dimensional gel strip into the second-dimensional gel and this has a favourable effect on resolution, as explained in Section 1.2. If the pH of the two gels is different, this difference disappears very soon after the second run is started, even before the RNA has left the first-dimensional strip. Indeed, the buffer ions are much more mobile than the RNA molecules and even the dye molecules. This rapid equilibration of pH can be clearly seen when an acid pH gel strip containing a bromophenol blue marker spot is embedded in a neutral pH gel; the bromophenol, which is also a pH indicator, turns from green to blue soon after the current is switched on, and before it has left the strip.

2.3 Variation and modifications of the procedure

Several variations of the basic procedure described above can be found in the literature. In one application of the urea shift method, the strip, excised from the first-dimensional gel, is immersed in water for several hours and then placed in 8 M urea before being polymerized into the second-dimensional gel, which has the same composition as the first gel except that it contains 8 M urea. Incubation in water removes buffer ions while the RNA remains in place due to its much lower diffusion rate. As a result the gel strip has a much lower conductivity and is subject to a much steeper voltage gradient than the surrounding second-dimensional gel slab. This is one way of obtaining band-sharpening in the case where the first- and second-dimensional gels have the same concentration. In another urea shift procedure, both the first- and second-dimensional gel slabs are covered with a 'stacking gel' of lower concentration and lower pH in an attempt to obtain band-sharpening. However, this complicates the preparation of the gels and it is doubtful whether any net improvement in resolution is obtained, since the amount of band-sharpening is ultimately determined by the difference in composition between the loading solution of the first-dimensional gel strip and the gel into which the material migrates.

An apparatus (10) slightly different in design from the one described previously is shown in *Figure 6*. Here the buffer containing the cathode is in direct contact with the upper edge of the gel slab via a notch cut out in the back glass plate and the front wall of the buffer reservoir. For running the first dimension, some authors prefer to use cylindrical gels in the apparatus described by Davis (11). Either the cylindrical gels used are thin enough to be squeezed between the plates used for casting the second-dimensional gel slab or they are sliced longitudinally.

2.4 Temperature control

In the types of apparatus described in *Figures 4* and *6*, little control can be exerted over the temperature of the gel. During electrophoresis the gel heats up due to its electrical resistance, and the heat is dissipated through the glass plates. The difference between the gel temperature and the ambient temperature remains small when a low voltage is applied, but it can reach tens of degrees when using steep voltage gradients.

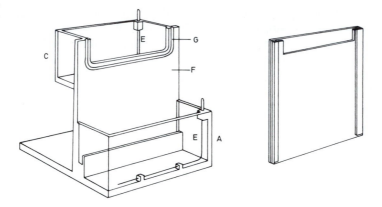

Figure 6. Modified electrophoresis apparatus. In this design the buffer in the upper reservoir is in direct contact with the top of the gel. This is achieved by making a notch in the front wall of the reservoir and in the rear glass plate of the gel mould, which is shown from the rear. C, cathodic buffer reservoir; A, anodic buffer reservoir; E, platinum electrodes; F, front plate; G, groove. This groove is fitted with a rubber tube before the gel mould containing the gel slab is fastened to the front plate with steel clips. Under pressure the rubber tube forms a leak-proof joint.

The geometry of the electrophoresis apparatus results in the heat being dissipated faster near the edges of the slab than in the middle. Since mobility increases with temperature, a sample loaded in the middle of the gel reaches the bottom faster than the same one loaded near the slides. For many applications this pattern distortion does not matter, but it can be avoided by working in an apparatus with liquid cooling. Some types of apparatus have plates containing cooling channels. In other types, one of which is described in Section 6.2, the mould containing the gel slab is entirely immersed in, or forms the boundary between, the buffer reservoirs. Temperature control is mandatory in some separations of large nucleic acid molecules, where temperature dependent conformational changes form the basis of the separation.

2.5 Autoradiography

Most two-dimensional separations of RNA mixtures have been carried out with ^{32}P-labelled RNA, for which autoradiography is the detection method of choice. Autoradiography can be carried out simply as described in *Protocol 4*.

Protocol 4. Detection of ^{32}P-RNA in gels by autoradiography

1. At the end of the second-dimensional electrophoresis run, dismantle the gel mould as after the first run.
2. After removal of the top glass plate, cover the gel slab with a sheet of thin plastic film (Saran Wrap from Dow Chemical or Glad Wrap from Union

Protocol 4. *continued*

Carbide are suitable) sufficiently large so that it can be folded back over the edges of the bottom glass plate. This prevents the gel from drying out while virtually no radiation is absorbed by the plastic film.

3. Stick small labels, marked with radioactive ink, on to the plastic film near the corners of the gel slab.

4. Take the gel to the dark-room and cover it with a suitable size of medical X-ray film. Ensure a good contact of the X-ray film with the gel surface by putting a thick glass plate on top.

It is difficult to give a rule for exposures times as a function of total radioactivity present in the gel. This depends not only on the thickness of the gel but also on such factors as the number of spots and the distribution of radioactivity in the different components, so it must be determined empirically. Preferably the radioactivity should be sufficient to allow exposure times of less than 24 h, since, otherwise, diffusion of the RNA spots, especially of short oligonucleotides, occurs. The time required for autoradiography can be reduced by the use of intensifying screens, which are placed on top of the X-ray film. Some of these, such as the rapid screen (CAWO) and the LH2 and SP screens (Kyokko) allow one to increase the sensitivity of detection by a factor of 5−15-fold, but only when autoradiography is carried out at −70°C. Others, such as the Trimax α16 screen (3M Company) improve the sensitivity by a factor of two- to three-fold even at room temperature, but these screens are much more expensive. Detection efficiency may also be increased by drying the gel using a gel slab drier prior to autoradiography, but this is advisable only if it is not necessary to recover RNA bands from the gel.

A detection method sufficiently sensitive for [3]H- and ^{14}C-labelled material is fluorography. This involves equilibrating the gel either in a solution of diphenyloxazole (PPO) in dimethylsulphoxide (12) or in an aqueous solution of salicylate (30) prior to autoradiography. Since the great majority of applications described in this chapter make use of ^{32}P-labelled nucleic acids, the interested reader is referred to Chapter 1, Section 6.5 for the detailed methodology of fluorography and the use of pre-exposed films to increase sensitivity.

2.6 Detection of non-radioactive RNA

This is not common practice in two-dimensional gel electrophoresis, since the amount of RNA that can be loaded on such gels is limited and autoradiography is by far the most sensitive detection method. Nevertheless, up to 500 μg of tRNA mixtures have been separated on two-dimensional gels (3) and the spots detected by methylene blue staining. Note that in this case no carrier RNA should be added to the sample. The procedure is as follows. Immerse the slab for 60 min in a solution containing 0.2% methylene blue in 0.2 M acetic acid, 0.2 M sodium acetate. Wash excess stain out of the gel by rinsing it with running tap water for several hours until the background is colourless, the RNA should then be visible as blue spots. The detection limit is approximately 0.05 A_{260} units. Stained gels can be kept for several weeks in the refrigerator if they are covered with Saran Wrap.

It is also possible to detect spots using UV light without any staining at all (13). In this case cover the gel slab on both sides with Saran Wrap and put on a fluorescent screen. Commercial pre-coated thin-layer chromatography plates containing a fluorescent indicator can be used for this purpose. Illuminate the gel slab lying on the screen from above with a short-wave UV lamp. RNA is visible as dark spots on the fluorescent background. The detection limit is approximately 0.1 A_{260} units.

2.7 Measurement of radioactivity in excised spots

Beta emissions of sufficient energy, such as those emitted by ^{32}P-phosphorus, give rise to Cerenkov radiation when emitted in a transparent medium such as an aqueous solution or a polyacrylamide gel. This radiation can be detected by the photomultipliers of a liquid scintillation counter without the need for any organic scintillator. Determination of the radioactivity of a spot excised from the gel may be required for a calculation of the radioactivity distribution in a fingerprint, or as a control of the recovery of RNA from the gel according to one of the methods described in Section 2.8. The gel disc can be put into a scintillation vial and counted as such, but it is important to remember that the optimal gain setting is much higher and

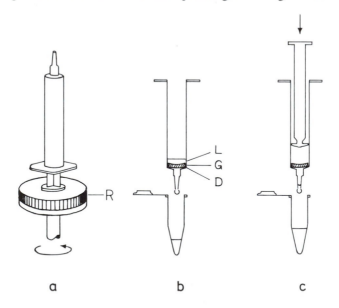

a   b   c

Figure 7. Extraction of RNA from polyacrylamide gel discs. (a) A filter disc is placed in the bottom of the syringe barrel, the piece of gel is introduced, and the plunger is inserted. The syringe plunger is then pressed against a rotating rubber disc (R) to grind the piece of gel. (b) The plunger is withdrawn, extraction liquid (L) is added to the ground gel cake (G) sticking to the filter disc (D). The liquid is collected in a disposable microcentrifuge tube. (c) The plunger is inserted in the syringe again to press the last drops out of solution.

the counting efficiency is lower than for counting ^{32}P-phosphorus in an organic scintillator. Usually, the ^{3}H channel is optimal for Cerenkov counting of ^{32}P-phosphorus but it is advisable to determine the optimal settings for each instrument. The counting efficiency can reach 40−50% but it may vary somewhat with the size of the piece of gel. To obviate this effect, Billeter et al. (9) recommend putting the gel disc in a small tube, covering it with 0.5 ml of 1 M NaCl, and then counting the tube in a standard vial. This ensures a constant geometrical factor and reproducible counting efficiencies. Liquid samples can be counted under the same conditions to simplify calculations.

2.8 Recovery of RNA from the gel

Areas of the gel containing RNA, located by autoradiography, can be cut out of the gel slab with a scalpel or with a cork-borer of suitable diameter. Many methods for the extraction of RNA from these discs have been described, most of them falling into two categories which could be called the soaking method and the grinding method.

Protocol 5. Extraction of RNA from the gel

Method A: soaking (14)

1. First put the gel disc, for example 0.4 mm thick and 5 mm in diameter, in a tube containing 200 μl of 0.001 M EDTA and 200 μg/ml of carrier RNA and incubate it at 37°C for 2 h.

2. Pipette the liquid off into another tube and add 20 μl of 3 M sodium acetate, 0.1 M magnesium acetate, 0.001 M EDTA (pH 6.0) followed by 700 μl of 95% ethanol.

3. Allow the RNA to precipitate in the cold (20 min at −70°C or 60 min at −15°C) and then pellet it by centrifugation at 13 000 g for 10 min.

Method B: grinding (8)

1. Make a disposable filter by removing the plunger from a 2 ml plastic syringe with a flat bottom into which a disc of Whatman 52 filter paper, fitting snugly in the syringe, is pressed against the bottom. The disc should be cut out with a cork-borer so that it is perfectly circular, otherwise gel particles may pass into the filtrate.

2. Put the piece of gel into the syringe and crush it between the filter disc and the plunger by turning the latter around under firm pressure. This can be done by inverting the syringe and pressing the head of the plunger against a slowly rotating rubber-covered disc, as illustrated in *Figure 7a*.

3. Then carefully withdraw the plunger while turning it in the opposite direction. In this way, the paper disc and gel cake remain on the syringe bottom.

4. Next, add 500 μl of 1 M NaCl and shake the syringe gently on a vortex mixer to bring the gel particles into suspension. It takes several seconds before liquid

Protocol 5. *continued*

    starts running through the filter disc, by which time the syringe should be positioned over a microcentrifuge tube (*Figure 7b*) for collection of the filtrate.

5. After a few minutes press out the last drops from the syringe by introduction of the plunger (*Figure 7c*).
6. Add a suitable amount of carrier RNA, 20 – 100 μg, to the filtrate, followed by an equal volume of isopropanol.
7. Allow the RNA to precipitate in the cold and recover it by centrifugation at 13 000 *g* for 10 min.

The soaking method is the simplest one and is probably best suited for the extraction from thin gel discs containing oligonucleotides, which diffuse relatively fast. The grinding method may be better for larger RNA fragments and thicker gel pieces. An alternative method that also can give good recoveries is electroelution; a protocol for this procedure is given in Section 2.9 of Chapter 2. If there is any doubt about the best procedure to be followed, it is best to measure the recovery in each case. This can be done very simply with ^{32}P-labelled RNA by Cerenkov counting of the gel before and after extraction, or even by measuring with a radiation monitor.

3. Examples of two-dimensional separations of RNA

3.1 Separation of oligonucleotides

Two-dimensional gel electrophoresis results in excellent fingerprints of T_1 RNase hydrolysates of large RNA molecules such as viral RNAs. The larger, unique oligonucleotides are especially well resolved. For isolating and characterizing these, gel electrophoresis has been found superior to fingerprinting methods based on cellulose acetate electrophoresis followed by DEAE-paper electrophoresis (15) or homochromatography (16).

In the first dimension, oligonucleotides are separated on a 10% gel (pH 3.3) containing 6 M urea. Mobility under these conditions is a function of base composition as well as chain length, and the presence of urea prevents the aggregation of longer complementary sequences. The second dimension consists of a 20 – 23% gel, buffered at pH 8.0 or pH 8.3, without urea. Mobility is then mainly a function of chain length. The recipes for the two gels according to the original procedure (8) is given in *Table 3*. Minor modifications in the gel composition have been introduced by several authors. The use of slightly more concentrated gels, buffered with Tris-borate buffer (pH 8.3) instead of with Tris-citrate (pH 8.0) has been advocated, but it is difficult to judge from the published results whether this actually improves resolution. Other differences in procedure reside in the size of the gel slabs, buffer concentration, voltage gradients applied, and the duration of electrophoresis. Comparative data for the main alternative procedures are summarized in *Table 4*.

Table 3. Original procedure for the preparation of gels for oligonucleotide fractionation

Stock solution	Volume (ml) required for 150 ml gel	
	first dimension	second dimension
Gel mixture (without catalysts)		
acrylamide (400 g/litre) bisacrylamide (13 g/litre)	37.5	75
9 M urea	100	–
1 M citric acid	3.75	–
1 M Tris, adjusted to pH 8.0 with citric acid	–	6
Catalysts		
$FeSO_4.7H_2O$ (25 g/litre)	0.6	–
ascorbic acid (100 g/litre)	0.6	–
H_2O_2 (300 g/litre)	0.06	–
TEMED	–	0.05
$(NH_4)_2S_2O_8$ (100 g/litre)	–	0.6

This table indicates the volume of each stock solution (*Table 2*) needed to prepare 150 ml of gel solution, which is the amount required to make 40 cm × 20 cm × 2 mm (first dimension) and 30 cm × 25 cm × 2 mm (second dimension) slabs. The catalyst is added to the mixture last, the volume adjusted to 150 ml and the gel cast immediately.

In general, a T_1 RNase digest of ^{32}P-labelled RNA containing some $1-20 \times 10^6$ cpm mixed with some 100 µg carrier RNA is loaded in a 5 mm × 2 mm well on the first-dimensional gel. If the bottom of the sample well has a larger or smaller area, the amount of carrier should be adjusted to approximately 20 µg per mm². The sample contains urea and sucrose, as mentioned in Section 2.1.2, and the coloured reference dyes bromophenol blue, xylene cyanol FF, and trypan red. The last of these dyes moves ahead of all nucleotide material on the first gel and so the electrophoresis is stopped when the red spot reaches the bottom of the gel. The bromophenol blue migrates slightly over half-way over this time period. A gel strip of 1 × 20 cm containing the radioactive zone is excised and the second gel prepared as described in Section 2.2. This gel is run until the bromophenol blue marker has migrated approximately half the length of the gel. For precise electrophoresis conditions according to alternative procedures, the reader is referred to *Table 4*. The time necessary to make a good autoradiograph of this type of gel depends not only on the amount of radioactivity applied, but also on the thickness of the gel slab. Special techniques for improving autoradiographic sensitivity can also be used (see Section 2.5). Most fingerprints are made with $1-15 \times 10^6$ cpm of ^{32}P and require less than 24 h autoradiography. This can be reduced to less than an hour if very thin gels are used.

An alternative procedure to fingerprinting RNase T_1 digests of uniformly ^{32}P-labelled RNA is to prepare unlabelled digests first and then to label the oligonucleotides at the 5′ terminus by means of polynucleotide kinase and [γ-^{32}P]ATP. Detailed

169

Table 4. Conditions for the electrophoresis of oligonucleotides: comparison of the original procedure and some modifications

Procedure: authors and reference	Dimension	Gel concentration (% w/v)[a]		Buffer	Gel size: height: width: thickness (mm)	Electrophoresis conditions		
		acrylamide	bisacrylamide			Voltage (V)	Time (h)	Marker distance[b] (cm)
1. De Wachter and Fiers (8)	first	10.0	0.325	0.025 M citric acid, 6 M urea	400:200:2	900	6	20
	second	20.0	0.65	0.04 M Tris-citrate (pH 8.0)	300:250:2	350	15	13.5
2. Coffin and Billeter (17)	first	10.0	0.33	0.025 M citric acid, 6 M urea	360:170:2	500	20	21
	second	21.8	0.65	0.04 M Tris-citrate (pH 8.0)	250:215:2	350	16	15
3. Frisby et al. (18)	first	10.0	0.35	0.025 M citric acid, 6 M urea	400:200:?	400		19
	second	20.0	0.66	0.1 M Tris-borate, 0.0025 M EDTA (pH 8.3)	400:400:?	400		19
4. Kennedy (19)	first	10.0	0.325	0.025 M citric acid, 6 M urea	380:170:3	500		21
	second	20.0	0.65	0.2 M Tris-borate, 0.005 M EDTA (pH 8.3)	380:380:3	400		18.5
5. Pedersen and Haseltine (14)	first	10.0	0.3	0.025 M citric acid, 6 M urea[c]	400:200:0.75	1700	2	18
	second	22.8	0.8	0.05 M Tris-borate (pH 8.3)	400:330:0.38	750	9	25

[a] Amounts of catalyst used were essentially the same in all procedures (see *Table 3*).
[b] Distance moved by the bromophenol blue marker when the electrophoresis was stopped.
[c] In this procedure, only the gel contained urea. The buffer compartments contained 0.025 M citric acid.

170

procedures are described elsewhere (Chapter 4, Section 2.6.2, Chapter 7, Section 3, and refs 14,20).

Oligonucleotide fingerprints of viral RNAs have been used to investigate sequence relationships between viruses belonging to different subgroups of the picornavirus family (18), sequence differences between distinct RNA molecules contained in a single virion (21) and between different strains of a virus (19), and to detect recombination between strains (22,23). The technique is illustrated in *Figure 8*, which shows a series of fingerprints of picornavirus RNAs uniformly labelled with ^{32}P-phosphorus (18). *Figure 9* shows fingerprints which have allowed the identification of some isolated influenza virus strains as mutants of a strain responsible for an epidemic in 1950 (23). In this case the oligonucleotides were obtained by T_1 RNase digestion of unlabelled RNA and then labelled *in vitro* at the 5' termini using polynucleotide kinase and [γ-^{32}P]ATP.

The most characteristic and useful part of fingerprints is the pattern formed by the longer oligonucleotides, which occur only once in the RNA molecule under study. In contrast, the shorter oligonucleotides completely fill the upper part of the graticule (*Figure 2*) since every sequence occurs once or more in any RNA of sufficient length. Differences between two viral mutants can be deduced simply by comparison of the fingerprint area consisting of unique oligonucleotides. If a difference is detected, the mutation can be identified by recovery of the material from the shifted spot and sequence analysis of the oligonucleotide. Another application of oligonucleotide fingerprinting is determination of the molecular weight or genome complexity of a viral RNA (9). When a unique oligonucleotide is selected from a fingerprint of a uniformly ^{32}P-labelled RNA, the chain length of the RNA can be computed from the relationship:

$$\frac{n}{N} = \frac{A_o}{A_t}$$

where n is the chain length of the oligonucleotide, N is the chain length of the RNA. A_o is the radioactivity of the oligonucleotide spot, and A_t is the radioactivity of the entire RNase T_1 digest. The value of n can be determined by sequence analysis of the oligonucleotide or by quantitative analysis of its pancreatic RNase breakdown products A_xC, A_yU and A_zG. The total radioactivity of the applied sample, A_t, and that of the excised oligonucleotide spot, A_o, can be measured by Cerenkov counting. The latter activity, however, must be multiplied by a correction factor to account for oligonucleotide losses during the RNase digestion and separation. Such losses are mainly due to a 'secondary splitting' by RNase T_1 preparations of phospho-diester bonds not adjacent to guanosine. The correction factor can be calculated from the recovery of known oligonucleotides from the fingerprints of an RNA of known molecular weight such as bacteriophage RNA or ribosomal RNA. This is a form of external standardization. Alternatively, a known oligonucleotide occupying a characteristic position in the map can be added to the unknown RNA before digestion and fingerprinting, in which case internal standardization can be used.

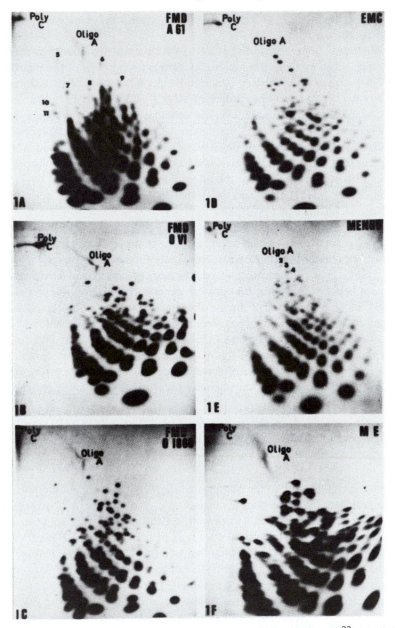

Figure 8. Fingerprints of T_1 RNase digests of picornavirus RNAs. Uniformly ^{32}P-labelled RNA obtained from virus grown on cell cultures was digested with RNase T_1 and fingerprinted according to procedure 3 in *Table 4* (18). Left, from top to bottom: three representatives of the foot-and-mouth disease subgroup. Right: three representatives of the cardiovirus subgroup, namely Encephalomyocarditis virus, Mengovirus and Maus Eberfeld virus. The origin is at the top left of each autoradiograph. Numbered oligonucleotides are those that have been extracted and further characterized by pancreatic RNase digestion. With permission from ref. 18.

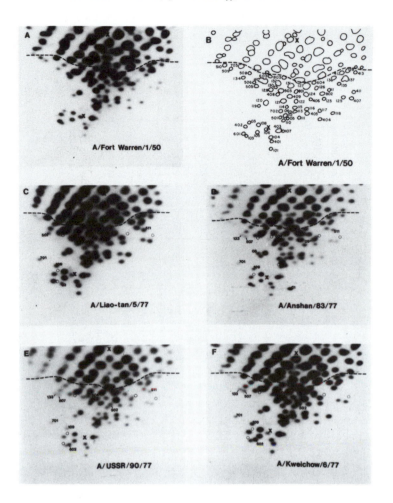

Figure 9. Fingerprints of T_1 RNase digests of influenza virus RNAs. The viral RNAs isolated from different H1N1 influenza A virus variants were digested with RNase T_1 and the resulting oligonucleotide mixture was 5'-end-labelled using [γ-^{32}P]ATP and polynucleotide kinase. Both labelling and fingerprinting (procedure 5 in *Table 4*) are as described elsewhere (14). Only the spots below the dashed line are used in comparing different viral variants, and these are characterized by gel sequencing methods. As compared to the reference strain (A, B) the variants (C to F) show additional spots indicated by new numbers and missing spots indicated by circles. The origin is at the bottom left of each autoradiograph. X's indicate the position of the dye spots xylene cyanol FF (bottom left) and bromophenol blue (top middle). With permission from ref. 23.

3.2 Separation of RNA fragments

The term RNA fragment is used here in the sense of a product obtained by partial RNase hydrolysis of an RNA, and oligonucleotide in the sense of a product obtained

Table 5. Conditions for the electrophoresis of of RNA fragments

Procedure: authors and reference; fragment length	Dimension	Gel concentration (% w/v)[a]		Buffer	Gel size: height: width: thickness (mm)	Electrophoresis conditions		Marker distance[a] (cm)
		acrylamide	bisacrylamide			Voltage (V)	Time (h)	
1. Vigne and Jordan (1); 5s RNA fragments 13≤N≤79	first	12.1	0.4	0.04 M Tris-acetate (pH 8.4)	180:130:3	300	3–4	
	second	12.1	0.4	0.04 M Tris-acetate, 8 M urea (pH 8.4)	400:200:4	400	14–18	
2. De Wachter and Fiers (8); bacteriophage RNA fragments 10≤N≤80	first	10.0	0.325	0.025 M citric acid, 6 M urea	400:200:2	900	6	20
	second	20.0	0.65	0.04 M Tris-citrate (pH 8.0)	300:250:2	350	15	13.5
3. De Wachter and Fiers (8); bacteriophage RNA fragments N>80	first	8.0	0.26	0.025 M citric acid, 6 M urea	400:200:2	900	4	19
	second	16.0	0.52	0.04 M Tris citrate (pH 8.0)	300:250:2	350	15	20

In the first procedure the gel–buffer mixtures for both dimensions were polymerized by the addition of 0.1 ml TEMED and 40 mg $(NH_4)_2S_2O_8$/100 ml of solution. The gels for the second procedure were prepared as described in *Table 3*. Preparation of gels for the third procedure was exactly the same except that the acrylamide and bisacrylamide concentrations were a factor of 0.8 lower. *N* designates the chain length of separated RNA fragments.

[a] Distance moved by the bromophenol blue when electrophoresis was stopped.

174

after complete hydrolysis with a base-specific RNase. In practice, oligonucleotides are seldom larger than 20 nucleotides, whereas fragments up to 200 nucleotides and more in size can be obtained by partial hydrolysis.

Vigne and Jordan (1) have used two-dimensional gel electrophoresis using the urea shift method to study interactions between *Escherichia coli* 5S RNA framents. Upon partial hydrolysis with T_1 RNase, only readily accessible guanosine-adjacent phosphodiester bonds in the RNA tertiary structure are split. Upon electrophoresis at neutral pH, some of the fragments thus obtained migrate together as a single larger fragment because of base-pairing between complementary segments. The hidden breaks can then be revealed by electrophoresis in the second dimension in the presence of 8 M urea, which disrupts hydrogen bonding. With a small molecule such as 5S RNA the number of fragments is limited (only five spots in the first dimension under the conditions of Vigne and Jordan) and it is possible to gain an insight into the secondary structure of the RNA. The gel composition and electrophoresis conditions used in this study are given in *Table 5*. The apparatus and procedure are slightly different from those described in Section 2. The first-dimensional gel was run in a commercial apparatus, the 'E.C. electrophoresis cell' (E.C. Apparatus Corporation), in which the slab is cooled by the surrounding buffer in the electrode compartments. The strip excised from this gel was washed in distilled water with gentle agitation for 4−6 h, and then in 8 M urea for 1−2 h. For the second dimension, a gel slab was cast in the type of apparatus shown in *Figure 3*, but with some space left at the top. After polymerisation, the strip from the first dimension is put on top and immersed in gel solution containing no buffer. The washed gel strip and the polymerized embedding solution both have a very low conductivity and hence the local voltage gradient is much steeper than in the surrounding gel. In this way, band-sharpening was obtained upon passage of the RNA from the first into the second-dimensional gel, although both have the same polyacrylamide concentration and pH.

Procedure 2 of *Table 5* is the pH-concentration-urea shift method which has also been used for oligonucleotide fingerprinting. Procedure 3 differs from it only in the gel concentration, which is slightly lower and is preferable for larger RNA fragments. Both procedures (8) have been extensively used for preparative separation of fragments from uniformly ^{32}P-labelled MS2 RNA during its sequence analysis. An example of such a separation is given in *Figure 10*.

3.3 Separation of small RNA molecules

RNA molecules ranging in size from approximately 80 to 400 nucleotides are separated most frequently by the concentration shift or by the urea shift methods. The essentials of the two procedures are summarized in *Table 6*. A few other methods have been described (6, 24), but the experimental protocols are poorly documented. A pH shift method is not so useful for molecules of this size. In contrast to the shorter oligonucleotides and fragments, they show only small variations in C+A/G+U ratio; hence they tend to remain close to the diagonal (see *Figure 2*).

Ikemura and Dahlberg (5) used a concentration shift method for separating a

Table 6. Conditions for the electrophoresis of small RNA molecules

Procedure: authors and reference	Dimension	Gel concentration (% w/v)[a]		Buffer	Gel size: height: width: thickness (mm)	Electrophoresis conditions	
		acrylamide	bisacrylamide			Voltage (V)	Time (h)
1. Ikemura and Dahlberg (5)[a]	first	9.5	0.5	0.089 M Tris base 0.0025 M EDTA 0.089 M H_3BO_3 (pH 8.3)	170:130:4	340	3
	second	19.0	1.0	0.089 M Tris base 0.0025 M EDTA 0.089 M H_3BO_3 (pH 8.3)	170:130:4	340	17
2. Varricchio and Ernst (3)[b]	first	15.2	0.8	0.089 M Tris base 0.0025 M EDTA 0.089 M H_3BO_3 7 M urea	150:ϕ = 6[c]	100	16
	second	15.2	0.8	0.089 M Tris base 0.0025 M EDTA 0.089 M H_3BO_3 (pH 8.3)	180:150:1.6	225	23

[a] Both the first- and second-dimensional gels were polymerized by the addition of 0.4 ml of diaminopropionitrile (DMAPN) and 40 mg $(NH_4)_2S_2O_8$/100 ml of acrylamide – buffer solution. Both runs were carried out using a commercial apparatus from E.C. Corporation.

[b] The first dimension was in tubes, the gel cylinders being 15 cm long and 6 mm in diameter. The second-dimensional slab gel was run in a 'Buchler' vertical starch gel electrophoresis apparatus. The catalyst for polymerization consists of 0.42 ml DMAPN and 56 mg $(NH_4)_2S_2O_8$/100 ml of gel. In both dimensions a layer of approximately 1.5 cm of 'stacking gel' was cast on top of the resolving gel. It contained 4.75 g acrylamide and 0.25 g bisacrylamide/100 ml, and 0.05 M Tris-HCl (pH 3.6). Immediately before use, 0.25 ml of 0.02% riboflavin was added to 5 ml of this solution, and the gels were polymerized by exposure to a fluorescent light source for 2 h. After the first dimension, the gel cylinder was sliced lengthwise and a 1.5-mm-thick slice was pushed between the glass plates containing the already prepared second-dimensional slab so that it laid horizontally on top of the stacking gel layer.

[c] Gel cylinder of 15 cm length and 0.6 cm diameter.

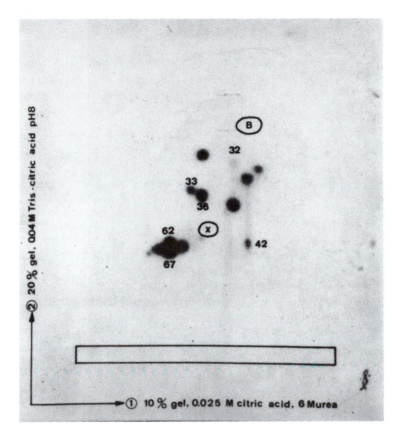

Figure 10. Separation of RNA fragments from bacteriophage MS2. Uniformly ^{32}P-labelled RNA from the bacteriophage MS2 was partially hydrolysed with RNase T_1 and fractionated on a non-denaturing slab gel at pH 8.0. One of the bands thus obtained was extracted and subjected to two-dimensional gel electrophoresis according to procedure 2 in *Table 5*. Total radioactivity on the gel was approximately 5×10^7 cpm and the autoradiograph was exposed for 5 min. The numbers near some spots indicate the fragment chain length as determined by sequencing analysis. The dye spots xylene cyanol FF and bromophenol blue are circled. The rectangle is the position of the strip excised from the first dimensional gel. With permission from ref. 8.

mixture of tRNAs, 5S RNA, 6S RNA, and some unknown RNA molecules of chain length up to 400 nucleotides isolated from *E. coli*. The same method was used in a study (25) of virus-coded molecules present in *E. coli* infected with DNA phages. In this case, a 5−10% acrylamide two-dimensional system was preferred for molecules of chain length longer than 150 nucleotides, the 10−20% system being used for RNAs of less than 200 nucleotides. The method has also been used in a study of small RNAs present in oncornaviruses, some of which are primers for RNA-directed DNA synthesis (26).

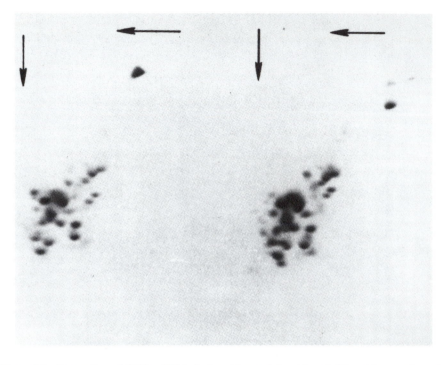

Figure 11. Separation of tRNAs. tRNAs isolated from adult rat liver (left) and from embryonic chicken liver (right), 400 μg in each case, were fractionated according to procedure 2 in *Table 6*. The intense spots at the top right, near the origin, correspond to 5S RNA. Staining was with methylene blue. With permission from ref. 3.

A urea shift method was found the most appropriate one for the fractionation of tRNA mixtures by Varricchio and Ernst (3) after they had tried several combinations (2,3), some of which include a small pH shift. *Figure 11* shows a separation of tRNA mixtures from chicken liver and from rat liver obtained by this procedure.

3.4 Separation of mRNAs

Burckhardt and Birnstiel (4) have used two-dimensional gel electrophoresis for purifying *Drosophila melanogaster* histone mRNAs from a polysome RNA preparation. Their method is not a general one for fractionating mRNAs. Rather, it is a specific method for purifying histone mRNAs, which are the main mRNA species synthesized during heat-shock treatment of *Drosophila* cells, from contaminating RNA of ribosomal origin. It is a good illustration of the urea shift method since it is based on the different behaviours of mRNAs and contaminating RNAs in the presence of the denaturing agent.

In a preliminary experiment, the polysomal RNA preparation is subjected to electrophoresis on a slab gel containing a urea concentration gradient perpendicular to the

5ᴍ   Urea gradient   Oᴍ

Figure 12. Analysis of *Drosophila melanogaster* total polysomal RNA on a urea-gradient slab gel. ³H-uridine labelled RNA, with an activity of 1.9×10^5 cpm, obtained from heat-shocked cells was loaded on to a 6% polyacrylamide gel slab containing a horizontal urea concentration gradient from no urea (right) to 5 M (left). Electrophoresis was from top to bottom, perpendicular to the urea gradient, at a constant temperature of 31 °C. Detection was by fluorography, using pre-fogged X-ray film and an exposure time of three weeks. Bands B, C, D, and E are histone mRNAs. With permission from ref. 4.

direction of separation. The RNA mixture is applied in a slot covering the entire width of the gel. In this way the influence of the urea concentration on the mobility of a component is reflected by a distortion of the band formed during the separation. In fact the band almost takes the shape of a melting curve, which would be obtained if electrophoretic mobility was plotted as a function of urea concentration. Since the melting of the RNA structure depends not only on the urea concentration but also on the temperature, the latter must be kept constant during the experiment. The result is shown in *Figure 12*. Most of the mRNA bands show a mobility transition in the urea concentration range up to 5 M, whereas the contaminating RNA bands do not. The interpretation of this pattern is that mRNAs start losing their tertiary structure under conditions where the background RNA (probably ribosomal RNA and its breakdown products) is not yet affected. The loss of tertiary structure results in an increased mobility. Those mRNAs that show a constant mobility under the conditions used (*Figure 12*) do show a mobility transition when the experiment is carried out at slightly lower temperature.

179

Table 7. Conditions for the electrophoresis of mRNA (4)

Dimension	Gel concentration (% w/v)[a]		Catalyst (amount/100 ml)	Buffer	Gel size: height: width: thickness (mm)	Electrophoresis conditions	
	acrylamide	bisacrylamide				Voltage (V)	Time (h)
first	6.0	0.16	$(NH_4)_2S_2O_8$, 200 mg; TEMED, 200 μl	0.04 M Tris base 0.02 M sodium acetate 0.001 M EDTA acetic acid to pH 7.2	$90:\phi = 2.2$[a]	60	4.5 – 9
second	6.0	0.16	$(NH_4)_2S_2O_8$, 200 mg; TEMED, 200 μl	0.04 M Tris base 0.02 M sodium acetate 0.001 M EDTA acetic acid to pH 7.2 5 M urea	90:170:2	60	4.5 – 9

The first dimensional gel was run in a cylindrical gel using the apparatus described by Davis (11). The 9-cm-long resolving gel was surmounted by a 1-cm-long 3% 'stacking gel' (half the amount of catalyst was used in casting the latter gel). After electrophoresis, the gel cylinder was immersed for 1 h in a solution containing half the buffer concentration used for electrophoresis, plus 5 M urea. The cylinder was then aligned horizontally near the top side between the glass plates assembled for casting the second-dimensional gel slab. The gel solution was poured between the plates to a height of 5 mm below the cylinder and overlayered with water during polymerization to obtain a flat surface. The space remaining between the slab and the cylinder was filled with 1% agarose solution containing half the buffer concentration used for electrophoresis, plus 0.2% SDS and 5 M urea. This solution was allowed to set at 4°C for 1 h. Electrophoresis was carried out in the apparatus shown in *Figure 6*. Temperature control (in the range 21°C – 35°C) was achieved by carrying out the electrophoresis in an incubator after equilibrium for 2 h.

[a] Cylinder height and diameter.

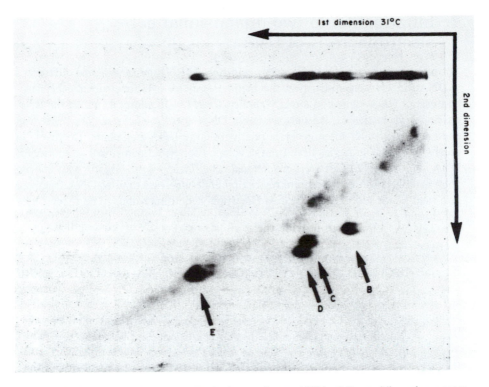

Figure 13. Separation of histone mRNAs from polysomal RNA of *Drosophila melanogaster* on two-dimensional gels. Poly(A)⁻, 6 – 14S RNA containing 2×10^4 cpm ^{3}H was electrophoresed for 8.5 h at 31°C in the first and for 9 h at 35°C in the second dimension, under the conditions described in *Table 7*. Spots B – E are histone mRNAs and correspond to bands B – E in *Figure 12*. Detection was by fluorography for 12 days. With permission from ref. 4.

The different behaviour of mRNAs and contaminating RNAs is then exploited in a two-dimensional gel system, the experimental details of which are summarised in *Table 7*. The second-dimensional gel contains urea whereas the first-dimensional one does not. Since urea has no influence on the mobility of the contaminating RNA, this remains on a diagonal. The mRNAs, on the other hand, have an increased mobility in the second-dimensional gel which contains urea, so they form spots below the diagonal; the resulting pattern is shown in *Figure 13*. Those mRNAs which do not show a mobility transition at the temperature of the experiment fall on the diagonal too, but they can be made to shift away from it by repeating the experiment at a different temperature. The mRNAs can be distinguished from background RNA on the urea-gradient gels because they are detectable by autoradiography but not by staining, as expected for a rapidly-labelled RNA species. They can also be identified on the two-dimensional gels by contact hybridization (4) with cloned recombinant DNA containing the appropriate histone genes.

181

4. Introduction to two-dimensional gel electrophoresis of DNA

The large size of most DNA molecules, compared to RNA molecules, and differences in DNA and RNA chemistry mean that methods used for two-dimensional gel electrophoresis of RNA cannot be directly applied to DNA. In addition, unlike the RNA base-specific nucleases, the only available DNA-specific nucleases are restriction endonucleases which often generate relatively large DNA fragments. However, in recent years two strategies have been developed for two-dimensional separation of DNA molecules. The first uses conventional gel electrophoresis (usually with agarose gels) and digestion with different restriction endonucleases to alter DNA fragment size (and hence mobility) in the two electrophoretic dimensions. The second uses denaturing gradient gels to separate DNA fragments according to base sequence.

5. Separation of DNA fragments on conventional gels

For most uses of these gels, large DNA molecules are digested with a restriction endonuclease and the resulting fragments are separated by gel electrophoresis. The fragments are then treated with a second restriction endonuclease and the resulting digestion products are separated by a second gel electrophoresis, with the electric field perpendicular to the original electrophoresis direction. This procedure generates a two-dimensional electrophoretic pattern of DNA fragments.

The experimental difficulty in two-dimensional electrophoresis of DNA on conventional gels has been in carrying out the second digestion. Three methods have been used. The first (and earliest) method was to excise or elute fragments from the first-dimensional gel into a number of fractions for subsequent digestion and electrophoresis (31−37). However, this procedure is tedious and resolution is limited by the number of fractions into which the gel can be divided. The second method has been to treat DNA fragments *in situ* in the first-dimensional gel with a second restriction endonuclease, then seal the treated gel on top of a slab gel and carry out electrophoresis in the second dimension.

To allow the second nuclease access to the DNA fragments, either the gel must be soaked in a nuclease solution (38−47) or the second nuclease is incorporated directly into the first gel (48). Although this second approach resolves problems associated with the first method, it introduces several new problems which limit its utility. These are the practical difficulties in exposing DNA in agarose gels to high specific activity nucleases, non-uniform exposure of DNA in the gel to nucleases, and inhibition of some nucleases by contaminants in agarose preparations which limits the choice of nucleases for the second digestion. The third method is to transfer the entire first-dimensional DNA fragment distribution from the gel to a DEAE cellulose membrane by electroblotting (49). This approach was developed by Poddar and Maniloff (49) to facilitate the second digestion: agarose contaminants are removed so all restriction endonucleases that have been examined can be used, and all fragments are uniformly exposed to the same high specific activity of nuclease. Hence, DNA transfer to DEAE cellulose membranes and subsequent restriction endonuclease treatment is currently the method of choice for carrying out the second digestion.

5.1 Basis of the separation

Genomic DNA, carefully isolated to reduce random breakage, is digested with a restriction endonuclease. The resulting fragments are separated by agarose gel electrophoresis, yielding a gel with a large number of DNA fragments, appearing almost as a continuum. The entire fragment distribution is transferred by electroblotting from the gel to the surface of a DEAE cellulose membrane, which is positively charged and binds anions such as DNA. This transfer facilitates subsequent nuclease digestion (since all fragments are uniformly exposed to added nuclease, and agarose contaminants that might interfere with digestion are absent) and improves final gel resolution (because of the fragment geometry on the membrane surface). The membrane-bound DNA fragments are then digested with a second restriction endonuclease, which cleaves those fragments containing recognition sites for that nuclease. After this, the membrane is embedded in a strip of agarose and soaked in high salt buffer, to break the ionic bonds between the DNA and DEAE, while holding the fragment distribution in place along the paper. Finally, the agarose strip containing the membrane is sealed on top of an agarose slab gel and electrophoresis is carried out perpendicular to the direction of the membrane. Fragments of DNA on the paper that are not cleaved by the second nuclease have the same mobility in the second electrophoretic dimension as they had in the first, and migrate to form a diagonal pattern in the second-dimensional gel. However, those fragments on the membrane that are cleaved by the second nuclease produce smaller fragments, which have greater mobility in the second electrophoresis than the original fragment had in the first, and form bands migrating below the diagonal in the second gel (49).

5.2 Experimental details

To prepare the sample for the first-dimensional separation digest DNA (30−40 µg) with the first restriction endonuclease in a total volume of 60 µl of reaction buffer, following reaction conditions recommended by the supplier. Depending on the type of analysis to be carried out (see Sections 5.3 and 5.4), this first digestion can be either a complete or a partial digestion. Following digestion, add 5 µl of gel loading solution (25% glycerol, 60 mM EDTA, and 0.1% bromophenol blue) to the mixture.

Protocol 6. First-dimensional electrophoresis

1. Cast a 0.7% agarose gel in 2 × TBE buffer on a glass plate (1 × TBE buffer is a 89 mM Tris-borate and 2.5 mM EDTA, pH 8.3). A typical gel is 15 cm wide, 19 cm long, and 2.5 mm thick, with sample wells 8.5 mm wide and 3 mm long.

Protocol 6. *continued*

2. Place the gel in a horizontal gel tank and load each well with a DNA digestion mixture. In addition, load one well with bacteriophage λ DNA digested with restriction endonuclease *Hin*dIII to provide a series of size markers.

3. Carry out electrophoresis at 1.5 V/cm for 16−18 h in 2 × TBE buffer.

A strip of gel containing the DNA separated in the first dimension is cut out and transferred on to DEAE-cellulose membranes by electroblotting (49) as described in *Protocol 7*.

Protocol 7. DNA transfer to DEAE-cellulose membranes by electroblotting

1. After the first-dimensional separation make two longitudinal cuts in the gel lanes to be used for second-dimensional separation. The cuts should be just inside each side of the lane.

2. Remove the strip of agarose containing the central 4.5 to 5.0 mm of each lane and set aside.

3. Stain the remaining gel with ethidium bromide (see Step 7 of *Protocol 8*) and visualize the DNA fragments in the trimmed edges with UV illumination, to determine the quality of the first-dimensional separation. Removing the band edges also improves subsequent electrophoretic separation, because the band edges tend to curve upward and decrease resolution in the second-dimensional separation.

4. Using the positions of the marker fragments of *Hin*dIII-digested bacteriophage λ DNA in the stained gel as a guide, the lower portion of each unstained gel strip, corresponding to DNA fragments <2.0 kb, can be removed. If these small fragments are not needed for the subsequent analysis, it is helpful to reduce the gel strip length to make it less fragile.

5. Mark DEAE-cellulose membranes (grade NA45 from Schleicher and Schuell, Keene, New Hampshire, USA) on one side using permanent ink and cut the membranes into strips 12 to 16 cm long and 2.5 to 3.0 mm wide. To maximize ion exchange capacity, soak the membrane strips in 10 mM EDTA (pH 7.6) for 10−15 min, then in 0.5 M NaOH for 5−10 min, and finally wash the strips several times in distilled water. Membrane strips can be stored for several weeks in water at 4 °C.

6. Place the agarose gel strips side-by-side on a glass plate, with adjacent gels separated from each other by a DEAE-cellulose membrane strip. The membrane strips should be the same width as the gel strips, and arranged with the unmarked side of each membrane against the gel to be blotted on to it.

7. Then place the glass plate in a horizontal gel box so that electrophoresis is perpendicular to the long axis of the strips. In this arrangement, for every gel-

Protocol 7. *continued*

membrane pair, the gel is closer to the cathode (negative electrode), and the unmarked membrane surface faces the gel and cathode.

8. Cover the gels with 2 × TBE buffer and, to ensure good contact between each gel and membrane, place long strips of Nylon scouring pad, slightly wider than the gel box, on the glass plate against the first and last gel strips to sandwich all the gel and membrane strips tightly together.

9. Carry out electrophoresis at 3.5 V/cm for 90 min. This is sufficient to transfer DNA bands from each gel to the adjacent unmarked membrane surface. Longer electroblotting times do not affect subsequent steps, but are not necessary.

10. Remove each membrane and gently wash them three times with buffer appropriate for the second nuclease digestion.

The DNA bound to the DEAE-cellulose membrane strip is then digested with a second restriction nuclease as follows. Insert each membrane strip containing bound DNA fragments into a tube, made by removing the ends of a 10 ml plastic disposable pipette. Fill the tube approximately half-full (3−4 ml is sufficient) with the appropriate digestion buffer containing 10 units of restriction endonuclease and 100 μl bovine serum albumin per millilitre, and tightly seal the ends with Parafilm. Seal the tubes in a plastic bag, lay them horizontally in a water bath at the appropriate temperature for the nuclease digestion and incubate them for about 4 h while gently shaking. Digestion of DNA bound to DEAE-cellulose membranes produces the same digestion products as digestion of the DNA in solution (49).

Protocol 8. Second-dimensional electrophoresis of DNA restriction fragments

1. Following the second restriction endonuclease digestion, remove each membrane strip from its tube and place it against a comb.

2. Position the comb near the top of a glass plate (so the direction of electrophoresis will be perpendicular to the length of the comb) and cast a slab gel, typically 15 cm wide and 19 cm long, of 0.7% agarose in 2 × TBE buffer. The membrane arrangement is such that its unmarked surface (containing the DNA fragments) faces the positive electrode (anode).

3. Remove the comb and then cut and remove a strip of agarose (5 mm wide and a few millimetres longer than the membrane), containing the membrane, from the slab gel.

4. Place the agarose strip in a solution of 2 × TBE buffer containing 1.5 M NaCl and 0.2% SDS and incubate at 60°C for 2 h, to break the ionic bonds between the DNA and DEAE-cellulose membrane while holding the DNA fragment distribution in place along the membrane.

5. Replace the agarose strip in the agarose slab gel and seal it in place with melted agarose; load bacteriophage λ DNA, digested with restriction endonuclease

Protocol 8. *continued*

*Hin*dIII, into a well at one or both ends of the gel, to serve as size markers (See Appendix 1).

6. Carry out electrophoresis in 2 × TBE buffer, at 3 V/cm for 30 min and then at 1.5 V/cm for 16−18 h. The use of DEAE-cellulose membranes for transferring DNA fragments does not affect DNA electrophoretic mobility in the second dimension (49).

7. Soak the gel in 1 μg/ml ethidium bromide for 2 h to stain the DNA fragments and destain the gel in water with occasional shaking. Visualize the DNA bands using UV light and photograph the gel using a Polaroid Type 55 P/N film (*Figure 14*).

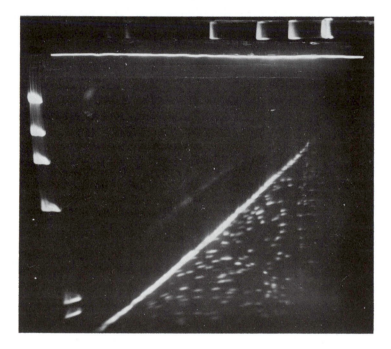

Figure 14. Two-dimensional electrophoresis pattern of *Acholeplasma laidlawii* strain K2 DNA (S. K. Poddar and J. Maniloff, unpublished data). The DNA was digested with *Bst*EII, and fragments separated by agarose gel electrophoresis, electroblotted on to a DEAE-cellulose membrane, and digested with *Hin*dIII. These fragments were separated by electrophoresis in an agarose gel with a well (on the left) containing *Hin*dIII-digested bacteriophage λ DNA. After electrophoresis the gel was stained with ethidium bromide. The sizes of the bacteriophage DNA fragments are 231, 9.4, 5.7, 4.4, 2.3, and 2.0 kb. The first-dimensional electrophoresis was from right to left and the second from top to bottom. The bright strip across the top is the DEAE-cellulose membrane. For this figure, a lane from the first dimension gel containing a *Hin*dIII digest has been laid across the top of the slab gel to show the first-dimensional separation.

If necessary, for a further analysis, the gel can be dried for autoradiography or transferred to a nitrocellulose filter and probed by Southern blotting (ref. 50 and Section 2.10 of Chapter 2).

5.3 Use of complete digests

5.3.1 Determination of genome size

The genetic complexity of an organism can be determined from an analysis of the DNA fragment distribution as a function of its size in a two-dimensional electrophoretic pattern, using equations derived independently by Yee and Inouye (42), and Poddar and Maniloff (49). In order to obtain the parameters necessary to calculate the genome size, experiments must be carried out using reciprocal restriction endonuclease digestions to generate two two-dimensional electrophoretic patterns. For one two-dimensional gel, the first digestion is by one restriction endonuclease (called nuclease 1) and the second digestion is by another restriction endonuclease (called nuclease 2). For the reciprocal two-dimensional gel, the first digestion is by nuclease 2 and the second digestion is by nuclease 1.

The relationship between the number of fragments under the diagonal in a two-dimensional electrophoretic pattern and DNA fragment size (42, 49) is:

$$\ln S_{12}(\geq L) = \ln S_{12}(0) - (L/M_0)$$

where $S_{12}(\geq L)$ is the total number of fragments under the diagonal of size $\geq L$, and M_0 is the number average molecular size of the fragments after the double digestion. For the reciprocal two-dimensional gel, the equation is:

$$\ln S_{21}(\geq L) = \ln S_{21}(0) - (L/M_0)$$

where $S_{21}(\geq L)$ is the total number of fragments under the diagonal size $\geq L$ for the reciprocal digest.

To calculate the genome size from two-dimensional gels, the number of DNA fragments under the diagonal larger than or equal to each marker DNA fragment [$S_{12}(\geq L)$ for one gel and $S_{21}(\geq L)$ for the reciprocal gel] is counted and plotted as a semilogarithmic function of fragment size (L). The data are fitted to the above equations by least-squares analysis and the slope and intercept [$S_{12}(0)$ for one gel and $S_{21}(0)$ the for the reciprocal gel] of each gel are calculated. The curves should be parallel and M_0 is taken as the mean of the two slopes.

With these three parameters [$S_{12}(0)$, $S_{21}(0)$, and M_0], the genome size is calculated (42, 49) using the equation:

$$M = (M_0/3) \{S_{12}(0) + S_{21}(0) + 2 [S_{12}(0)^2 + S_{21}(0)^2 - S_{12}(0) S_{21}(0)]^{1/2}\}$$

The number average molecular size of the DNA after digestion by each restriction endonuclease alone, M_1 and M_2, can also be calculated from these data:

$$M_1 = M_0 \{M/[M - M_0 S_{12}(0)]\}^{1/2}$$

$$M_2 = M_0 \{M/[M - M_0 S_{21}(0)]\}^{1/2}$$

and the total number of cleavage sites for each restriction endonuclease, n_1 and n_2, is given by:

$$n_1 = M/M_1$$

$$n_2 = M/M_2.$$

Genome size determination by this method does not require DNA fragment size measurements, because the sizes of the marker fragments that are used are known from sequence data, and it is only necessary to count the number of genomic DNA fragments of size equal to or greater than each marker DNA fragment. The genome sizes measured using two-dimensional conventional gel electrophoresis are *Escherichia coli*, 3520−4399 kb (42, 44, 49, 64); *Myxococcus xanthus*, 5690 kb (42); *Acholeplasma laidlawii*, 1680−1719 kb (49); and *Azotobacter chroococcum*, 1784−1920 kb (44). The standard deviation for genome sizes determined this way has been estimated to be ±10−20% (42, 44, 49).

5.3.2 Genome organization

The two-dimensional gel pattern of an organism's DNA is a genomic fingerprint. These have been determined, using a variety of methods for carrying out the second restriction endonuclease digestion, for λ (35, 48), *E.coli* (31, 35, 42, 44, 49), *B.subtilis* (32), *M.xanthus* (42), *Acholeplasma laidlawii* 1500-1700 kb (49), *Azotobacter chroococcum* 1800 kb (44), *Drosophila* (34), mouse (cited in ref. 34), rat (48), and human DNA (37). In addition, two-dimensional electrophoresis of mouse repetitive DNA has allowed the relationship between repetitive DNA bands to be determined (45), *B.subtilis* rDNA has been located using rRNA to probe two-dimensional gel patterns of *B.subtilis* DNA (32), and DNA fragments near the *M.xanthus* DNA replication origin have been identified by pulse-labelling followed by two-dimensional electrophoresis of *M.xanthus* DNA (46).

Specific methylation of DNA can be identified by two-dimensional gel electrophoresis using restriction endonuclease isoschizomers, such that the first nuclease does not cleave if the recognition site is methylated and the second nuclease cleaves whether or not the recognition site is methylated (42). Only DNA fragments containing uncleaved methylated recognition sites are cleaved by the second digestion to produce fragments that migrate under the diagonal during the second-dimensional separation. This protocol was developed by Yee and Inouye (42) and used to study DNA methylation during *M.xanthus* development.

Repetitive DNA sequences and genome rearrangements can be observed by denaturing and renaturing the DNA after the first restriction endonuclease digestion and before the first-dimensional separation, and using nuclease S1 (which is specific for single-stranded DNA) for the second digestion (43). The DNA for this procedure can be from a single organism (to look for repetitive sequences) or from two variants (to look for genome rearrangements). After the first digestion, denaturation, and renaturation, unique DNA regions re-anneal to form perfect DNA duplexes, but regions with different flanking sequences form partial duplexes with single-stranded tails. The latter are cleaved by the second digestion using nuclease S1, to produce DNA fragments that migrate under the diagonal during the second-dimensional separation. This method was devised by Yee and Inouye (43) and used to identify repetitive elements in *E. coli*, and genome rearrangements in *E. coli* and *M. xanthus*.

5.3.3 Genome mapping

DNA fragments in a two-dimensional gel, following digestion with two different restriction endonucleases, represent the double digestion products for that particular genome, and their positions contain information about fragment locations on the genome restriction endonuclease cleavage map. Additional map positional information can be obtained by comparing reciprocal two-dimensional electrophoretic patterns of a particular genome. This approach has been used by Rosenvold and Honigman (38) to map the 49 kb bacteriophage λ genome, with restriction endonucleases having 3 to 8 cleavage sites, and by Hildebrand *et al.* (47) to map a 158 kb chloroplast genome, using restriction endonucleases which cleave the DNA at 11 to 13 sites. However, even for the relatively simple chloroplast genome, there was some ambiguity in the DNA fragment map order determined from two-dimensional gels and it was necessary to use supplemental map data from cloned DNA fragments.

For large DNA molecules the two-dimensional gel pattern becomes too complex for straightforward analysis. Part of the problem is that there are two types of DNA fragments under the diagonal; those digestion products arising from termini of first-dimension DNA fragments, with a site for the first nuclease at one end and a site for the second nuclease at the other end, and those digestion products arising from within the first dimension DNA fragments, with sites for the second nuclease at both ends. These two types of fragments can be differentiated by end-labelling the fragments after the first digestion. The fragments from termini in the final two-dimensional gel pattern can then be identified by autoradiography. This method was used by O'Farrell *et al.* (41) to map the 166 kb bacteriophage T4 genome, using reciprocal digests with restriction endonucleases that cleaved the DNA at 14 to 17 sites. Amibiguities were resolved by including an additional restriction endonuclease with many (about 70) cleavage sites in the digestion prior to the second-dimensional separation.

5.4 Use of partial digests

Another approach to genome mapping is to use a partial restriction endonuclease digestion for the first dimension and a complete restriction endonuclease digestion (with either the same or a different restriction endonuclease) for the second digestion. Villems *et al.* (39) mapped the five *Eco*RI cleavage sites in the 49 kb bacteriophage λ genome this way. However, the number of partial digest products increases as the square of the number of cleavage sites (50) and so the number of fragments becomes too large to allow this method to be used for larger DNA molecules.

The complexity of the two-dimensional gel pattern, arising from a partial digestion for the first dimension followed by a complete digestion for the second dimension, can be reduced by end-labelling the DNA termini before the first (partial) digestion and examining the two-dimensional gel by autoradiography. This is a two-dimensional version of the Smith and Birnstiel method (51). In a two-dimensional gel, partial digestion products from each labelled terminus form a row of fragments in the second dimension, and the distance of each fragment in the row from the diagonal is a function of the distance of a restriction endonuclease cleavage site from the labelled terminus. This method, together with data from complete digestions and two-dimensional gel electrophoresis, has been used by Kovacic and Wang (40) to map the 7 *Hind*III and 17*Hae*III cleavage sites in bacteriophage PM2 DNA (about 10 kb). For the 166 kb bacteriophage T4 genome, O'Farrell *et al.* (41) first digested the genome with *Sal*I to produce eight large DNA fragments, then end-labelled these fragments and used the mixture of fragments for two-dimensional mapping using partial digestion. This allowed mapping of some of the sites for restriction endonucleases with many (30 to 80) cleavage sites in the T4 genome. Two-dimensional gel electrophoresis with partial digestion has also been used by Chen *et al.* (36) to map the eight *Bst*EII sites in 59 kb *Physarum* rDNA.

5.5 Other variations

In the protocol discussed in the preceding section, digestion with restriction endo-nucleases was used to change DNA mobility in the two electrophoretic dimensions. Other methods have been used to produce mobility changes based on DNA properties other than size.

5.5.1 Use of different gel matrix concentrations
Derynck and Fiers (27) separated DNA restriction endonuclease digests using an agarose−acrylamide composite gel (without crosslinker) in the first dimension, and a conventional polyacrylamide gel run at 50°C in the second dimension. The physical basis of DNA fragment separation in this system is not clear and, while resolution of DNA fragments in the range of 50−1000 bp is satisfactory, the fragments migrate close to the diagonal. This method has been used to separate DNA fragments of bacteriophage λ, SV40, and adenovirus (27).

Serwer (52) has developed a procedure for separating DNA molecules based on DNA conformation by two-dimensional electrophoresis in different agarose concentrations: 0.15% agarose for the first dimension and 0.7% agarose for the second dimension. With this method, open circular DNA molecules migrate above the diagonal formed by linear DNA molecules.

5.5.2 Use of different temperatures
Mizuno (53) has described a method for identifying DNA fragments with bend sequences, based on the observation that bent DNA migrates relatively normally at high temperature but slower than would be expected at low temperature. A restriction endonuclease digest of the DNA is separated by first-dimensional electrophoresis in a tube polyacrylamide gel at 60°C. The gel is then laid on a

polyacrylamide slab gel and the second-dimensional separation is at 10°C. Bent DNA fragments migrate above the diagonal in this procedure, and this has been used to identify and clone *E.coli* bent DNA fragments (53).

6. Separation of DNA fragments on denaturing gradient gels

The electrophoretic mobility of a double-stranded DNA molecule is markedly reduced if the duplex is partially denatured without strand separation, so that the molecule consists of both double-stranded and melted single-stranded regions (28, 29, 54). DNA melting is a co-operative process and, therefore, is observed as a sharp reduction in electrophoretic mobility as a function of the denaturing conditions. Melting and the consequent transition in electrophoretic mobility is related to base sequence, because melted single-stranded loops form co-operatively within a DNA molecule at regions (called domains) that are slightly richer in AT base-pairs than neighbouring regions. Therefore electrophoretic retardation of a DNA molecule is determined by the melting of its least stable domain and is related in a complex way to base sequence (54,55). Lerman and co-workers (28, 54) have described two experimental electro-phoretic configurations for using denaturing gradient gels to separate DNAs: perpendicular gradient gels in which the electric field is perpendicular to the direction of the denaturing gradient, and parallel gradient gels in which the directions of the electric field and denaturing gradient are the same. In both cases, denaturing gradients are preformed using urea and formamide and electrophoresis is at 60°C in poly-acrylamide slab gels.

6.1 Basis of separation

6.1.1 Perpendicular denaturing gradient gels

For this method DNA is applied in a continuous line along the top of a slab gel near the cathode, and electrophoresis is carried out so that DNA molecules migrate perpendicular to a linear gradient of denaturant (28, 54). The DNA molecules at any point migrate into the gel at a constant denaturant concentration, with mobility determined by the equilibrium DNA structure at that denaturant concentration and temperature.

DNA molecules migrate as an S-shaped band in perpendicular denaturing gradient gel (*Figure 15*). At low denaturant concentrations electrophoresis is conventional and DNA mobility is a function of molecular size. However, for each DNA fragment there is a denaturant concentration at which its least stable domain melts, causing a significant reduction in mobility. A similar DNA profile was observed by Thatcher and Hodson (56) using electrophoresis perpendicular to thermal gradients.

Electrophoresis perpendicular to linear urea−formamide concentration gradients is useful for studying the effects of DNA sequence on helix stability and for detection of single base changes in DNA (28, 54, 56−63).

20%   denaturing gradient ⟶   67%

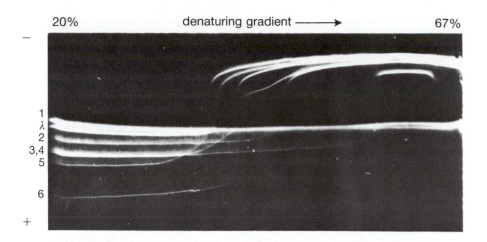

Figure 15. Electrophoresis of DNA in a perpendicular denaturing gradient gel using the Fisher and Lerman procedure (28). A mixture of 50 μg of intact bacteriophage λ DNA and 400 μg of an *Eco*RI digest of λ DNA were loaded in a continuous line across the top of a polyacrylamide gel containing a perpendicular linear gradient of denaturant from 20 to 67% of the denaturant stock solution. Electrophoresis (from top to bottom) was at 60°C and 5 V/cm for 17 h. After electrophoresis the gel was stained with ethidium bromide. The symbols on the left identify intact DNA (48.5 kb) and the six *Eco*RI fragments; 1, 21.2 kb; 2, 7.4 kb; 3, 5.8 kb; 4, 5.6 kb; 5, 4.9 kb; and 6, 3.5 kb. With permission from ref. 28.

6.1.2 Parallel denaturing gradient gels

For two-dimensional parallel gradient gels, a DNA restriction endonuclease digest is separated according to size in the first dimension by conventional agarose gel electrophoresis, and then according to a property related to base sequence in the second dimension by polyacrylamide gel electrophoresis in a parallel denaturing gradient (28, 29, 54). For this procedure, the agarose gel lane containing separated DNA fragments is placed at the top of a polyacrylamide slab gel, such that the linear denaturant gradient is parallel to the direction of the second electrophoresis. Each DNA fragment has an initial mobility determined by its size, as in conventional electrophoresis, but then migrates into higher denaturant concentrations. At the gradient depth at which its least stable domain melts, the fragment's mobility is sharply reduced. This produces a two-dimensional DNA pattern in which fragment positions are almost time-independent (*Figure 16*).

Agarose gel electrophoresis followed by parallel denaturing gradient gel electrophoresis separates DNA according to different properties in each dimension and produces two-dimensional patterns that can be used to observe genomic changes (28). A one-dimensional variation of this technique, in which similar DNA molecules migrate in adjacent lanes in a parallel denaturing gradient gel, can be used to analyse wild-type and mutant DNA fragments (54, 55, 57–63).

R. De Wachter, J. Maniloff, and W. Fiers

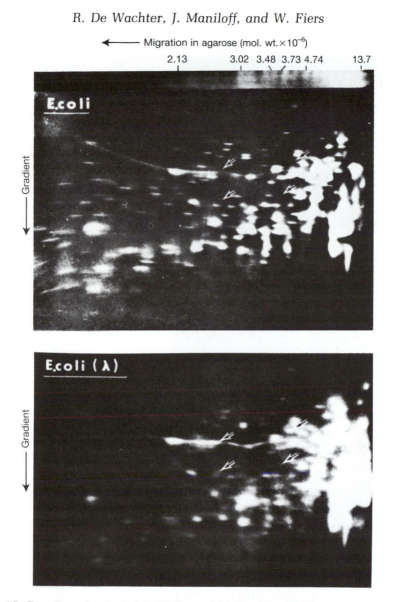

Figure 16. Two-dimensional gel of *Eco*RI-digested *E.coli* and *E.coli* (λ) DNAs using the Fischer and Lerman procedure (28). A mixture of 60 μg of *E.coli* DNA and 0.5 μg of ^{32}P-labelled *E.coli* (λ) DNA was digested with restriction endonuclease *Eco*RI and separated by electrophoresis on a 1% agarose gel. The agarose lane containing the DNA fragments was laid across the top of a polyacrylamide gel containing a parallel linear denaturant gradient from 10 to 75% of the denaturant stock solution. Electrophoresis (from top to bottom) was at 60°C and 9 – 12 V/cm for 20 h. After electrophoresis the gel was stained with ethidium bromide (upper gel) to show *E.coli* fragments, and autoradiography was carried out (lower gel) to show *E.coli* (λ) fragments. The arrows in the lower gel mark the positions of fragments found in *E.coli* (λ) DNA but not in *E.coli* DNA. An ethidium bromide stained first-dimensional agarose gel is shown at the top, together with molecular weight markers. With permission from ref. 28.

193

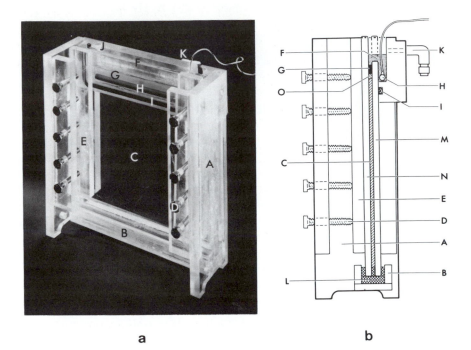

a   b

Figure 17. Apparatus for two-dimensional electrophoresis in denaturing gradient gels developed by Fischer and Lerman (28, 29). Front view (a) and side view (b). The parts are: A, acrylic frame; B, trough; C, gradient gel; D, stainless-steel screws; E, screw force spreader; F, cathode buffer chamber; 6, agarose gel strip; H, carbon cathode; I, rubber gasket; J, cathode buffer chamber inlet; K, cathode buffer chamber outlet; L, polyacrylamide or agarose plug; M, rear glass plate with notch at top; N, front glass plate; O, 1% agarose seal. With permission from ref. 29.

6.2 Apparatus

Fischer and Lerman (28, 39) developed an apparatus for running denaturing gradient gels at high temperatures. The polyacrylamide denaturing slab gels are formed in an acrylic plastic apparatus that holds two glass plates, such that the top of the plates forms part of the cathode chamber and the bottom of the plates sits in a trough which contains a gel plug in contact with the anode buffer reservoir (*Figure 17*). The front plate (N in *Figure 17*) is 20 × 17.7 cm, and the rear plate (M in *Figure 17*) is 17.7 × 17.7 cm with two 2.3 × 1.5 cm tabs at the sides of the top edge to form a notch across the top and allow contact with the cathode buffer chamber (F in *Figure 17*). The plates are separated by 2.5 mm × 1 cm acrylic spacers along the sides (not shown in *Figure 17*) and sealed against the cathode chamber with 4.5 mm silicon rubber tubing (I in *Figure 17*) and stainless steel screws (D in *Figure 17*).

The cathode (H in *Figure 17*) is a 6 mm diameter graphite rod. The entire apparatus (after the gel has been poured) is set in a 25-litre aquarium which is filled with buffer (about 14 litres) to a height equal to the top of the slab gel. This forms the anode reservoir: the anode is a platinum wire parallel to the trough at the bottom of the plates. The anode reservoir is maintained at 60°C by a thermostatically-controlled heater and stirred, and serves to keep the gel at constant temperature during electrophoresis by heat exchange through the glass plates. The buffer is recirculated between the cathode chamber and the anode reservoir by a peristaltic pump at about 10 ml/min.

6.3 Experimental details

For the analysis of genomic DNA by parallel gradient gel electrophoresis, first digest the DNA with an appropriate restriction endonuclease and separate the fragments by conventional agarose gel electrophoresis (e.g. see Section 5.2). For the first dimension, use a 1% agarose gel, TAE buffer (40 mM Tris, 20 mM sodium acetate, and 1 mM EDTA, pH 8.0), and electrophorese the gel at 2 V/cm for 20 h (28). Following electrophoresis, cut out a strip of agarose (about 5 mm wide and 13 cm long) of the DNA-containing lane, transfer it on a flexible plastic ruler to the top of a second-dimensional parallel denaturing gel and carry out the second-dimensional separation as described in *Protocol 9*.

Protocol 9. Second-dimensional electrophoresis of DNA using a denaturing gradient gel

1. Degassed solutions are used for gel preparation (29). Fill the trough (B in *Figure 17*) with 40 ml of either 1% agarose in TAE buffer (58) or 9% acrylamide (from a stock solution containing 30% acrylamide and 0.8% bis-acrylamide, both electrophoresis grades) in TAE buffer, then add 0.2 ml of a 200 mg/ml ammonium persulphate solution and 40 μl of TEMED (29).

2. After the gel plug (L in *Figure 17*) has set, pipette hot 1% agarose in TAE buffer along the outer edges of the acrylic spacers and into the corners around the plug to prevent leakage of the denaturing gradient gel solution.

3. Prepare the denaturing gradient gel solutions from an acrylamide stock solution (29). The low density solution (0% denaturant) is the acrylamide stock solution diluted to 4% acrylamide in the TAE buffer. The high density solution (100% denaturant) used is 4% acrylamide, 7 M urea, and 40% (v/v) formamide in TAE buffer. Other denaturant concentrations can be made in an analogous manner as needed. Immediately before use degas the solutions and add 0.2 ml of a 200 mg/ml ammonium persulphate solution and 4 μl of TEMED; this allows 30 min before the onset of gel polymerization.

4. Mix the low and high density solutions and pump the mixture between the plates, either into the side of the plates for perpendicular denaturing gradient gels or into the top of the plates for parallel denaturing gradient gels. Conventional gravity-balanced mixing chambers or a digitally-controlled gradient pump can be used to generate the desired linear gradient of denaturant. Use a long syringe needle reaching the bottom of the gel mould to put the mixed solutions with

195

Protocol 9. *continued*

the low density solution pumped in first at the bottom and displaced upward to about 1 cm below the top edge of the rear plate.

5. Spray the top edge of the gel with 0.1% SDS to ensure a smooth flat surface and allow the gels to set for at least 2 h.

6. Then pour off the SDS solution. Do not begin electrophoresis until at least 3.5 h after the onset of gel polymerization.

7. *Either*
    (a) For perpendicular denaturing gels, after the gel is poured, plug the top of the gel with a Teflon spacer to form a single large well after polymerization (58). The amount of DNA that can be loaded into the well and the electrophoresis conditions required vary with the experiment. For example, in *Figure 15* each band corresponds to about 50 μg DNA and electrophoresis was at 5 V/cm for 17 h (28).
    or
    (b) For parallel denaturing gels, either lay the first-dimensional agarose gel containing DNA fragments across the polyacrylamide gel (29) or form wells for loading DNA solutions (54, 59). The first-dimensional agarose gel containing DNA fragments should be the same thickness as the polyacrylamide gel so that it can be gently pushed into the space between the plates. Leave a 3 mm space between the bottom of the agarose strip and the top of the polyacrylamide gel, and fill this with hot 1% agarose. Allow this agarose seal to set for about an hour, and carry out electrophoresis using appropriate conditions. For example, the two-dimensional gel in *Figure 16* contains 60 μg DNA and electrophoresis was at 9−12 V/cm for 20 h.

8. For denaturing gradient gel electrophoresis, immerse the apparatus (*Figure 17*) in TAE buffer up to the top of the slab polyacrylamide gel (29). Maintain the temperature of the anode reservoir at 60°C, keep the buffer stirred, and re-circulate the cathode reservoir buffer. Electrophorese the gel at sufficient electrical potential and time for the slowest, most denaturation-resistant fragment to reach its mobility transition point.

9. Stain the gels with ethidium bromide and photograph under UV light or dry them down for autoradiography (e.g. see Step 7 of *Protocol 8*).

6.4 Use of perpendicular gradient gels

These gels have been used to analyse melting profiles of DNA molecules as a function of base sequence, which has allowed comparisons between experimental data and theoretical results from the statistical mechanisms of helix-coil transitions as a function of base sequence (54, 56). It was found that almost all single base changes in the least stable melting domain of a DNA molecule can be detected, but base changes in more stable, higher melting domains cannot be detected. Two methods have been developed to enable a single base change almost anywhere within a DNA fragment to be detected. In the first method, Myers and co-workers (58, 59) attached a GC-rich sequence (called a GC-clamp) to the DNA fragment being studied. The GC-clamp alters the melting profile within the DNA fragment, allowing nearly all single base changes to be observed as an alteration in electrophoretic mobility transition point. Both perpendicular and parallel denaturing gradient gels have been used to characterize the properties of DNA sequences attached to a GC-clamp (58, 59). The GC-clamp is used in a general method for saturation mutagenesis (57) and to detect single base changes in human DNA (60–62). In the second method, Noll and Collins (53) hybridized the DNA fragment of interest to labelled single-stranded probe DNA. Analysis of the hybrid electrophoretic mobility transition point in parallel denaturing gradient gels can be used to identify human DNA polymorphisms (63).

6.5 Use of parallel gradient gels

6.5.1 Genome size determination

Poddar and Maniloff (64) have applied the equations derived for genome size determination from two-dimensional conventional gels to two-dimensional denaturing gradient gels. In the latter case, there is only one restriction endonuclease digestion, and the relationship between $S(\geq L)$, the number of fragments $\geq L$, and the genome size, M, is:

$$\ln S(\geq L) = \ln (M/m) - (L/m)$$

where L is the fragment size and m is the number average molecular weight of the fragments after digestion.

To determine the genome size from two-dimensional parallel denaturing gradient gels, the number of DNA fragments $[S(\geq L)]$ larger than or equal to marker DNA fragments (run in a lane in the first-dimensional gel next to the genomic DNA digest) is counted and plotted as a semilogarithmic function of fragments size L (64). The data are fitted to the above equation by least squares analysis and the slope $(-1/m)$ and intercept $[\ln(M/m)]$ are calculated. From these values, m and M are calculated.

As for two-dimensional conventional gels, genome size determination by this method does not require DNA fragment size measurements, because the sizes of the marker fragments that are used are known from sequence data, and it is only necessary to count the number of genomic DNA fragments of size equal to or greater than each marker DNA fragment. The genome sizes that have been measured using two-dimensional denaturing gradient electrophoresis are *Haemophilus influenzae*, 1833 kb; *Acholeplasma laidlawii*, 1646 kb; and *Mycoplasma capricolum*, 724 kb (64).

6.5.2 Genome organization

The only application of these two-dimensional denaturing gradient gels to genome analysis has been in *E.coli*, where several DNA fragments were identified in a λ lysogen that were not in the parental strain (*Figure 16*) (28). Hence, these gels can resolve (at least) a 49 kb insertion in a 4700 kb chromosome.

One-dimensional parallel denaturing gradient gels are particularly useful for comparing the electrophoretic mobility transition points of similar DNA molecules (54, 56), and for identifying DNA molecules with base changes (58–63).

Acknowledgements

Research in the authors' laboratories has been supported by grants from the Fonds voor Kollektief Fundamenteel Onderzoek, the Fonds voor Geneeskundig Wetenschappelijk Onderzoek, and the Geconcerteerde Onderzoeksakties of Belgium (to R.D.W. and W.F.) and by United States Public Health Service, National Institutes of Health, Grant GM32442 and University of Rochester Biomedical Research Support Grant (to J.M.).

References

1. Vigne, R. and Jordan, B. R. (1971). *Biochimie*, **53**, 981.
2. Stein, M. and Varricchio, F. (1974). *Anal. Biochem.*, **61**, 112.
3. Varricchio, F. and Ernst, H. J. (1975). *Anal. Biochem.*, **68**, 485.
4. Burckhardt, J. and Birnstiel, M. L. (1978). *J. Mol. Biol.*, **118**, 61.
5. Ikemura, T. and Dahlberg, J. E. (1973). *J. Biol. Chem.*, **248**, 5024.
6. Fradin, A., Gruhl, H., and Feldmann, H. (1975). *FEBS Lett.*, **50**, 185.
7. De Wachter, R., Merregaert, J., Vandenberghe, A., Contreras, R., and Fiers, W. (1971). *Eur. J. Biochem.*, **22**, 400.
8. De Wachter, R. and Fiers, W. (1972). *Anal. Biochem.*, **49**, 184.
9. Billeter, M. A., Parsons, J. T., and Coffin, J. M. (1974). *Proc. Nat. Acad. Sci. (USA)*, **71**, 3560.
10. Studier, W. (1973). *J. Mol. Biol.*, **79**, 237.
11. Davis, B. J. (1964). *Ann. N. Y. Acad. Sci.*, **121**, 404.
12. Laskey, R. A. and Mills, A. D. (1975). *Eur. J. Biochem.*, **56**, 335.
13. Hassur, S. M. and Whitlock, H. W. (1974). *Anal. Biochem.*, **59**, 162.
14. Pedersen, F. S. and Haseltine, W. A. (1980). In *Methods in Enzymology* (Grossman, L. and Moldave, K. eds). Academic Press, Inc. New York and London, Vol. 65, p. 680.
15. Sanger, F., Brownlee, G. G., and Barrell, B. G. (1965). *J. Mol. Biol.*, **13**, 373.
16. Brownlee, G. G. and Sanger, F. (1969). *Eur J. Biochem.*, **11**, 395.
17. Coffin, J. M. and Billeter, M. A. (1976). *J. Mol. Biol.*, **100**, 293.
18. Frisby, D. P., Newton, C., Carey, N. H., Fellner, P., Newman, J. F. E., Harris, T. J. R., and Brown, F. (1976). *Virology*, **71**, 379.
19. Kennedy, S. I. T. (1976). *J. Mol. Biol.*, **108**, 491.
20. Frisby, D. P. (1977). *Nucl. Acids Res.*, **4**, 2975.
21. Clewley, J., Gentsch, J., and Bishop, D. H. L. (1977). *J. Virol.*, **22**, 459.
22. Rommelaere, J., Faller, D. V., and Hopkins, N. (1978). *Proc. Nat. Acad. Sci (USA)*, **75**, 495.
23. Young, J. F., Desselberger, U., and Palese, P. (1979). *Cell*, **18**, 73.
24. Reddy, R., Sitz, T. O., Ro-Choi, T. S., and Busch, H. (1974). *Biochem. Biophys. Res. Commun.* **56**, 1017.

25. Ikemura, T. and Ozeki, H. (1975). *Eur. J. Biochem.*, **51**, 117.
26. Peters, G., Harada, F., Dahlberg, J. E., Panet, A., Haseltine, W. A., and Baltimore, D. (1977). *J. Virol.*, **21**, 1031.
27. Derynck, R. and Fiers, W. (1977). *J. Mol. Biol.*, **110**, 387.
28. Fischer, S. G. and Lerman, L. S. (1979). *Cell*, **16**, 191.
29. Fischer, S. G. and Lerman, L. S. (1980). In *Methods in Enzymology* (Wu, R., ed.). Academic Press Inc., New York and London, vol. 68, p. 183.
30. Chamberlain, J. P. (1979). *Anal. Biochem.*, **98**, 132.
31. Potter, S. S. and Newbold, J. E. (1976). *Anal. Biochem.*, **71**, 452.
32. Potter, S. S., Bott, K. F., and Newbold, J. E. (1977). *J. Bacteriol.*, **129**, 492.
33. Parker, R. C. and Seed, B. (1980). In *Methods in Enzymology* Grossman, L. and Moldave, K. (eds.) Academic Press, New York, Vol. 65, p.358.
34. Smith, S. C. and Thomas, C. A. (1981). *Gene*, **13**, 395.
35. Chen, C. W. and Thomas, C. A. (1983). *Anal. Biochem.*, **131**, 397.
36. Chen, C. W., Braun, R., and Thomas, C. A. (1984). *Experientia*, **40**, 921.
37. Guerin, P. and Lucotte, G. (1985). *Electrophoresis*, **6**, 339.
38. Rosenvold, E. C. and Honigman, A. (1977). *Gene*, **2**, 273.
39. Villems, R., Duggleby, C. J., and Broda, P. (1978). *FEBS Lett.*, **89**, 267.
40. Kovacic, R. T. and Wang, J. C. (1979). *Plasmid*, **2**, 394.
41. O'Farrell, P. H., Kutter, E., and Nakanishi, M. (1980). *Mol. Gen. Genet.*, **179**, 421.
42. Yee, T. and Inouye, M. (1982). *J. Mol. Biol.*, **154**, 181.
43. Yee, T. and Inouye, M. (1984). *Proc. Natl. Acad. Sci. (USA)*, **81**, 2723.
44. Robson, R. L., Chessshyre, J. A., Wheeler, C., Jones, R., Woodley, P. R., and Postgate, J. R. (1984). *J. Gen. Microbiol.*, **130**, 1603.
45. Boehm, T. L. J. and Drahovsky, D. (1984). *J. Biochem. Biophys. Met.*, **9**, 153.
46. Komano, T., Inouye, S., and Inouye, M. (1985). *J. Bacteriol.*, **162**, 124.
47. Hildebrand, M, Jurgenson, J. E., Ramage, R. T., and Burque, D. P. (1985). *Plasmid*, **14**, 64.
48. Peacock, A. C., Bunting, S. L., Cole, S. P. C., and Seidman, M. (1985). *Anal. Biochem.*, **149**, 177.
49. Poddar, S. K. and Maniloff, J. (1986). *Gene*, **49**, 93.
50. Maniatis, T., Fritsch, E. F., and Sambrook, J. (1982). *Molecular Cloning: A Laboratory Manual*. Cold Spring Harbor Laboratory Press, NY.
51. Smith, H. O. and Birnstiel, M. L. (1976). *Nucleic Acids Res.*, **3**, 2387.
52. Serwer, P. (1985). *Anal. Biochem.*, **144**, 172.
53. Mizuno, T. (1986). *Nucleic Acids Res.*, **15**, 6827.
54. Lerman, L. S., Fischer, S. G., Hurley, I., Silverstein, K., and Lumalsky, N. (1984). *Ann. Rev. Biophys. Bioeng.*, **13**, 399.
55. Fischer, S. G. and Lerman, L. S. (1983). *Proc. Natl. Acad. Sci. (USA)*, **80**, 1579.
56. Thatcher, D. R. and Hodson, B. (1981). *Biochem. J.*, **197**, 105.
57. Myers, R. M., Lerman, L. S., and Maniatis, T. (1985). *Science*, **229**, 242.
58. Myers, R. M., Fischer, S. G., Maniatis, T., and Lerman, L. S. (1985). *Nucleic Acids Res.*, **13**, 3111.
59. Myers, R. M., Fischer, S. G., Lerman, L. S., and Maniatis, T. (1985). *Nucleic Acids Res.*, **13**, 3131.
60. Myers, R. M., Lumelsky, N., Lerman, L. S., and Maniatis, T. (1985). *Nature (London)*, **313**, 495.

61. Lerman, L. S., Silverstein, K., and Grinfeld, E. (1986). *Cold Spring Harbor Symp. Quant. Biol.*, **51**, 384.
62. Lerman, L. S. (1987). *Somatic Cell Mol. Genet.*, **13**, 419.
63. Noll, W. W. and Collins, M. (1987). *Proc. Natl. Acad. Sci. (USA)*, **81**, 3330.
64. Poddar, S. K. and Maniloff, J. (1989). *Nucleic Acids Res.*, **17**, 2889.

6

Gel retardation analysis of nucleic acid – protein interactions

MARK M. GARNER and ARNOLD REVZIN

1. Introduction

To understand gene expression at the molecular level requires study of the inter-actions of nucleic acids with a wide variety of regulatory proteins. As always, progress in such an area depends on the availability of appropriate techniques. A method often applied to quantifying DNA−protein interactions is the nitrocellulose filter binding assay (1), which relies on the retention of protein-bound (but not free) double-stranded DNA on a nitrocellulose filter. While this approach has been used successfully in many systems, it does not work well for all DNA-binding proteins, perhaps because some proteins do not adhere well to nitrocellulose filters. The filter assay also has limited applicability if multiple protein interactions are involved, since one cannot discriminate between different types of complexes. Assays which depend on the protection (by bound proteins) of the DNA from DNase cleavage (2) or chemical modification (3,4) can provide quantitative information (5) but are most often used qualitatively, to localize the DNA binding site. In some cases binding studies utilize an enzymatic activity inherent in the DNA−protein interactions for example if RNA polymerase (6) or DNA gyrase (7) is involved. Gel filtration, centrifugation, and optical techniques have also been used to study site-specific nucleic acid−protein interactions.

In this chapter rapid, simple electrophoretic methods for detecting and quantifying nucleic acid−protein interactions are described. Electrophoresis has been used for many years for characterizing complicated systems such as nucleosome core particles (8). Complexes of 5S RNA with specific binding proteins have been resolved on polyacrylamide gels (9). Agarose gel electrophoresis has been used to identify promoter-containing DNA fragments by running quenched transcription reactions from a mixture of restriction fragments on gels, and identifying which bands contained ternary DNA−RNA polymerase−RNA complexes (10); the mobility of a DNA fragment is not appreciably altered by being in such a complex. Co-operative complexes of the *Escherichia coli* recA protein with DNA have also been separated from free DNA on agarose gels (11). However, a variation of these electrophoretic procedures has found wide use and this technique is based on the observation that, in an acrylamide gel, the mobility of a specific DNA—protein complex may differ substantially from that of the free DNA (12,13). This procedure is widely known as gel retardation analysis and it allows one to separate and quantify the free and complexed components of a binding reaction. In addition, one can distinguish between complexes involving multiple proteins or multiple classes of proteins, which allows the method to be applied to solutions containing a mixture of binding activities. Alternative terms used to describe this type of method are 'gel shift' or 'gel mobility' analysis.

In principle, electrophoresis techniques can be used with any nucleic acid—protein system. This chapter emphasizes the authors' own studies with proteins which bind to linear double-stranded DNA, but the protocols may be modified for work with other types of nucleic acids as well.

2. Characteristics of nucleic acid – protein interactions

To optimize the use of this gel electrophoresis method it is necessary to consider several features of protein—nucleic acid interactions. These binding phenomena have been modelled as exchange reactions between DNA-bound or RNA-bound ions and free protein (reviewed in ref. 14). Hence, they are exceptionally sensitive to salt concentration. All site-specific binding proteins also bind non-specifically, generally with lower affinity, to essentially any DNA or RNA template. The ratio of the strengths of the specific versus non-specific binding will depend on the solution conditions. Indeed, it is possible that in certain situations the specific binding may be weaker than the non-specific binding. Thus, no universal buffer system is likely to be found for the study of all nucleic acid—protein interactions; the salt composition of the buffer needs to be assessed for each system of interest.

An additional complicating factor is that preparations of nucleic acid binding proteins are normally only fractionally active in sequence-specific binding reactions, even though usually every molecule can bind non-specifically. The apparent fractional activity can vary from 5% to 75%, depending on the particular protein, the individual sample, and the method used to assay for activity. Hence, even under conditions of very tight binding it is necessary to add a greater than stoichiometric amount of protein in order to obtain complete formation of complexes. The molecular basis for the observed lack of protein activity is not known. On the other hand, the DNA fragments used are often isolated from recombinant plasmids, and are totally active if handled with reasonable care.

3. Principles of the gel retardation assay

The principles of the technique are illustrated in *Figure 1*. When a solution containing a mixture of protein—nucleic acid complexes and unbound molecules is subjected to gel electrophoresis, the free DNA will rapidly enter the gel as a band, physically removed from the other components. The highly negatively-charged DNA will pull bound proteins into the gel, but the DNA—protein complexes are retarded in moving

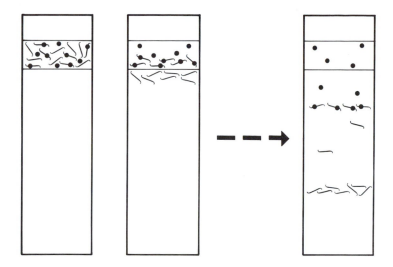

Figure 1. Schematic diagram of the gel retardation method. The filled circles represent protein, curved lines represent linear DNA fragments. The left-hand panel illustrates the DNA – protein solution loaded on to the gel. The middle panel shows free DNA entering the gel just after the power is turned on. The right-hand panel depicts the situation later in the run; bands of complexes and of free DNA are seen. If the complexes dissociate during electrophoresis, the DNA released never catches up with the main band of free DNA.

through the gel matrix. The amount of DNA in the 'free' band is not affected by any subsequent re-equilibration, either in the solution loaded on to the gel, or during electrophoresis. Any DNA released by dissociation during the run will trail behind the free DNA band as a smear, and is usually not detectable. Hence the assay yields an accurate value for the amount of uncomplexed DNA in the original solution, provided that no significant association or dissociation occurs in the 'dead time' between when the solution is loaded on to the gel and when the free nucleic acid enters the gel. In practice, this is not a serious limitation, since specific DNA–protein complexes are usually quite stable. From the concentration of free DNA and the known level of input DNA, the amount of DNA in complex can be calculated. Furthermore, in many cases the complexes themselves are stable during electrophoresis, and hence can be quantified, or studied for their stoichiometry.

This chapter uses as examples of the application of the gel retardation method the authors' own studies of the binding of the *E.coli* catabolite activator protein (CAP; also known as the cAMP receptor protein, CRP) and RNA polymerase to the lactose and galactose promoter regions. CAP stimulates transcription by binding to the *lac* promoter and enhancing the ability of RNA polymerase to form 'open' (pre-initiation) complexes. Assaying the equilibrium properties of proteins bound to cognate sites on a single fragment is discussed, as well as measurement of association and dissociation kinetics. Several useful variations on the basic method are given together with a review of the experimental details which need to be taken into account. The latter part of this chapter considers the applications of gel electrophoresis assays in conjunction with other techniques for probing DNA−protein interactions, and in measuring binding activities in nuclear or cellular extracts.

4. Basic methodology

The basic method uses slab gels. Here an example of this technique is demonstrated by measurement of the site-specific binding of CAP to the wild type *E.coli* lactose promoter region.

4.1 Preparation of the gel

The reader is referred to Chapters 2 and 4 in this volume for detailed protocols for preparing polyacrylamide gels.

It is important to take precautions to avoid contact with acrylamide solutions and do not breathe acrylamide dust since this is a dangerous chemical. The authors have found that the vertical gel apparatus supplied by Watson Products (Pasadena, California) works quite well. This device can accommodate gels of different lengths. Typically the gels used are 10 cm long by 16 cm wide by 0.1 cm thick.

Protocol 1. Preparation of the slab gel

1. Assemble the plates and spacers as usual, sealing the sides with Vaseline or plastic tape.

2. Place the plates in the vertical support stand, the bottom of the sandwich resting on the stand.

3. Seal the plates at the bottom as follows:

    (a) Mix 5 ml of 5% stock acrylamide:bisacrylamide (30:1) in 1 × TBE buffer (89 mM Tris-borate, 2.5 mM EDTA, pH 8.3) with 200 μl 10% ammonium persulphate.

    (b) Add 20 μl of TEMED (*N,N*-tetramethylethylenediamine) and vortex immediately.

    (c) Using a Pasteur pipette, allow this solution to run down the inside edges of the plates, until about 1 cm of gel mixture is at the bottom.

    (d) The solution polymerizes rapidly to form a plug (within a minute or so).

4. When the plug is solid, degas 25 ml of the 5% acrylamide solution in 1 × TBE buffer under vacuum.

5. Then add 200 μl of 10% ammonium persulphate and 20 μl of TEMED, and swirl the mixture gently.

6. Pour the solution into the gel plate/spacer assembly, insert a 10-, 12- or 16-tooth sample well-former (comb), and allow the gel to polymerize for about 60 min.

Protocol 1. *continued*

7. After polymerization, remove the comb, and clamp the gel into place on the electrophoresis apparatus. Add 1 × TBE buffer to the reservoirs, and remove any bubbles. Flush out the wells with reservoir buffer using a Pasteur pipette.

8. Prior to loading the samples, pre-electrophorese the gel for 60 min at 120 V, in order to remove any excess persulphate and unpolymerized acrylamide.

4.2 Preparation of samples

Protocol 2. Binding reaction and sample preparation

1. Prepare the reaction mixtures by mixing on ice:
   - 2 μl 10 × reaction buffer, (1× = 0.1 M KCl, 0.02 M Tris-HCl, pH 8.0, 3 mM $MgCl_2$, 0.1 mM EDTA, 0.1 mM dithiothreitol)
   - 2 μl 0.2 mM 3′,5′-cyclic AMP (cAMP)
   - 6 μl 67 μg/ml *lac* promoter DNA (203 bp fragment)
   - 9 μl water.

2. Add 1 μl of 0.47 mg/ml CAP in 'storage buffer' (0.2 M NaCl, 0.02 M Tris-HCl, pH 8.0, 0.1 mM EDTA, 50% glycerol) to the reaction mixture. Mix gently (vigorous vortexing may inactivate the protein), and incubate the sample for 10 min at 37°C.

3. Next, add 2 μl of '10 × loading buffer', containing 25% Ficoll, 0.05% xylene cyanol, 0.05% bromophenol blue, and load the sample on to the previously prepared gel, using a 100 μl Hamilton syringe. When multiple samples are used, be sure to rinse the syringe with electrophoresis buffer between loadings.

4.3 Electrophoresis

Protocol 3. Electrophoresis of the DNA-protein complexes

1. After all the samples have been loaded, run the gel at 200V, 20 V/cm (positive electrode at bottom of the gel), until all of the dye has entered the gel (approx. 5 min); then decrease the voltage to 120 V and continue electrophoresis for about 1.5 h, or until the bromophenol blue reaches the bottom of the gel. The xylene cyanol will then be about half-way down the gel, and the 203 bp DNA fragment runs somewhat ahead of the xylene cyanol in a 5% polyacrylamide gel.

2. Remove one plate and the spacers, and immerse the remaining plate plus gel (gel uppermost) in 0.5 μg/ml ethidium bromide in 1 × TBE buffer, and gently agitate for 30 min. Note that ethidium bromide is a dangerous carcinogen and

Protocol 3. *continued*

mutagen. Protective gloves (plastic) must always be worn when handling solutions of this chemical.

3. Photograph the gel on a transilluminator which produces UV light at 302 nm (see Section 2.8.2 of Chapter 2).

Figure 2 shows that much of the DNA in the wild-type *lac* promoter reaction is shifted into a band which migrates more slowly (lane B). This not seen in the reaction with DNA containing the catabolite-insensitive L8/UV5 promoter (lane D), indicating that the decrease in mobility in lane B is indeed due to promoter-specific binding of the CAP protein. This conclusion is substantiated by the observation that solutions containing a higher ratio of CAP to DNA show the same pattern on a gel. Note that CAP shows a significant co-operativity in its non-specific binding to DNA (15), hence non-specific complexes would contain several CAP per DNA fragment and do not migrate very far into the gel.

4.4 Technical aspects of gel retardation assays

4.4.1 Electrophoresis buffers

An important experimental consideration is the electrophoresis buffer itself. The authors have found that for many binding reactions the 'standard' nucleic acid electro-

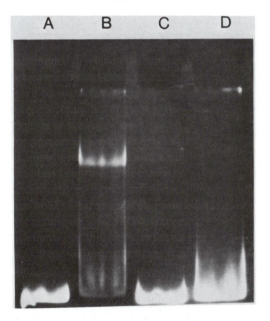

Figure 2. Gel stained with ethidium bromide, showing the specific binding of CAP to 203 bp *lac* DNA promoter fragments. Lane A: wild-type *lac* DNA alone; lane B: wild-type *lac* DNA plus CAP; lane C: mutant L8 *lac* DNA alone; lane D: L8 *lac* DNA plus CAP. The L8 mutation prevents specific CAP binding, so no band of complexes is seen in lane D. The somewhat broadened DNA band in lane D may reflect a small degree of non-specific binding.

phoresis buffers work quite well (i.e. 1 × TBE buffer as above, or Tris-acetate-EDTA buffer, etc.). Note, though, that while TBE buffer has been used extensively, some caution is dictated since borate-containing electrophoresis buffers can give artefactual separations of native proteins causing them to migrate as doublets, presumably due to borate−protein interactions (16). Moreover, for some systems it is necessary to have essential co-factors present in the buffer and gel during electrophoresis. For example, DNA gyrase−DNA complexes are not stable during electrophoresis in the absence of magnesium ions. Hence, to observe a complex, 3 mM $MgCl_2$ must be included in the buffer. For CAP binding to the wild-type galactose promoter, a band of complexes is seen when the buffer contains 200 μM cAMP, but only a smearing of the free DNA occurs at lower cAMP levels (e.g. 20 μM). Electrophoresis of nucleic acids has been performed in buffers ranging from 10 mM Tris to 2 M NaCl, though at higher salt concentrations the mobility is reduced and gel heating can be a problem.

Interpretation of data is easiest if the electrophoresis buffer is the same as that normally used for other experiments with the DNA−protein system under consideration. In practice this will require that low salt buffers are used. High salt concentrations lead to a reduction in the stability of DNA−protein complexes. This, coupled with a longer electrophoresis time results in smearing of the bands making it difficult to resolve bands of the complex under high salt conditions. The advantages of working with the same solution and gel buffers are illustrated by experiments described in Section 7, in which it is shown that the gel matrix does not affect either transcription by RNA polymerase, or the dissociation of CAP−DNA complexes.

It is important to note that the gel electrophoresis assay does seem to reflect the properties of the complexes in the buffer which is loaded on to the gel. Consider the interactions of CAP with DNA fragments containing the *lac* promoter. If the protein and nucleic acid are mixed in 1 × TBE buffer, then in both sedimentation and gel electrophoresis experiments the authors have observed the co-operative, non-specific binding characteristically seen at low salt concentrations. However, as indicated above, when CAP-*lac* DNA complexes are made in 0.1 M KCl buffer, only specific binding and no non-specific interactions are observed in the gel. Thus, any changes in ionic conditions which occur in the solution layered on to the gel as electrophoresis begins do not affect the migration of the bands. The gel assay yields information about the composition of the solution *in the reaction buffer*.

4.4.2 Loading buffers

There are, however, situations in which the ingredients in the sample buffer can affect the results. In particular, problems may arise from components of the dense dye-containing loading buffer, added to facilitate layering of the solution on to the gel. It is known that bromophenol blue can rapidly disrupt some protein−DNA complexes. Glycerol is a chaotropic solvent that has substantial effects on water structure and protein stability and so at high concentrations it may affect RNA polymerase−DNA interactions. Finally, the authors have found that it is preferable to have Ficoll rather than glycerol used in loading buffers. When Ficoll is used, bands in the gel are straight, while the presence of glycerol tends to cause significant trailing at the edges of the bands.

4.4.3 Gel heating

It is very important to control the temperature of the gel during electrophoresis. Excess heating can perturb the equilibrium reaction, or even can cause protein denaturation. Most commercial electrophoresis devices have temperature control mechanisms which are only marginally acceptable. Hence, the major limitation on the rate at which the electrophoresis can be performed is the ability of the gel to dissipate heat. The Watson apparatus is supplied with a small cooling fan, which is reasonably effective. Electrophoresis can be carried out in a temperature-controlled room if desired. Since specific DNA−protein interactions generally have high affinities, the electrophoresis rate can usually be slow enough so that heating is not a problem.

4.4.4 Quantifying the DNA

While analysis of the level of protein in a band of complexes can be achieved under some circumstances, the gel assay usually also involves quantifying the amount of DNA. If gels are stained with ethidium bromide as in *Figure 2*, it is easiest to perform densitometry on the negative of the gel photograph, taken with the gel on a UV transilluminator. The photographic negative can be scanned in any commercially-available laser scanning device. Care must be exercised to be sure that the DNA band is not so intense that saturation of the film image occurs. As controls, lanes containing known amounts of DNA should be run and scanned to ensure that the sample being assayed is in the linear response range of the film and scanner.

Generally, gel assays are performed using slab gels. If rod gels (typically 5 mm diam. See Section 5.2) are used, these may be scanned directly in a spectrophotometer equipped with a gel scanning attachment. In this case, the gel is removed from the tube and placed in a quartz trough, which rests on the transporter. A motor drive slowly moves the gel past the light beam, and the absorbance at 260 nm at each position of the gel is displayed on a dedicated chart recorder. If the acrylamide is pure, and the gels are pre-run to remove unpolymerized acrylamide, the background absorbance is virtually nil.

Staining with ethidium bromide is a simple procedure, but it does not allow a very wide range of DNA concentrations to be used. In addition, the fluorescence signal obtained may be affected when protein is bound to the DNA and so quantifying the amounts of complexes could be inaccurate. Radioactive labelling of the DNA is, therefore, often the method of choice. There are a number of ways to label at DNA at either the 3′ or 5′ termini. Protocols for labelling DNA are given in Section 2.6.2 of Chapter 4 and Section 3 of Chapter 7. Alternatively one can find similar protocols in most cloning manuals (e.g. see ref. 17).

If possible use sufficient radioactive DNA so that the aliquot of solution run in a lane contains about 2000 cpm. Following electrophoresis, autoradiography can be done either with the wet gel (if the DNA is to be extracted for additional work; see below) or a dried gel (which is convenient if bands are to be cut out and counted).

To autoradiograph a wet gel, take the gel sandwich from the apparatus and remove one plate. Carefully lay one thickness of plastic wrap on the gel and around the remaining glass plate. Put a piece of Kodak XAR-5 film on to the plastic and expose overnight; exposure for 16 h provides a clear picture if the DNA bands contain in the order of 1000 cpm. If fewer counts are available, then use an intensifying screen (Dupont Cronex-Lightning Plus). In this case exposure must be carried out at -70 to $-80°C$, in order to allow the screen to function. The screen increases the sensitivity of the autoradiography by about a factor of ten, although this is at the expense of some sharpness of the bands. After exposure develop the film and, if required, the autoradiogram can be scanned using a laser scanner to quantify the intensity of each band, just as would be done with the photographic negative of an ethidium bromide stained gel as discussed previously. Again, care must be taken to ensure that the bands are not so dark that the film image is saturated.

If it is desired to get more accurate values for the amounts of DNA in the free and complex bands, the gel can be dried, the bands cut out and the radioactivity of each measured in a scintillation counter. To do this, transfer the gel from the glass plate to a piece of Whatman 3MM filter paper, simply by laying the paper on to the gel and peeling it off the plate. Make alignment spots at 3 or 4 positions around the gel by placing $1-2$ μl of radioactive ink on to the filter paper. The amount of radioactivity in each ink dot should be about equal to that in the DNA bands of interest. Then smoothly cover the gel (but not the back of the filter paper) with plastic wrap, and dry on a slab gel dryer. Autoradiography is carried out as previously described. To identify the location of bands, place the gel over the autoradiogram on a light box, line up the ink marks, and cut out the bands with a scalpel. Add each band to 5 ml of toluene-based scintillation fluid and count in a liquid scintillation counter. The quenching by the gel when ^{32}P is used is 50% or less, so adequate counts are easily obtained. It is also possible to solubilize the gel, release the DNA into the scintillation fluid, and minimize quenching by adding 0.3 ml of Protosol or other tissue solubilizer to each vial, and measure the radioactivity after incubating the samples at 37°C overnight in the dark.

A second means for locating the bands of DNA and of complexes involves staining the gel with ethidium bromide before drying. This works well if the bands contain 1 μg or more of DNA. Remove one plate from the gel sandwich after a run, and immerse the gel and plate (plate down) in 1 × TBE buffer containing 0.5 μg/ml ethidium bromide. After 45 min of gentle agitation, carefully remove the plate and gel from the tray, transfer the gel to Whatman 3MM paper and dry, as described previously. Visualize the bands by holding a UV light above the dried gel; the bands can then be marked, cut out, and counted.

5. Variations on the basic methodology

5.1 Determination of equilibrium constants

Because the gel assay yields concentrations of unbound DNA and of complexes in the DNA−protein solution, it can readily be used to evaluate the affinity constant for the reaction. This involves choosing a reaction buffer such that varying the nucleic acid and protein concentrations over convenient ranges leads to detectable changes in the relative amounts of free DNA and complexes. For the CAP-*lac* promoter system, Fried and Crothers (18) used a low salt buffer containing 10 mM Tris-HCl (pH 8.0 at 20°C), 1 mM EDTA, and 10% glycerol, plus 0.2−10.0 μM cAMP. These authors prepared 1:1 complexes of CAP and ^{32}P-DNA at concentrations typically of 1×10^{-8} M, then made a series of threefold dilutions. After suitable incubation times to allow equilibration to occur, solutions were loaded directly on to polyacrylamide gels and electrophoresed. The data were analysed on a double logarithmic plot by plotting, for each reaction lane, log([DNA in complex]/[free DNA]) against log([CAP]). At the mid-point of the reaction, where half the DNA is in complexes and half is free the equilibrium constant, K, is equal to the concentration of CAP.

Fried and Crothers used the same buffer for their reactions and for the electro-phoresis itself. As discussed in Section 4.4.1, it appears that the level of free DNA in the gel accurately reflects the amount in the solution put on to the gel, regardless of the reaction buffer used. For each system, however, it is desirable to check the value obtained for the binding constant if other means are available. For example, if association and dissociation rate constants can be derived from kinetics experiments (see Section 6 below), their ratio provides another estimate of the equilibrium constant.

5.2 Determining the stoichiometry of complexes

A unique advantage of the gel assay is that when bands of complexes are found, these are separated from other components in the solution, including protein molecules which are not active in specific DNA binding. Thus analysis of the amounts of DNA and protein in the complex band yields the stoichiometry of the complexes. As discussed earlier, DNA can be quantified by radioactivity, by ethidium bromide fluorescence, or by directly scanning a rod gel for UV absorbance. If the protein can be radioactively labelled, it too can be readily quantified. The protein can be made radioactive by iodination with ^{125}I-iodine, by reductive methylation with ^{14}C-formaldehyde or with ^{3}H-sodium borohydride (28), or it can be isolated from cells grown in the presence of ^{35}S-methionine. The last of these options is the safest method, since modification of the protein by chemical reaction runs the risk of inactivating or denaturing it.

Another approach involves locating the complex bands either by ethidium bromide staining or by autoradiography if using ^{32}P-DNA, then cutting out the complex band from the wet gel and running the protein on a second-dimensional SDS−polyacrylamide gel. This procedure is similar to that used for analysing nucleosomes (see Section 4 of Chapter 8). In this case an SDS-containing running gel is poured to within 1−2 cm of the top of the gel plate. After polymerization occurs, the slice containing the DNA−protein band is soaked in SDS-containing buffer and is placed horizontally between the gel plates above the running gel. The stacking gel is carefully poured around the gel slice and electrophoresis performed as usual (19). The protein can be identified after staining the gel with Coomassie blue, or silver staining (see Section 4.3 of Chapter 8). In principle, this is a very useful approach for determining whether one or several proteins are specifically bound to the DNA. Known amounts of protein, run in parallel lanes on the same SDS gel, can be used to calibrate the level of protein in the DNA−protein complex. However, in practice, it is difficult to achieve sufficient accuracy for meaningful data because often the protein recovered from the complex does not run as a tight band and so is not easily quantified.

Alternatively, if the protein binds a ligand tightly which does not interfere with (or enhance) nucleic acid binding, then the radioactive ligand can be used to evaluate the stoichiometry of protein binding, provided that the protein−ligand interaction has been characterized. As an example, consider the CAP−*lac* DNA system (20). In this case gel shift experiments were performed using rod gels. Such gels, 10 cm long, can be made in glass tubes, 4−5 mm i.d. and about 12 cm long, and run in a rod gel apparatus after layering samples on top using a Hamilton syringe. For most applications rod gels are more cumbersome than slab gels, but for stoichiometry determinations, which do not involve many repetitive experiments, tubes can be advantageous. Following electrophoresis, the gels must be removed from the tube.

The stoichiometry of cAMP−CAP−*lac* DNA complexes was assayed using solutions similar to those described in Section 4.2. Here, 30 μl of sample solution containing 2.0×10^{-7} M *lac* DNA molecules (203 bp long), 7.7×10^{-7} M CAP, and 2.0×10^{-5} M ^{3}H-cAMP (the cAMP was 31 GBq/mmole, 833 mCi/mmole) was applied to 10 cm rod gels and electrophoresed for 90 min at 2 mA/gel. Each gel was removed from its tube and scanned in a spectrophotometer at 260 nm to locate the complex band and to quantify the DNA (see *Figure 3*). The gel was then cut into 0.25 cm slices which were placed in scintillation vials together with 0.3 ml of tissue solubilizer and 10 ml of scintillation fluid, and the radioactivity of each was measured. The most reliable data will be obtained if the solutions are left overnight in the dark before counting. To determine the absolute amount of ligand, it is necessary to account for quenching of the tritium. To do this, known amounts of ^{3}H-cAMP were run for a short time in parallel gels. The small cAMP molecule gives somewhat diffuse bands due to diffusion, but these can be located, excised, and the radioactivity measured. Results of stoichiometry experiments are shown in *Figure 3*. The position of the band of DNA−CAP complexes coincides with the location of a peak in the radioactivity profile. In this instance, the data show that there is one cAMP molecule bound to each *lac* DNA fragment. Other experiments indicate that a single cAMP molecule is bound to each CAP molecule at the *lac* promoter. Hence one concludes that the stoichiometry of the complex is one molecule each of cAMP, CAP, and *lac* DNA (20).

The data shown in *Figure 3* were obtained without addition of cAMP to the gel or electrophoresis buffer. Thus the stability of cAMP binding to the complexes is such that little dissociation occurs during the run. If significant dissociation of ligand does take place during electrophoresis, it may be necessary to add ligand to the buffer. The amount added will depend on the system involved. The binding affinity of ligand

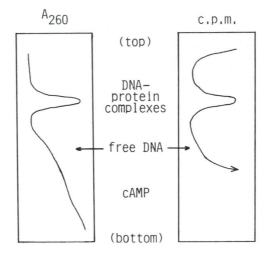

Figure 3. Absorbance and radioactivity profiles of ^{3}H-cAMP-CAP-*lac* DNA complexes, after gel electrophoresis. Peaks of absorbance and radioactivity are seen at the position in the gel corresponding to the position of the complexes. The location of free DNA is given by the arrow. The increases in A_{260} and radioactivity near the bottom of the gel are due to unbound cAMP, which has a higher mobility than do the macromolecular components of the system.

must be high enough so that the amount of radioactivity in the complexes is well above the background radioactivity of free ligand in the buffer.

5.3 Analysis of multiple protein molecules binding a DNA fragment

An important advantage of the gel electrophoresis assay is that it allows the study of systems in which several proteins interact simultaneously with the same DNA fragment. *Figure 4* shows the electrophoresis patterns for CAP and RNA polymerase binding to two similar, but different, DNA fragments. The reactions were prepared as described previously for the CAP binding experiments, and contained the amounts of proteins and DNA indicated in the figure legend. After reactions had been incubated for 10 min, they were layered on to a 4% (30:1) polyacrylamide slab gel and electrophoresed as described in Section 4.3. Comparing lanes 2, 3, and 4, or lanes 6, 7, and 8, it can be seen that complexes containing both proteins (upper bands, lane 4 and 8) show lower mobility than do those with only CAP or only RNA polymerase.

5.4 Studies of DNA bending

A novel application of the gel method using CAP-promoter DNA complexes demonstrated that CAP bends *lac* DNA (21). This phenomenon is illustrated in *Figure 4*, lanes 2 and 6, which show that the location of the CAP binding site on the fragment determines the extent to which bound protein retards the complex. The

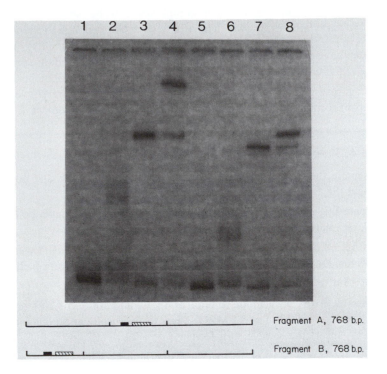

Figure 4. Autoradiogram of a gel electrophoresis assay of CAP and RNA polymerase binding to the DNA fragments A and B. The filled box shows the CAP binding site in each fragment, the cross-hatched box is the RNA polymerase site. Lanes 1 – 4: fragment A alone, with CAP, with RNA polymerase, and with both proteins, respectively. Lanes 5 – 8 are the same for fragment B. Each reaction contained 5.0×10^{-8} M ^{32}P-DNA; CAP, when present, was at 2.5×10^{-7} M; RNA polymerase, when present, was at 1.0×10^{-7} M. The buffer was 20 mM Tris-HCl (pH 8.0 at 23°C), 100 mM KCl, 3 mM MgCl$_2$, 0.1 mM DTT, 0.1 mM EDTA.

migration of the DNA in the complex is most affected when the bend is at the centre of the DNA fragment, least when it is at one end. Similar conclusions have been deduced from studies of DNA fragments which are thought to have a natural curvature. Comparing lanes 3 and 7 in *Figure 4* indicates that RNA polymerase also may bend the DNA, at least to a small extent. Finally, comparing mobilities of the slowest moving bands in lanes 4 and 8 shows that the bending of DNA caused by CAP binding is largely retained in complexes where RNA polymerase is bound as well. Thus these simple gel mobility assays can be quite revealing as to fundamental properties of the complexes.

5.5 Co-operative, non-specific DNA – protein complexes

In the absence of cAMP, CAP forms highly co-operative complexes with non-specific DNA (15). Co-operativity is also seen in the presence of cAMP. With DNA fragments

213

from 100 – 1500 bp in length, such complexes can be studied using mixed agarose – polyacrylamide gels (which have pore sizes large enough to allow the large complexes to migrate at a convenient rate; the agarose provides stability). Typically, the authors use 0.5% agarose, 2.5% acrylamide (30:1; acrylamide:bisacrylamide) gels, made in 1 × TBE buffer. With larger proteins or DNA fragments, different levels of agarose and acrylamide, or only agarose, may be more convenient.

To prepare a mixed agarose – polyacrylamide gel, follow *Protocol 4.*

Protocol 4. Preparation of an agarose-acrylamide slab gel

1. Assemble the plates and spacers for the slab gel apparatus, and pour a plug to seal the bottom of the gel mould (Section 4.1).

2. Prepare 1% agarose in 1 × TBE buffer and heat in an autoclave, boiling water bath or microwave oven, until the agarose dissolves.

3. Mix 5.2 ml of 15% acrylamide stock solution in 1 × TBE buffer with 10.0 ml of water. Degas and allow the solution to equilibrate in a water bath at 48°C.

4. Add 15.4 ml of the agarose solution and keep at 48°C.

5. To the agarose – acrylamide solution at 48°C, add 200 μl of 10% ammonium persulphate and 20 μl of TEMED, swirl to mix well, and pour into the gel mould. Insert the sample well-former.

6. Leave overnight at room temperature (the authors have been more satisfied with the results when gels have been left overnight before use).

7. Remove the comb, place the gel on to the slab gel apparatus, rinse wells with buffer, and pre-electrophorese for at least an hour at 30 mA ($\sim$215 V), using 1 × TBE buffer in the reservoirs.

Figure 5 shows the data obtained using a 203 bp *lac* DNA fragment with the L8 mutation (this fragment does not show specific binding of CAP). For this experiment, no cAMP was present in the gel or electrophoresis buffers. In the right-hand side lane one sees a sharp band of complexes, corresponding to DNA fragments saturated with non-specifically bound CAP, plus a faint smear, which reflects DNA molecules with fewer bound proteins, or from which one or more CAP molecules have dissociated during the run.

5.6 Use of gel retardation to assess the degree of DNA – protein crosslinking

The gel assay can provide a sensitive means for determining the degree to which a DNA – protein complex has been covalently linked by a crosslinking agent. To do this, the complexes are prepared, the agent is added, and the crosslinking reaction is allowed to proceed for the appropriate length of time. Following quenching of the reaction, two aliquots of reacted complexes are electrophoresed on a

Figure 5. Non-specific binding of CAP to 203 bp long DNA fragments. The solution loaded on to the left-hand lane contained DNA at 6.6×10^{-5} M, in 1 $\times$ TBE buffer. That run in the right-hand lane was the same but also had CAP at 3.3×10^{-6} M. This picture is the negative from an ethidium bromide stained gel.

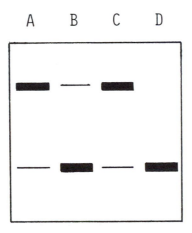

Figure 6. Idealized result of a gel assay for crosslinked DNA – protein complexes. Lanes A and B represent complexes exposed to crosslinker, lanes C and D show complexes not exposed to the reagent. Lanes B and D illustrate the result of raising the salt concentration of the solutions before electrophoresis, which causes dissociation of uncrosslinked complexes.

polyacrylamide gel; one aliquot is run without further treatment (in, say, lane A of *Figure 6*), while the second is run (in lane B) after addition of salt to dissociate any complexes which are not crosslinked. As controls, unreacted complexes without (lane C) and with (lane D) salt, are also electrophoresed. In such an experiment, then comparing lanes A and C should indicate that the crosslinking reagent has not grossly affected the complexes (i.e. that a moderate degree of reaction has occurred, without severe inter-complex crosslinking). Comparing lanes B and D will then reveal the extent to which covalent DNA-protein links have been made; the crosslinked complexes will not dissociate at high salt, and will be found at the location of the complex. Since this experiment involves quantifying the appearance of a band, it is quite sensitive. Using radioactively labelled DNA, it should be possible to detect as little as a few percent of crosslinked complexes easily.

6. Measurement of binding kinetics

The gel retardation assay can be used to measure on-rates or off-rates for nucleic acid – protein interactions. To measure kinetics in either direction a 'scavenger' molecule must be available. Such a scavenger is used to quench association reactions at various times by sequestering unbound protein; it is used in dissociation rate experiments to prevent proteins which come off of the DNA from rebinding. However, it is important to establish that the agent does not perturb the complexes themselves.

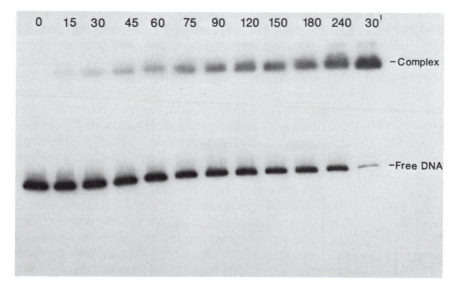

Figure 7. Autoradiogram of a gel binding experiment used to assay the association of RNA polymerase with the bacteriophage lambda P_R promoter. Solutions and electrophoresis conditions are described in the text. Times at which samples were quenched are at the top of the gel, in seconds, except for the right-most lane, which shows a solution which was incubated for 30 min before the reaction was stopped.

It is often convenient to use a large excess of DNA fragments containing the specific site under study as scavenger. If radioactive DNA is to be used, non-labelled DNA can be used as the scavenger, since it will not be detected by autoradiography. It is also possible to use a competitor DNA of different size than the DNA fragment being studied and this avoids interference when analysing gels stained with ethidium bromide. For experiments with *E. coli* RNA polymerase, heparin is often used as a sequestering agent. This negatively-charged polysaccharide has a high affinity for RNA polymerase. In the presence of heparin, non-specific RNA polymerase−DNA binding is quickly lost, but there is little effect on specific complexes. The effect, if any, of a scavenger on specific complexes can be tested by performing assays at several concentrations of the scavenger molecule.

6.1 Measurement of rates of association

Results of an association rate experiment using *E. coli* RNA polymerase and a 890 bp DNA fragment containing the P_R promoter of bacteriophage lambda are shown in *Figure 7*. The protocol for this experiment is as follows:

Protocol 5. Measurement of the kinetics of association of RNA polymerase with promoter DNA

1. Prepare the following solutions:
   (a) 125 µl of a solution of radioactive 890 bp DNA fragment giving a concentration of bacteriophage lambda P_R promoters of 2.0 nM. The binding buffer was 40 mM Tris-HCl (pH 8.0 at 22°C), 10 mM $MgCl_2$, 120 mM KCl, 1 mM DTT. In this experiment, 10 µl of DNA solution yielded about 6000 cpm when electrophoresed and cut out of the gel.
   (b) 187.5 µl of RNA polymerase at 1.33 nM, in the same binding buffer, except that BSA was also present, at 0.167 mg/ml. The BSA was 'RIA grade', from Sigma Chemical Company.
   (c) Heparin at 472 µg/ml, in binding buffer.
   (d) '10 × loading buffer' (25% w/v, Ficoll, 0.05% xylene cyanol, 0.05% bromophenol blue).

2. Prewarm the individual DNA and RNA polymerase solutions at 37°C for 5 min.

3. Gently mix 10 µl of DNA solution and 15 µl of RNA polymerase (no vortexing) and incubate at 37°C for the indicated time. Final concentrations are: 0.8 nM promoter DNA, 8 nM RNA polymerase, so the RNA polymerase is in a tenfold excess.

4. Quench the reaction at the indicated times by the addition of 30 µl of 472 µg/ml heparin solution, with brief vortexing.

5. Next, add 3 µl of 10 × loading buffer and briefly vortex the mixture. Load immediately on to a 10 cm, 12 lane, 5% (30:1) polyacrylamide gel (made with 1 × TBE buffer).

Protocol 5. *continued*

  6. Electrophorese the samples at 150 V for about 60 min, until the bromophenol blue runs off the bottom of the gel.

  7. Dry the gel, expose overnight, cut out, and measure the radioactivity of the bands as described previously.

  8. Plot data as the log[fraction of free DNA] against time. The graph in this case is linear as expected for a pseudo-first-order reaction (RNA polymerase excess).

As can be seen from *Figure 7*, as the reaction proceeds, the amount of free DNA diminishes, while the amount of DNA in complexes increases. The samples in different lanes were loaded at different times. The experiment can be modified so that all samples are available for loading simultaneously after quenching. This involves preparing those DNA-protein solutions which will be incubated for the longest times first. The advantage of this approach is that it minimizes any artefacts which might arise from time-dependent partial dissociation of complexes prior to electrophoresis. This is not a problem in the case of the very stable RNA polymerase-promoter complexes. In any case, the problem is obviated by focusing measurements on the free DNA band (see Section 3).

Another variation on the association experiment would be to mix the RNA polymerase and DNA together, then remove 25 μl aliquots at various times and quench. This has the advantage that all solutions run on the gel contain identical concentrations of reactants. The limitation here occurs if the early time points are necessarily quite close together so that one cannot perform the manipulations rapidly enough. To slow the reaction to manageable rates, the temperature can be lowered, the concentrations of reactants reduced, or the salt concentration altered.

6.2 Measurement of rates of dissociation

To ascertain the dissociation rate of complexes one should prepare about 325 μl of DNA – protein solution at the desired concentration and incubate it at the temperature of interest for a time long enough to ensure that the reaction is complete. For RNA polymerase – promoter interactions, this is typically 10 min at 37°C. At 'time zero', add heparin to 50 – 100 μg/ml. As polymerase dissociates from the DNA it is prevented from rebinding by the heparin. A convenient dissociation rate is established empirically for the particular system being studied by choosing an appropriate salt concentration. Remove aliquots (25 μl) at various times, add loading buffer, electrophorese the samples, and determine the amounts of free and complexed DNA. This protocol leads to different lanes being loaded at different times. However, if individual reactions are prepared in separate 25 μl aliquots, it is possible to arrange that all are completed and loaded at the same time.

In order to increase the number of samples that can be assayed at one time conveniently, Hendrickson and Schleif (22) used the following procedure. They placed a 4.5-mm-thick glass plate into a 15 × 15 cm, 0.8 cm Perspex (Plexiglas) tray, and made a lid for the tray containing 30 'teeth', 6 × 6 × 3 mm, in two suitably positioned rows. To make a gel, 90 ml of acrylamide solution is poured on to the plate and the lid carefully put in place so that the teeth are in the gel solution and thus form sample wells. After polymerization, the plate and gel are removed and used for electrophoresis in a 'submarine gel' device (see Section 2.1 of Chapter 2). In this device, the gel is run submerged, such that it is covered by about 1 mm of electrophoresis buffer.

7. Characterizing complexes in gels

Once a band of complexes is isolated in a gel, it can be assayed for appropriate chemical activity. One might ask, for example, whether RNA polymerase−promoter complexes can initiate transcription, or whether DNA polymerase−DNA complexes are capable of DNA replication. These are important questions and if the stoichiometry of complexes is to be determined by gel electrophoresis, it is desirable to establish that the complexes retain their biological activity.

One approach to this problem is to carry out an electrophoresis experiment in the usual way, using radioactive DNA, to locate the complexes in the wet gel by auto-radiography, and then to cut out that part of the gel containing the complexes. This piece of gel may be immersed in an appropriate reaction buffer and, after the reactants have diffused into the gel, the products are eluted from the gel slice and assayed in an appropriate manner. If RNA is the product, α-^{32}P-labelled nucleoside triphosphates can be used, and the RNA measured on denaturing gels (17). To avoid confusion between the labelled DNA and RNA, it may be useful to run parallel lanes of complexes first, one with ^{32}P-DNA to locate the band, the other without label which can be cut out and used for the RNA assay.

In some cases it is feasible to characterize complexes in the gel as part of the electro-phoresis run. For example, transcription reactions have been studied in the following way (23). Complexes between a tenfold excess of *E.coli* RNA polymerase and a 203 bp DNA fragment containing the *lac* UV5 promoter were made as described in Section 4.2, except that the buffer used was 0.014 M Tris-HCl (pH 7.9), 0.013 M NaCl, 0.01 M MgCl$_2$. This buffer, which is known to support transcription, was also used in making the gel, and as the reservoir buffer. The complexes were incubated at 37°C for 10 min and then heparin was added to 100 μg/ml. The sample was loaded onto a 5% polyacrylamide gel and electrophoresis begun. After 30 min the power was turned off and 6 μl of a solution containing 1.6 mM each of ATP, GTP, and CTP, plus 80 μM [α-^{32}P]-UTP (4.6 GBq/μmole; 125 mCi/μmole) was loaded on top of the gel lane; electrophoresis was resumed. The nucleotides have a much higher mobility than the complexes so they catch up with and pass through the band of complexes. However, during the time that the bands of nucleotides and complexes coincide, transcription can occur. After an additional 3 h of electro-phoresis, the experiment was stopped and the wet gel autoradiographed. A band of radioactivity was observed, which coincides more or less with the location of the complexes (determined by a separate control lane, using ^{32}P-DNA). This band was cut out of the gel and the RNA eluted by the 'crush and soak' method (see Section 2.8 of Chapter 5). The RNA was analysed on a denaturing gel. It was found that the RNA products made in the gel are the same as those made in a separate reaction done in solution. Thus the polymerase−promoter complexes behave in the gel precisely as they do in solution.

A second type of experiment involved studies of dissociation of CAP from wild type *lac* promoter complexes during electrophoresis in a polyacrylamide gel (23). Binding reactions (110 μl) were prepared containing 5×10^{-8} M ^{32}P-*lac* DNA fragments (203-bp long) and 1.25×10^{-7} M CAP in 20 mM Tris-HCl (pH 7.9), 3 mM MgCl$_2$, 1 mM DTT, 0.1 mM EDTA, 100 mM KCl, 20×10^{-5} M cAMP. At this concentration of KCl, CAP does not show non-specific binding to DNA. A 20-cm-long 5% polyacrylamide gel was made in $1 \times$ TBE buffer containing 5×10^{-6} M cAMP. The CAP−DNA solution was incubated for 5 min at 37°C, then 25 μl was loaded onto the gel and electrophoresis performed at 15 V/cm. A second 25 μl sample was loaded 30 min later, a third 30 min after that, and a fourth aliquot 30 min later still. After 30 min more of electrophoresis the run was ended, the gel was dried and autoradiographed, and the bands of complexes identified, cut out, and the radioactivity of each measured. It was found that those samples in the gel for longer periods had fewer CAP−DNA complexes remaining at the end of the experiment. That is, dissociation had occurred during the run, and the number of complexes declined with time. Analysis of the data was consistent with a first-order kinetic process, as expected for a simple dissociation reaction. This experiment yields the dissociation rate constant in the gel, in $1 \times$ TBE buffer containing 5 μM cAMP. The dissociation rate in free solution in this buffer, determined by another approach, was the same as that in the gel.

Thus for the two systems reported here, the presence of the gel does not affect DNA−protein interactions. This may not be true in every case. The polyacrylamide matrix can act as a volume excluding agent, and may also cause structuring of layers of water molecules near it, which could alter the properties of the complexes.

8. Use of the gel retardation assay in conjunction with other methods for studying DNA – protein interactions

Electrophoresis has been used to separate DNA−protein complexes from other components in nuclease or chemical protection experiments, and in 'interference' experiments (24). A detailed description of these techniques is given in Chapter 7.

8.1 Protection experiments

Basically, in a typical protection experiment, DNA −protein complexes are allowed to form, then are exposed to a reagent such as DNase I which will attack the DNA. On average the reaction is limited to less than one nick by the nuclease per DNA molecule. It is convenient to use about 10^6 cpm of radioactive DNA in such an experiment. After incubation with nuclease for the desired time, the solution is loaded on to a slab gel and electrophoresed. The band of complexes is identified by auto-radiography of the wet gel. The large amount of radioactivity used allows exposure times to be as short as 10 min. The complexes are excised from the gel, the DNA eluted by electroelution (Section 2.9 of Chapter 4) or the crush−soak method and electrophoresed on a denaturing gel. Control lanes contain samples in which nuclease attack is performed on the DNA in the absence of the binding protein. Autoradiography allows comparison of the DNase I digestion patterns of DNA alone and of DNA-protein complexes, revealing which nucleotides are protected by the bound protein. Nuclease protection experiments can be done without isolating complexes by gel electrophoresis. However, isolating complexes leads to a reduction of background and a substantial improvement in quality of the data.

8.2 Interference experiments

'Interference' experiments *require* the separation of DNA-protein complexes from unbound DNA. In this approach, the DNA is subjected to limited reaction with dimethyl sulphate or ethylnitrosourea. Alkylation of critical nucleotides of the DNA will prevent the protein from binding. Modifications of non-essential bases do not prevent binding. Complexes can be separated from the free DNA using nitrocellulose filters, but gel electrophoresis is the method of choice. Comparison is made of DNA eluted from complex and free DNA bands, following cleavage of the DNA by heating in NaOH solution and denaturing gel electrophoresis. This allows determination of those bases or phosphate linkages which, when alkylated, inhibit the DNA−protein interactions.

9. Detection of specific DNA-binding proteins in crude extracts

As described in detail in Chapter 7, the gel shift assay is now being widely applied to identify specific nucleic acid binding activities in extracts of whole cells or of nuclei (see also ref. 25). Radioactive DNA fragments are incubated with the extract, followed by gel electrophoresis and autoradiography. A band of radioactivity with a slower mobility than that of free DNA indicates the presence of a specific binding protein. This approach is quite useful for the detection of factors which may influence transcription; in this case the DNA used to probe the extract contains all or parts of a promoter region. If specific binding is seen only to, say, certain sequences in the upstream region of the promoter segment, then these sequences are tentatively identified as sites of action of the factor and can be further investigated by other techniques.

To do gel assays using cell extracts does not require extraordinary preparations. The labelled DNA fragment (conveniently about 5000 cpm per lane) is added directly to aliquots of extract which in turn are loaded on to the gel. The only complicating feature is the presence of the many other proteins in the extract, some of which will bind to the DNA probe in a non-specific manner. This, in general, prevents the probe from migrating very far into the gel. To circumvent this problem, one needs to add an excess of unlabelled competitor DNA. This competitor DNA ideally will bind only the non-specific proteins and will not interfere with the specific binding. Thus, *E.coli* DNA would be useful for eukaryotic extracts and calf thymus DNA for prokaryotic extracts. Any natural DNA, however, may contain adventitious tight-binding sites for the specific protein of interest. Therefore some investigators have turned to using poly(dI-dC)·poly(dI-dC) as competitor (26). The optimal amount of non-specific DNA to be added must be determined empirically for each extract. Typically, several micrograms of competitor are added to a reaction mixture containing less than a nanogram of probe DNA. An appropriate level of non-specific DNA results in appearance of a DNA–protein band which migrates into the gel, though at a slower rate than does the free DNA. The specificity of the reaction can be proved if this band disappears when excess unlabelled probe DNA is used as a competitor in the extract. The sensitivity of the gel electrophoresis assay in detecting specific DNA-binding factors may be enhanced by using short fragments (100 bp or less) as probes, since this will minimize interference from proteins in the extract binding non-specifically to extra DNA sequences on the probe itself. The type of gel to be used must also be evaluated for each system. In many cases 4–5% polyacrylamide gels work well. In another situation, 1% agarose was used, even with small DNA fragments (27).

   In principle, the band of specific DNA–protein complexes can be analysed further. SDS–polyacrylamide gel electrophoresis in a second dimension followed by silver staining of protein can reveal the molecular weight of the factor polypeptide(s). This has not yet proved to be a very fruitful approach, perhaps due to the low level of factor in the extracts. In addition, other proteins in the extract may by chance migrate to the same position in the gel as the complexes, thus obscuring analysis as to which proteins are truly bound to the DNA. It is also conceivable that thermodynamics and kinetics studies as described above can be applied to specific proteins in crude extracts (e.g. one might follow the rate at which DNA-factor complexes form by doing gel assays at various times after adding probe to the extract). Rigorous interpretation of such data will be hindered by the fact that many other proteins are present in the solution, but estimates of relevant parameters may be obtained. At present, the gel assay is finding much use in studies of the DNA sequence specificity of the reaction. The DNA probe is easily modified by site-directed mutagenesis techniques, hence the effects of particular base changes can readily be investigated.

Acknowledgements

We thank John Ceglarek for *Figure 4*, Stephanie Shanblatt for *Figure 7*, and Donald Lorimer and Stephanie Shanblatt for helpful discussions and suggestions. This work was supported by NIH grant GM 25498.

References

1. Jones, O. W. and Berg, P. (1966). *J. Mol. Biol.*, **22**, 199.
2. Galas, D. J. and Schmitz, A. (1978). *Nucleic Acids Res.*, **5**, 3157.
3. Gilbert, W., Maxam, A., and Mirzabekov, A. (1976). Control of ribosome synthesis. In *Alfred Benzon Symposium XII* (ed. N. O. Kjelgaard and O. Maaloe), p. 139. Munksgaard, Copenhagen.
4. Siebenlist, U., Simpson, R. B., and Gilbert, W. (1980). *Cell*, **20**, 269.
5. Brenowitz, M., Senear, D. F., Shea, M. A., and Ackers, G. K. (1986). *Proc. Natl. Acad. Sci. (USA)*, **83**, 8462.
6. McClure, W. R. (1980). *Proc. Natl. Acad. Sci. (USA)*, **77**, 5634.
7. Maxwell, A. and Gellert, M. (1984). *J. Biol. Chem.*, **259**, 14472.
8. Varshavsky, A. J., Bakayev, V. V., and Georgiev, G. P. (1976). *Nucleic Acids Res.*, **3**, 477.
9. Douthwaite, S., Garrett, R. A., Wanger, R., and Feunteun, J. (1979). *Nucleic Acids Res.*, **6**, 2453.
10. Chelm, B. K. and Geiduschek, E. P. (1979). *Nucleic Acids Res.*, **7**, 1851.
11. McEntee, K., Weinstock, G. M., and Lehman,I. R. (1981). *J. Biol. Chem.*, **8835**, 256.
12. Garner, M. M. and Revzin, A. (1981). *Nucleic Acids Res.*, **9**, 3047.
13. Fried, M. and Crothers, D. M. (1981). *Nucleic Acids Res.*, **9**, 6505.
14. Record, M. T., Jr. and Mossing, M. C. (1986). In *RNA Polymerase and the Regulation of Transcription* (ed. W. S. Reznikoff, R. R. Burgess, J. E. Dahlberg, C. A. Gross, M. T. Record,Jr, and M. P. Wickens), p. 61. Elsevier, New York.
15. Saxe, S. A. and Revzin, A. (1979). *Biochemistry*, **18**, 255.
16. Cann, J. R. (1966). *Biochemistry*, **5**, 1108.
17. Maniatis, T., Fritsch, E. F., and Sambrook, J. (1982). *Molecular Cloning: A Laboratory Manual*. Cold Spring Harbor Laboratory, Cold Spring Harbor, NY.
18. Fried, M. G. and Crothers, D. M. (1984). *J. Mol. Biol.*, **172**, 241.
19. Laemmli, U. K. (1970). *Nature (London)*, **227**, 680.
20. Garner, M. M. and Revzin, A. (1982). *Biochemistry*, **21**, 6032.
21. Wu, H.-M. and Crothers, D. M. (1984). *Nature*, **308**, 509.
22. Hendrickson, W. and Schleif, R. F. (1984). *J. Mol. Biol.*, **174**, 611.
23. Revzin, A., Ceglarek, J. A., and Garner, M. M. (1986). *Anal. Biochem.*, **153**, 172.
24. Hendrickson, W. and Schleif, R. (1985). *Proc. Natl. Acad. Sci. (USA)*, **82**, 3129.
25. Strauss, F. and Varshavsky, A. (1984). *Cell*, **37**, 889.
26. Singh, H., Sen, R., Baltimore, D., and Sharp, P. A. (1986). *Nature*, **319**, 154.
27. Green, P. J., Kay, S. A., and Chua, N.-H. (1987). *EMBO J.*, **6**, 2543.
28. Hames, B. D. and Rickwood, D. (1990). *Gel Electrophoresis of Proteins: A Practical Approach*, 2nd Edition, Appendix III.

7

The analysis of sequence-specific DNA-binding proteins in cell extracts

GRAHAM H. GOODWIN

1. Introduction

Gene transcription in eukaryotes is controlled by the binding of nuclear proteins to DNA sequence elements termed promoters and enhancers (1). Proteins binding to specific sequences within such elements probably interact with one another and with RNA polymerase and its associated transcription factors to potentiate the initiation of transcription. In order to purify and characterize such sequence-specific DNA-binding proteins, techniques have been developed to detect their presence in nuclear or whole cell extracts. The simplest and most sensitive is the gel retardation or mobility shift method (2,3) in which a labelled DNA fragment containing the sequence of interest is mixed with the protein(s) and the mixture then analysed by polyacrylamide gel electrophoresis. As described in the previous chapter, the protein − DNA complexes are seen as discrete bands migrating slower than the free DNA. The protein component may be either a purified protein or a complex mixture of proteins. The sequence within the fragment bound by a protein can be determined by a methylation interference assay, using a DNA fragment that is partially methylated at guanine (G) bases; if methylation of specific G residues within the binding site inhibits protein binding then fragments containing these methylated bases will be depleted in the retarded band and these can be detected on a sequencing gel following piperidine cleavage of the DNA extracted from the band. Alternatively, the less sensitive footprinting method (4) can be used in which the DNA fragment, labelled on one strand at one end, is mixed with excess protein, digested with DNase I or a chemical cleaving agent, and the resulting DNA fragments analysed on a sequencing gel (*Figure 1*). The presence of a sequence-specific DNA-binding protein in the protein extract is revealed by a protected region on the sequencing gel. Exonucleases can similarly be used to map the boundaries of protein binding sites (5).

This chapter describes these techniques, using as an example the detection of proteins binding to the enhancer of the avian sarcoma virus DNA. The long terminal repeat (LTR) of this virus contains typical eukaryotic enhancer and promoter elements

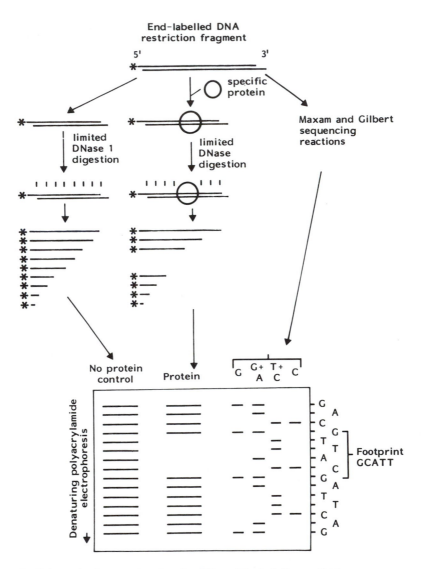

Figure 1. Schematic diagram showing the DNase I footprinting method.

within the U3 region (*Figure 2*). Upstream of the TATA-box promoter sequence there are two enhancer elements, termed B and C (6−8) which have been shown to bind at least three protein factors present in nuclear extracts (9,10). Two of the factors bind to CAAT-box motifs, the third binds to an inverted repeat sequence (10).

When searching for proteins that bind to a particular sequence, both the gel retardation and the footprinting techniques should be employed, since there have

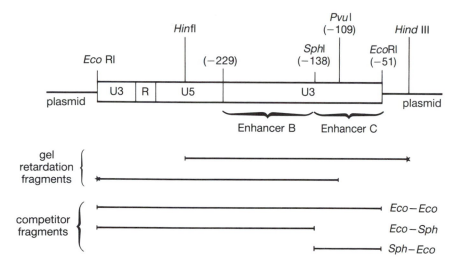

Figure 2. Structure of the subcloned Rous sarcoma viral LTR in the plasmid pAT153. The 0.32 kb *Eco*RI fragment of the duplicated LTR in the plasmid pSAR2 (which contains the complete Schmidt-Ruppin A2 RSV DNA [12]) subcloned into pAT153 is shown as a box containing the U3, R, and U5 sequences. The U3 region contains the two enhancer elements B and C which are upstream of a TATA box (at -30 bp) in the complete proviral DNA. The numbers in brackets refer to distances upstream of the start of transcription of the proviral DNA. Also shown below are the DNA fragments that were ^{32}P-labelled for gel retardation and footprinting analyses and DNA fragments that were used for competition experiments.

been instances when a factor has been detected by footprinting but not by gel retardation and *vice versa*. For example, the avian TGGCA-binding protein (or Nuclear Factor 1) in erythroid nuclear fractions gives a clear footprint on the inverted repeat that it binds to the β^H-globin promoter (11). However, due to heterogeneity or aggregation, it gives a rather ill-defined smear in gel retardation assays. Also, it is advisable to analyse cytoplasmic supernatants as well as nuclear extracts, since some factors may be washed out of the nucleus during the isolation of nuclei.

2. Preparation of protein extracts

Nuclei are prepared from tissues or tissue culture cells as described in *Protocol 1* by lysing the cells with a detergent, pelleting the nuclei, and washing the nuclei one or more times. The combined cytoplasmic supernatants are retained for analysis. The nuclei are extracted with 0.3−0.35 M NaCl since at this ionic strength the bulk of the non-histone nuclear proteins are extracted leaving the chromatin complex of DNA and histone as an insoluble residue. The proteins in the nuclear extract can then be precipitated with ammonium sulphate (to concentrate the protein and remove nucleic acid and high mobility group proteins (HMG proteins) which are soluble at high concentrations of ammonium sulphate), redissolved and dialysed. Alternatively, the extract can simply be dialysed versus buffer containing 100 mM NaCl. Detergents (e.g. Brij) are sometimes included in the final dialysis to minimize aggregation and precipitation of proteins. The protein extracts, stabilized with glycerol, are finally stored frozen at $-70°C$ in small aliquots. The protein concentration of the nuclear extracts should be about 10 mg/ml.

Many of the DNA-binding proteins are very susceptible to proteolytic enzymes and it is essential to include a mixture of inhibitors in all the buffers throughout (*Protocol 1*); phenylmethylsulphonyl fluoride (PMSF) alone is often insufficient.

Whilst $0.3-0.35$ M NaCl is sufficiently high an ionic strength to elute many sequence-specific factors from nuclei, higher ionic strengths have been used (e.g. ref. 9). However, with ionic strengths greater than 0.4 M, the histones are progressively solubilized and, unless they are removed, will introduce a large amount of strong but non-specific binding activities into the extracts which will interfere with the DNA-binding assays. Histones are best removed by precipitating the non-histone proteins in the extracts with ammonium sulphate.

Protocol 1. Preparation of nuclear and cytoplasmic protein extracts from tissue culture cells[a]

1. Collect cells grown in suspension by centrifugation and wash with phosphate-buffered saline (PBS) by gentle resuspension and recentrifugation. If cells are grown as monolayers attached to dishes or roller bottles, decant the medium and rinse the cells with PBS. Scrape the cells off into PBS and collect by centrifugation.

2. Wash the cells once with RSB buffer[b] (10 mM Tris-HCl, pH 7.6, 10 mM NaCl, 3 mM MgCl$_2$).

3. Resuspend the cells in $10-20$ volumes of RSB buffer containing $0.1-0.5\%$ Triton X-100 and homogenize in a Dounce homogenizer[a] until all the cells have lysed.

4. Pellet the nuclei by centrifugation at 2000 g for 10 min at 4°C. Collect the cytoplasmic supernatant. Wash the nuclei with RSB buffer by resuspending with a Dounce homogenizer and centrifuging at 2000 g for 10 min. Combine the cytoplasmic supernatants, dialyse against E buffer (20 mM Hepes−NaOH, pH 7.6, 5 mM MgCl$_2$, 0.1 mM EDTA, 1 mM DTT, 20% glycerol) containing 100 mM NaCl and freeze at $-70°C$. The pellet of crude nuclei contains approximately 5 mg of DNA from 10^9 cells.

5. Resuspend the nuclei in RSB buffer using a homogenizer (1 ml per nuclei from 2×10^9 cells). Add 2 M NaCl slowly with stirring to give a final concentration of $0.3-0.35$ M NaCl as required. Stir for 30 min at 4°C and then centrifuge at 100 000 g for 45 min at 4°C.

6. *Either*, dialyse the supernatant versus E buffer[c] containing 100 mM NaCl, centrifuge at 100 000 g for 30 min and freeze supernatant at $-70°C$ in aliquots (protein concentration $5-10$ mg/ml).

Protocol 1. *continued*

Or, precipitate the proteins from the 0.3 M NaCl extract supernatant by adding solid $(NH_4)_2SO_4$ (enzyme grade). For each 7.5 ml of supernatant add 3.5 g of $(NH_4)_2SO_4$. Stir at 4°C for 30 min and then centrifuge at 5000 *g* for 30 min. Redissolve the pellet in E buffer[c] containing 100 mM NaCl but containing no glycerol, and dialyse overnight against E buffer[c] containing 100 mM NaCl and 20% glycerol. Centrifuge at 100 000 *g* for 30 min and freeze the supernatant at −70°C in aliquots.

[a] This procedure can also be used for isolating nuclear proteins from animal tissues but in this case it is necessary to homogenize the tissue using either a Potter-Elvejhem homogenizer with a motor-driven pestle or, for tougher tissues, a blender with rotating blades.

[b] All buffers used throughout this procedure *must* contain the following proteolytic inhibitors: 0.25 mM phenylmethylsulphonyl fluoride (from a 50 mM stock solution in dried isopropanol), 0.5 mM benzamidine (from a 50 mM stock solution), 5 µg/ml leupeptin (from a 0.5 mg/ml stock solution), 5 µg/ml aprotinin (from a 0.5 mg/ml stock solution) and 2 µg/ml pepstatin (from a 0.2 mg/ml stock solution in ethanol, it is sonicated briefly to dissolve).

[c] The solubility of the non-histone protein in this buffer may be improved if a detergent is present (e.g. 0.1% w/v Brij or 0.025% w/v *n*-octyl D-glucopyranoside) but it is not essential.

---

3. End-labelling of DNA fragments

DNA fragments containing the sequences of interest are normally subcloned into a suitable plasmid vector (e.g. pUC). For gel retardation analysis, the DNA fragment should be less than 300 bp. Footprinting can be carried out on longer fragments, but it must be remembered that the region of interest should not be more than about 200 bp from the labelled end due to the limited resolution of DNA fragments larger than this size on linear sequencing gels. The plasmid subclone is cut at a suitable restriction site near one end of the insert (e.g. in the polylinker region), the 5′- or 3′-ends labelled with ^{32}P-phosphate and then cut at a site near the other end of the insert. The labelled fragment is then isolated by polyacrylamide gel electrophoresis (*Protocols 2* and *3*). Polynucleotide kinase is used to label the 5′-hydroxyl at restriction sites giving 5′-overhangs (*Protocol 2*). The other strand (i.e. the 3′-recessed end) can be labelled with Klenow polymerase or, more efficiently, with reverse transcriptase (*Protocol 4*). A 3′-overhang is labelled using the terminal transferase enzyme (*Protocol 5*). Blunt ends can be labelled with polynucleotide kinase, but the reaction is not very efficient and is best avoided.

Protocol 2. Labelling a 5′-overhang end

1. Digest 50 µg of the plasmid DNA with the first restriction enzyme in 100 µl of buffer.

2. Then add 20 units of calf intestinal alkaline phosphatase to the incubation and continue to incubate at 37°C for 30 min.

3. Add 10 µl of 100 mM EDTA (pH 8.0) and 10 µl of 10% SDS.

Protocol 2. *continued*

4. Extract twice with 100 μl of phenol, 0.1% hydroxyquinoline and then once with chloroform:iso-amyl alcohol (25:1 v/v).

5. Add two volumes of ethanol and precipitate the DNA at $-80°C$ for 60 min.

6. Centrifuge at 13 000 g for 10 min at 5°C and wash the pellet with 70% ethanol, then ethanol, and dry under vacuum.

7. Redissolve the DNA in 10 μl of 10 mM Tris-HCl (pH 7.6), 1 mM EDTA (TE buffer) and store frozen at $-20°C$.

8. To label the DNA with ^{32}P-phosphate mix the following:
   - 2 μl of the cut and phosphatased DNA
   - 5 μl of 10 × kinase buffer (500 mM Tris-HCl, pH 7.6, 100 mM MgCl$_2$, 50 mM DTT, 1 mM spermidine, 1 mM EDTA)
   - 10 μl [γ-^{32}P]-ATP (3.7 MBq, 180 TBq/mmol; 100 μCi, 5000 Ci/mmol)
   - 32 μl of water
   - 1 μl (10 units) polynucleotide kinase
   Incubate at 37°C for 45 min.

9. Add EDTA, SDS, and carry out phenol and chloroform:iso-amyl alcohol extraction as described in Steps 3 and 4.

10. Precipitate the DNA with two volumes of ethanol at $-80°C$ (see Step 5).

11. Centrifuge (see Step 6) and redissolve the pellet in 20 μl of TE buffer.

12. Add 2 μl of 3 M sodium acetate (pH 7.0) and reprecipitate the DNA with two volumes of ethanol.

13. Centrifuge as in Step 6, wash the pellet first with 70% ethanol then with absolute ethanol and dry.

14. Redissolve the DNA in 10 μl TE buffer and digest with the second restriction enzyme in 50 μl buffer.

15. Add 5 μl of 100 mM EDTA (pH 8.0), 5 μl of 10% SDS and 12 μl of 20% Ficoll, 0.1% bromophenol blue.

16. Load into two wells of a 15 × 15 × 0.3 cm 5% polyacrylamide gel containing 1 × TBE (see gel retardation electrophoresis, *Protocol 6*) and electrophorese at 150 V until the bromophenol blue reaches the bottom ($\sim$3 h).

17. Remove from the apparatus and take one glass plate off the gel.

18. Cover the gel and the remaining glass plate with Cling-film. Place in a light-proof box. Place a film (Kodak XAR-5) over the gel and expose the film for 3 min.

19. Develop the film, locate the radioactive band of interest and excise with a scalpel.

20. Extract the DNA from the polyacrylamide strip as described in *Protocol 3*.

Protocol 3. Isolation of DNA from polyacrylamide gels

1. On a clean glass plate, chop the acrylamide strip containing the DNA band into approximately 1 mm cubes and place in a 10 ml centrifuge tube with 2 ml of LS buffer (0.2 M NaCl, 20 mM Tris-HCl, pH 7.6, 1 mM EDTA).

2. Cap the tube and incubate with shaking at 37°C overnight.

3. Pipette off the buffer and add a further 0.5 ml of LS buffer. Vortex, remove the buffer and combine the two extracts.

4. Centrifuge the combined extracts at 2000 *g* for 10 min and collect the supernatant.

5. Equilibrate a Schleicher & Schuell Elutip D-column by passing 1 ml of HS buffer (1.0 M LiCl, 20 mM Tris-HCl, pH 7.6, 1 mM EDTA) through the column using a syringe, then pass through 5 ml of LS buffer.

6. With a 5 ml syringe attached to the Elutip, pipette the centrifuged extract from the acrylamide gel into the syringe, fit the plunger, and pass the extract through the column.

7. Wash the column with 5 ml of LS buffer.

8. Elute the radioactive DNA fragment from the Elutip with 0.4 ml of HS buffer using a 1 ml syringe.

9. Precipitate the DNA from the aqueous phase by adding 2.5 volumes of ethanol and leaving it at $-80°C$ for an hour. Centrifuge in a microcentrifuge at 13 000 *g* for 1 h.

10. Wash the pellet with 70% ethanol, then with absolute ethanol and redissolve in $100-200$ μl of TE buffer (see *Protocol 2*) containing 50 mM NaCl, giving a solution of approximately 20 000 cpm per microlitre.

Protocol 4. Labelling a 3′-recessed end

1. Cut 50 μg of plasmid DNA with the first restriction enzyme and extract the DNA with phenol and chloroform:iso-amyl alcohol and precipitate with ethanol as described in Steps $4-6$ of *Protocol 2*.

2. Redissolve the DNA in 10 μl TE buffer (see *Protocol 2*). Label 2 μl of the DNA solution as follows.
   Mix:
   - 2 μl DNA
   - 5 μl 10 × RT buffer (100 mM Tris-HCl, pH 8.3, 800 mM KCl, 100 mM MgCl$_2$)
   - 2 μl 300 mM 2-mercaptoethanol
   - 31 μl water

231

Protocol 4. *continued*

- 10 μl [α-^{32}P]-deoxynucleoside triphosphate^a
  (3.7 MBq, 200 TBq/mmol; 100 μCi, 6000 Ci/mmol)
- 2 μl reverse transcriptase (40 units)
  Incubate at 37°C for 1 h.

3. Add 5 μl of 100 mM EDTA (pH 8.0). Extract with phenol and chloroform: iso-amyl alcohol and ethanol precipitate and reprecipitate as described in Steps 10−13 of *Protocol 2*.

4. Cut the plasmid with a second restriction enzyme and isolate the fragment from polyacrylamide gel as for *Protocol 3*.

---

^{a} The radioactive deoxynucleotide used depends on the restriction enzyme used. Thus, use [α-^{32}P]-dATP for *Hind*III cut DNA.

Protocol 5. Labelling a 3′-overhang end

1. Digest 50 μg of plasmid DNA to give a 3′-overhang. Phenol extract, ethanol precipitate and redissolve the DNA in 50 μl of TE buffer as described in *Protocol 2*.

2. DNA containing 10 pmoles of 3′-OH groups is end-labelled in a single reaction. For example, a 3200 bp plasmid with a single restriction cut contains ∼2 pmoles of 3′-OH groups/μg of DNA. Thus mix:
   - 5 μl (5 μg) of DNA
   - 5 μl of 10 × TdT buffer (1 M potassium cacodylate, pH 7.2, 20 mM CaCl$_2$, 2 mM DTT)
   - 30 μl of water
   - 5 μl of aqueous [α-^{32}P]-dideoxyATP
     (180 TBq/mmol, 370 MBq/ml; 5000 Ci/mmol, 10 mCi/ml)
   - 10 units of terminal deoxynucleotidyl transferase.

   Incubate the solution at 37°C for 60 min. Phenol extract and ethanol precipitate and reprecipitate as described in Steps 10−13 of *Protocol 2*.

3. Digest the DNA with the second restriction enzyme and purify the DNA fragment by polyacrylamide gel electrophoresis as described above.

---

4. Gel retardation assay

In this assay, described in *Protocol 6*, the labelled DNA fragment is incubated with an excess of nuclear (or cytoplasmic) protein extract and the mixture analysed by polyacrylamide gel electrophoresis. Nuclear extracts contain a large number of DNA-binding proteins which will bind non-specifically to the labelled DNA fragment.

Therefore, in order to detect the specific binding of a minor component it is necessary to include an excess of a non-specific competitor DNA in the incubation to bind non-specific proteins. Poly(dI-dC)·poly(dI-dC) is most commonly used and is preferable to sheared *E. coli* DNA (the latter appears to bind tightly some sequence-specific proteins, possibly through single-stranded regions in the DNA), but with some unfractionated extracts, poly(dI-dC)·poly(dI-dC) may not bind all the non-specific DNA-binding proteins present. In such cases, the addition of the plasmid pBR322 as additional competitor DNA often helps. However, it is important to note that this plasmid has been found to contain specific sequences which bind some eukaryotic factors (e.g. the factor binding to the enhancer 'core' sequence) and therefore initially the assays should be carried out with and without plasmid DNA.

Protocol 6. Gel retardation assay

Preparation of polyacrylamide gel

1. Prepare a gel containing 0.2 × TBE buffer by mixing:
   - 12.5 ml stock acrylamide solution (29% acrylamide, 1% bisacrylamide, deionized with mixed bed ion-exchange resin, e.g. BioRad AG-501-X8, and filtered)
   - 3.0 ml 5 × TBE buffer (108 g Tris base, 55 g boric acid, 9.3 g disodium EDTA dihydrate made up to 2 litres, final pH 8.3)
   - 59.5 ml water

2. Degas the solution and add 400 μl of 10% ammonium persulphate and 30 μl of TEMED.

3. Pour the gel (15 × 15 × 0.3 cm) and allow to polymerize ($\sim$1 h).

4. Pre-electrophorese the gel at 4°C for 90 min at 150 V ($\sim$10 mA) circulating the buffer (0.2 × TBE buffer) between the reservoirs.

Protein−DNA incubation

1. Mix the following in a final volume of 20 μl, adding the nuclear protein last.
   - 9 μl 2 × retard buffer (40 mM Hepes−NaOH, pH 7.6, 8% Ficoll, 10 mM MgCl$_2$, 80 mM NaCl, 0.2 mM EDTA, 1 mM DTT)
   - Poly(dI-dC)·poly(dI-dC)[a,b]
   - 1−3 μl 1 mg/ml pBR322 plasmid DNA[a]
   - 1 μl ^{32}P DNA fragment ($\sim$5 fmol, $\sim$20 000 cpm)
   - 1−5 μl nuclear protein extract and water to give a final volume of 20 μl.

2. Leave on ice for 1 h.

3. Load into one well of the polyacrylamide gel and electrophorese at 4°C at 150 V with circulation of the buffer (0.2 × TBE). For a 250 bp labelled DNA fragment electrophoresis takes about 3 h.

4. Place the gel on a sheet of Whatman 3MM chromatography paper and then cover with Saran Wrap or Cling-film. Place this on to a second sheet of Whatman 3MM paper and dry under vacuum using a gel drier heated at 80°C.

Protocol 6. *continued*

5. When dry, place the gel attached to the first sheet of paper in a cassette and detect the bands of radioactive DNA by autoradiography at $-70°C$ (2−15 h).

---

[a] The amount of poly(dI-dC)·poly(dI-dC) used as a competitor to abolish non-specific binding of protein to labelled probe must be determined by titration but is usually in the range of 0.5−10 μg depending on the nuclear protein preparation. Addition of pBR322 plasmid DNA assists in abolishing non-specific binding when using crude nuclear extracts and again the amount added has to be determined empirically. With proteins that have been partially purified (e.g. by HPLC chromatography) pBR322 DNA is no longer required.

[b] Poly(dI-dC)·poly(dI-dC) should be dissolved in 50 mM NaCl, 10 mM Tris−HCl (pH 7.6), 1 mM EDTA.

In order to determine the optimum concentration of poly(dI-dC)·poly(dI-dC) a series of incubations is set up with a constant amount of labelled fragment and protein, and increasing amounts of poly(dI-dC)·poly(dI-dC). After electrophoresis and autoradiography, most of the labelled fragment will be seen migrating as free DNA (the fastest band in *Figure 3*) and minor DNA−protein complexes migrating more slowly. If most of the DNA migrates as a smear near the top of the gel, this indicates insufficient poly(dI-dC)·poly(dI-dC) has been used. If, even with increased amounts of poly(dI-dC)·poly(dI-dC) (<20 μg), substantial amounts of radioactivity remain at the very top of the gel plasmid pBR322 should be tried in the incubation. Polyacrylamide gels 3 mm thick are easier to load and give sharper bands than 1-mm-thick gels. *Figure 3* shows an example of an analysis of a total nuclear extract from the avian erythroid cell line HD3 binding to the avian sarcoma viral enhancer. In order to determine which of the retarded bands are due to proteins binding to specific sequences, it is necessary to carry out competition experiments with varying excess amounts of unlabelled DNA fragments (10- to 100-fold molar excess). In the example shown in *Figure 3*, the labelled *Hin*fI − *Hin*dIII fragment containing the LTR enhancer was mixed with poly(dI-dC)·poly(dI-dC) and a 100-fold molar excess of the unlabelled fragments shown in *Figure 2* prior to the addition of protein. It can be seen, for example, that the *Eco*RI − *Eco*RI fragment containing the whole enhancer competes for the proteins forming retarded bands BI, BII, and BIII, whilst an equivalent amount of pAT153 plasmid DNA does not compete. The *Eco*RI − *Sph*I fragment competes for proteins producing gel retardation bands BI and BIII, whilst the *Sph*I − *Eco*RI fragment competes for the band BII protein. These three bands are therefore due to the specific binding of three proteins. The other bands are due to non-specific binding of proteins; the band migrating just behind the free DNA is probably a non-specific HMG protein−DNA complex.

If no specific retarded bands are detected, this could be due to protein degradation, denaturation, loss of the protein during nuclear isolation, too dilute a protein extract, incorrect structure of the labelled DNA fragment, or incorrect amounts or type of non-specific competitor DNA. Protein degradation is often manifested by the appearance of smeared gel retarded bands, in which case it is essential to ensure

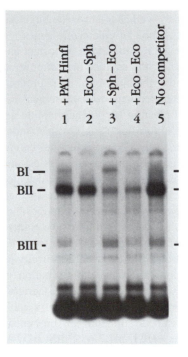

BI —
BII —

BIII -

Figure 3. Gel retardation analysis of binding to the LTR of HD3 proteins extracted from nuclei with 0.3 M NaCl.

The *Hin*fI – *Hin*dIII labelled fragment (2 ng) was incubated with 0.3 M NaCl extracted HD3 nuclear protein (15 μg), poly(dI-dC) · poly(dI-dC) (9 μg) and pBR322 (3 μg) together with various unlabelled competitor DNA fragments (see *Figure 2*) prior to electrophoresis: lane 1: 250 ng plasmid pAT153 digested with *Hin*fI. lane 2: 150 ng 0.23 kb *Eco – Sph* LTR fragment. lane 3: 150 ng 0.09 kb *Sph – Eco* LTR fragment. lane 4: 250 ng 0.032 kb *Eco – Eco* LTR fragment. lane 5: no additional competitor DNA.

that all the proteolytic inhibitors are freshly prepared and added to all the cold buffers just prior to the start of the preparation of the protein extracts and that the preparation time is kept as short as possible. Changing the cell type or tissue can also help; tissues are generally worse than tissue culture cells in terms of the amounts of proteolytic enzymes present. It is likely that many of the DNA-binding factors have zinc-fingers or other metal-binding domains and it is therefore possible that the use of chelators (e.g. EDTA), high salt and/or reducing agents could remove essential metal ions resulting in the loss of DNA-binding activity, as has been reported to occur with the RNA polymerase III factor TFIIIA (13). Therefore, it is worth trying the preparation and gel retardation analyses in the absence of EDTA and dithiothreitol. Factors can be eluted from the nuclei during nuclear preparations; for example, large amounts of the CCAAT-box factor are found in the cytoplasmic supernatants, presumably due to leaching out of the nucleus. It is therefore very worthwhile retaining and analysing cytoplasmic supernatants.

The protein concentrations of the nuclear extract, and from how much nuclear DNA it was extracted, must be ascertained to ensure that sufficient protein is being added to the assay. It must be remembered that most of the transcription factors are not very abundant (approximately 10^4 molecules per cell). Therefore it is important to calculate that one is putting into the DNA-binding reaction sufficient molecules of a putative factor to give a detectable gel retarded band. One way to check that the nuclear extract has been correctly prepared is to test the extract on a labelled DNA fragment that is known to bind a fairly well characterized ubiquitous factor such as the protein that binds to the CCAAT-box of eukaryotic genes.

The size of the labelled DNA fragment used for gel retardation is important and should be in the $30-300$ bp range; as a general rule, the smaller the better so as to minimize non-specific protein binding and to detect the binding of large proteins more readily. The labelled and unlabelled DNA fragments used for these analyses are prepared from the restriction enzyme cleaved plasmid by polyacrylamide gel electrophoresis followed by removal of acrylamide impurities using Elutip columns (*Protocols 2* and *3*). Alternatively, defined synthetic oligodeoxynucleotides can be used. The DNA fragments are usually finally ethanol precipitated, dried and redissolved. The presence of denatured or otherwise altered DNA can be detected by the uncomplexed DNA on the gel retardation electrophoresis running as a doublet. However, a cryptically altered DNA fragment may be revealed only when it binds protein artefactually as described by Svaren *et al.* (14). These authors have pointed out the possible artefacts that can be generated by using organic solvents to purify and precipitate small DNA fragments for gel retardation analyses. In particular, drying ethanol precipitates of DNA at room temperature can cause the formation of hairpin-like structures which may bind protein in a non-specific fashion. It is therefore recommended that after ethanol precipitation and centrifugation, as much as possible of the ethanol supernatant be removed with a pipette tip and the precipitate immediately redissolved in TE buffer (10 mM Tris−HCl, pH 7.6, 1 mM EDTA) containing 50 mM NaCl without prior drying down. Alternatively, one can avoid ethanol precipitation entirely by dialysis or desalting on a column of Trisacryl GF05 or by gel filtration using the spun column technique described by Maniatis *et al.* (15).

5. Methylation interference assay

Having detected the presence of one or more putative sequence-specific DNA binding proteins by the gel retardation method, further proof for sequence-specific binding and the location of the binding site is carried out by a methylation interference assay (16). In this assay the end-labelled DNA fragment is partially methylated at the N7 of guanine, using dimethyl sulphate as described in *Protocol 7*. The length of reaction time with dimethyl sulphate to obtain optimum methylation ($\sim 50\%$ of the DNA molecules should have a methylated G) is determined in a preliminary experiment in which the labelled DNA is methylated for varying lengths of time followed by cleavage with piperidine and analysis on a sequencing gel (*Protocols 8* and *9*). In the methylation interference assay (*Protocol 8*), the methylated DNA is incubated with poly(dI-dC)·poly(dI-dC) and protein and separated by electrophoresis as for a gel retardation analysis. In order to obtain sufficient radioactive counts in the gel retarded band for subsequent analysis on a sequencing gel, the analytical gel retardation analysis must be scaled up to such a level that the gel retarded band can be detected in a wet gel after autoradiographic exposure for approximately 4 h. Thus for the methylation interference experiment shown in *Figure 4*, 30 μl of HD3 nuclear

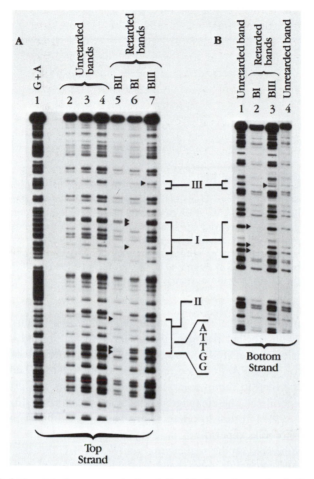

Figure 4. Methylation interference analysis of the binding of proteins I, II, and III. The *Hin*fI − *Hin*dIII LTR fragment labelled at the *Hin*dIII site on either strand was methylated with dimethyl sulphate and used in scale-up gel retardation analyses with MonoQ FPLC partially purified proteins I and III and the total HD3 0.3 M NaCl nuclear extract (30 μl). 90 μg poly(dI-dC)·poly(dI-dC) and 30 μg pBR322 were included in the incubation. After autoradiographic exposure of the electrophoretic gel, the retarded bands BI, BII, and BIII together with the unretarded counterparts were excised from the gel. The DNA was purified, cleaved with piperidine, and analysed on sequencing gels.

(A) Top strand. Lane 1 − G plus A cleavage reaction; lanes 2, 5 − DNA from unretarded band and retarded band BII respectively, produced by incubation with HD3 nuclear extract; lanes 3, 6 − DNA from unretarded band and retarded band BI respectively, produced by incubation with protein I; lanes 4, 7 − DNA from unretarded band and retarded band BIII respectively, produced by incubation with protein III.

(B) Bottom strand. Lanes 1, 3 − DNA from unretarded band, retarded band BIII respectively, produced by incubation with protein III; lanes 4, 2 − DNA from unretarded band and retarded band BI respectively, produced by incubation with protein I.

Arrowheads show G residues exhibiting reduced cleavage.

237

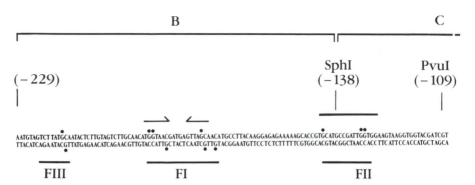

Figure 5. Sequence of the enhancer elements B and part of C showing the binding sites for the three proteins I, II, and III. The thick horizontal lines show the footprints of FI, FII, and FIII obtained in **Figure 6**. (Note that the exact boundaries of the footprints are not definite due to the presence of sequences which are poorly cut by DNase I in the no protein controls.) The filled circles mark the G bases where methylation inhibits protein binding as detected by the gel retardation analysis of **Figure 4**. Note that FIII has the GCAAT motif on the top strand, FII has the CCAAT on the bottom strand, and FI has the inverted repeat shown by the arrows.

extract (compare this with the $2-3$ μl used in the analytical gel retardation of *Figure 3*) was mixed with labelled fragment (4×10^5 cpm) and loaded into three wells of a 3-mm-thick gel prepared in $0.2 \times$ TBE buffer. After electrophoresis and autoradiography, the DNA from the retarded and non-retarded bands is extracted from the gel, purified to remove acrylamide contaminants and protein, and then cleaved with piperidine. The cleavage products are then analysed on a sequencing gel (see *Protocol 9*). In order to obtain sharp bands, it is important that the DNA samples are free of salts, piperidine, and acrylamide contaminants. For gel retarded band BII, it can be seen on the sequencing gel (*Figure 4A*) that the DNA has three G cleavage bands (shown by arrowheads) which are reduced in intensity relative to the uncomplexed DNA (compare lanes 2 and 5). This shows that DNA fragments containing methylated G residues at these positions will not bind the proteins strongly and hence these methylated bases are under-represented in the gel retarded band. In order to locate these G residues within the DNA sequence, it is necessary to run a G plus A sequencing reaction in parallel if the sequence is known (if not, the other sequencing reactions must be carried out also). It can be seen that in the example given in *Figure 4A*, the G residues affected lie within and flanking the CCAAT-box element (*Figure 5*). The positions of the G-methylation that inhibit the binding of proteins forming retardation bands BI and BIII are also shown in *Figure 5*.

Protocol 7. Limited methylation of guanine with dimethyl sulphate for methylation interference assays[a]

1. Mix 200 μl 50 mM sodium cacodylate (pH 8.0), 1 mM EDTA with end-labelled DNA in a 1.6 ml microcentrifuge screw-cap tube.

Protocol 7. *continued*

2. Chill to 0°C on ice, and add 1 μl of dimethyl sulphate (reagent grade). Mix and incubate at 20°C for 3 min[b].

3. Add 25 μl of 3 M sodium acetate (pH 7.0) and 600 μl of ethanol (0°C). Chill at −70°C for 4−5 min. Centrifuge at 12 000 *g* for 45 min.

4. Remove the supernatant and transfer to a waste bottle containing 5 M NaOH.

5. Wash the pellet with 70% ethanol, then with absolute ethanol and dry. Redissolve in 10 μl of TE buffer or water.

[a] **Caution:** all operations must be carried out in a fume-hood since dimethyl sulphate is a carcinogen.
[b] The optimum time of reaction must be determined empirically.

Protocol 8. Methylation interference method

1. Mix the following in 120−200 μl of gel retardation buffer (see *Protocol 6*):
   - Labelled and methylated DNA fragment (50−100 fmol; 4×10^5 cpm)
   - Poly(dI-dC)·poly(dI-dC)
   - pBR322 (if necessary)
   - Nuclear extract

   The quantities used are about 10× that used for gel retardation analysis (*Protocol 6*).

2. Incubate on ice for an hour and load into 3−5 wells of a 3-mm-thick polyacrylamide gel in 0.2 × TBE buffer.

3. After electrophoresis remove one glass plate from gel. Cover the gel on the other plate with Saran Wrap or Cling-film and place in a lightproof box with an X-ray film on top of the gel. Expose for approximately 4 h.

4. Excise the retarded and unretarded bands and extract the DNA as described in *Protocol 3*. Purify the DNA by passing the acrylamide gel extract through an Elutip column (see *Protocol 3*). Add tRNA (10 μg) to the 1 M LiCl eluate, phenol extract, ethanol precipitate and dry the DNA.

5. Redissolve the DNA in 100 μl of 1.0 M piperidine (freshly diluted) in a screw-cap microcentrifuge tube. Heat at 90°C for 30 min.

6. Cool on ice and then centrifuge for a few seconds in a microcentrifuge. Punch holes in the tube cap with a needle. Evaporate to dryness for 2 h using a SpeediVac centrifugal desiccator.

7. Redissolve the DNA in 50 μl of water and dry down again for 1 h. Redissolve the DNA in 40 μl of water and dry again for 1 h.

8. Redissolve in 5 μl of sequencing sample-solvent (95% v/v deionized formamide, 0.02% w/v xylene cyanol FF, 0.025% w/v bromophenol blue). Heat at 90°C for 3 min and quick freeze until ready to load on to the sequencing gel (see *Protocol 9*). Note that to obtain sharp bands on the sequencing gel the DNA samples must be free of salts, residual piperidine, and acrylamide contaminants.

Protocol 9. Polyacrylamide DNA sequencing gel electrophoresis

The DNase I-digested or piperidine-cleaved DNA is resolved on a sequencing gel (34 × 40 × 0.4 mm). One glass plate should be siliconized to aid pouring and subsequent removal at the end of the electrophoresis.

1. Dissolve 50 g urea, 5.0 g acrylamide, 0.2 g bisacrylamide and 20 ml of 5 × TBE buffer in a final volume of 100 ml (with water). Warm to about 20°C to help the urea dissolve but do not overheat. Filter the solution under vacuum through a Nalgene sterilization filter, and leave under vacuum for 1−2 min to degas.

2. Add 200 μl of 10% ammonium persulphate and 70 μl of TEMED. Pour the gel using a 50 ml syringe. Polymerization should occur in 5−10 min at room temperature. Allow the gel to polymerize for at least an hour (gels can be stored for up to 36 h at room temperature).

3. Pre-electrophorese for 40−60 min at 40 mA using 1 × TBE buffer in the electrode chambers. This removes persulphate ions and heats the gel. The surface temperature should reach 50−60°C.

4. Load the samples (2−5 μl) and electrophorese for 90−180 min at 40 mA (∼1500 V). Bromophenol blue migrates at the position of a DNA fragment of about 30 nucleotides, and xylene cyanol at about 130 nucleotide fragments. To resolve fragments 20−250 bases long on one gel, it is possible to do two loadings; that is, to dissolve the samples in 8 μl of sample buffer, load 4 μl and after 90 min of electrophoresis, load the remainder in separate wells and electrophorese for a further 90 min. The first and second loadings will resolve fragments ∼100−250 and ∼20−100 bases from the end-label, respectively.

5. After electrophoresis, remove the siliconized plate, leaving the gel attached to the other plate. Place a sheet of Whatman 3MM paper over the gel, press it gently down on the gel and peel it off together with the gel attached. Cover the gel surface with Saran Wrap or Cling-film and dry the gel under vacuum at 80°C for 60 min. Expose against pre-flashed Kodak XAR5 film at −70°C in a cassette containing Lightning-Plus (Dupont) screens.

6. Footprinting techniques

A number of techniques have been described for identifying a DNA binding site based on the premise that a protein bound tightly to a specific sequence will inhibit or alter access of reagents to the bases or the sugar phosphate backbone of that sequence. *Table 1* lists some of the more commonly used reagents, and *Protocols 10* and *11* describe footprinting using DNase I and methidium propyl-EDTA (MPE). One of the disadvantages of DNase I as a cleaving agent is that some sequences, especially tracts of A and T residues, are not readily digested by DNase I due to

Table 1. Footprinting and related techniques.

DNA cleaving or modifying agent	Comments
DNase I (4)	Bound protein inhibits DNase I binding in the minor groove. Region protected from cutting usually larger than the actual protein binding site. Some sequences are not cut by DNase I and therefore binding sites or part of binding site may not be detected. However, DNase I can give information on changes in DNA conformation induced by protein binding. Method can be made more sensitive by combining with gel retardation technique.
Dimethyl sulphate (DMS) (19)	Bound protein inhibits N-7 methylation of G bases. Gives information on which G residues are in intimate contact with the protein. Useful for comparing *in vitro* protection patterns with those obtained with nuclei or whole cells.
Methidium propyl-EDTA (MPE). hydroxyl radical (17, 18)	Bound protein inhibits scission of DNA by hydroxyl radicals. It is a better agent than DNase I since cutting is not sequence-dependent and the protein protected region is smaller, thereby giving better definition of the binding site. Sensitivity can be increased by combining with gel retardation technique.
UV light (20, 21)	Bound protein alters DNA double-helix geometry resulting in altered reaction of photons with bases. Can be used to detect proteins bound to specific sequences in whole cells. Non-specifically bound proteins such as histones do not interfere with reaction.

the conformation of the minor groove. Chemical cleavage agents such as MPE and hydroxyl radicals do not suffer from this disadvantage (17,18). Also, since they are small molecules, the footprints obtained are smaller than DNase I footprints and define more precisely the binding site of the protein.

In order to detect a protein binding to a specific sequence using these methods (see *Figure 1*), it is necessary that all the DNA fragment molecules are covered with the protein. Thus much more protein (approximately 10- to 100-fold more) is required in the footprinting assay compared with that required to detect a band gel retardation. Again poly(dI-dC)·poly(dI-dC) is required to compete out non-specific DNA binding proteins when using crude extracts. The amount of poly(dI-dC)·poly(dI-dC) used must be determined empirically.

6.1 DNase footprinting

After preincubation of the labelled DNA fragment with varying amounts of poly(dI-dC)·poly(dI-dC) and nuclear extracts (0−250 µg protein) (*Protocol 10*) the samples are digested briefly with DNase I; again the amounts of DNase I required will depend on the concentrations of components in the mixture and must therefore be determined by trial and error. It is advisable to include $1-2$ mM $CaCl_2$ in the incubation to optimize DNase I enzyme activity. The optimum ionic strength for protein binding to the DNA is usually in the range $40-100$ mM monovalent cations, usually NaCl. Normally, $1-5$ mM $MgCl_2$ is also required. After purification of the DNA by phenol extraction, the cleaved DNA fragments can be analysed on a sequencing gel. Comparison of the zero protein control digestions with those containing nuclear proteins reveals the presence of protected region(s).

Protocol 10. DNase I footprinting

1. Mix in a final volume of 100 μl^a of 20 mM Hepes-NaOH (pH 7.6) $40-50$ mM NaCl, 2mM $CaCl_2$ 5 mM $MgCl_2$, 1 mM DTT, 10% glycerol, the following:
   - ~ 5 fmol end-labelled DNA fragment ($\sim 20\ 000$ cpm)
   - Poly(dI-dC)·poly(dI-dC)
   - $0-50$ μl nuclear protein ($0.5-10$ mg/ml protein)

   Make several such mixtures with varying amounts of protein and poly(dI-dC)· poly(dI-dC) in duplicate (e.g. 8 samples total). Incubate on ice for 1 h.

2. Digest samples with 1 μl of DNase I (usually about 20 ng/μl) in above buffer for 15 sec and 30 sec at room temperature. Stop the reaction by rapid addition of 100 μl of stop buffer (2% SDS, 10 mM EDTA, pH 8.0) containing 0.1 mg/ml tRNA).

3. Extract twice with equal volume of phenol 0.1% hydroxyquinoline and once with chloroform:iso-amyl alcohol.

4. Precipitate with 2.5 volumes of ethanol, wash and dry.

5. Redissolve in $5-10$ μl of sequencing sample solvent (step 8 *Protocol 8*) and heat at 90°C for 3 min. Freeze rapidly at -80°C. Thaw just prior to loading on to a sequencing gel (see *Protocol 9*).

[a] The volume can be reduced to 20 μl when using partially purified protein fractions in which the DNA-binding activity is a major component; for example, using DNA-cellulose purified material from chicken erythrocytes in which the CCAAT-box binding protein is very abundant (see *Figure 6B*).

Examples of this technique are shown in *Figure 6*, in which a total 0.3 M NaCl nuclear extract from HD3 cells gives two footprints FI and FIII (*Figure 6A*) over the regions detected by the gel retardation−methylation interference technique (*Figure 5*). Note that a footprint over the CCAAT-box was not detected under these conditions, presumably due to inhibitory proteins or to incorrect amounts of poly(dI-dC)·poly(dI-dC), despite the fact that the CCAAT-box binding factor gives the most prominent gel retard band (BII) see *Figure 3*). However, once the protein from

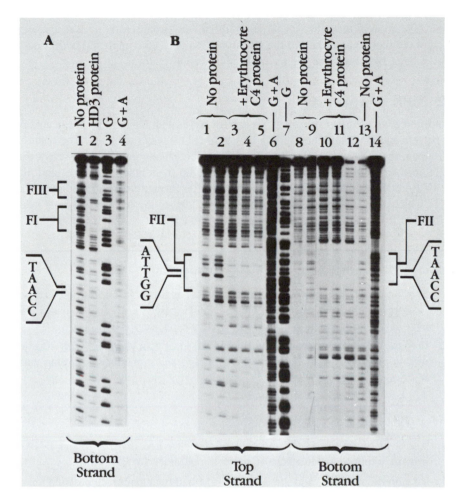

Figure 6. DNase I footprint analyses of erythroid nuclear proteins.

(A) The 0.25 kb HinfI – HindIII fragment 5'-end-labelled at the HindIII site was incubated with no protein (lane 1) or 250 μg of 0.3 M NaCl extracted protein from HD3 nuclei (lane 2) together with 3 μg poly(dI-dC)·poly(dI-dC), digested with DNase I (20 ng), and analysed on a sequencing gel. Lanes 3 and 4 show G and G plus A cleavage reactions. The position of the CCAAT-box next to the SphI site is shown.

(B) The HinfI – HindIII fragment 5'- and 3'-labelled at the HindIII site was incubated with 5 μg Poly(dI-dC) and no protein (lanes 1, 2, 8, 9 and 13) or with DNA-cellulose partially purified chicken erythrocyte protein II (lanes 3 and 12: 0.5 μg protein; lanes 4 and 10: 1 μg protein; lanes 5 and 11: 2 μg protein). Lanes 6, 7 and 13 show G and G plus A cleavage reactions. The position of the CCAAT-box is shown.

chicken erythrocytes had been partially purified by DNA-cellulose chromatography, a footprint could be obtained (*Figure 6B* and *Figure 5*). This illustrates the importance of using both techniques for detecting sequence-specific proteins. It is also important to note that the footprint results of *Figure 6B* were obtained using sufficient protein in the incubation so that when an aliquot of the lowest protein mixture was analysed by a gel retardation assay, all of the DNA was found to be complexed with protein to give band BII and some higher molecular weight material.

6.2 MPE footprinting

Footprinting with MPE (*Protocol 11*) is carried out as with DNase I by preincubating the protein and DNA and then adding the chemical cleavage reagents. The reaction is stopped, the DNA purified and analysed on a sequencing gel. It is important that the ferrous ammonium sulphate solution is made up fresh just before use, and that the MPE-Fe(II) complex is diluted immediately after formation. The reducing agent, DTT, which activates the cleavage reagent, is added to the MPE-Fe(II) just prior to addition to the DNA−protein complex.

Protocol 11. Methidium-propyl-EDTA footprinting

1. Mix labelled DNA, protein, competitor, etc., as for DNase I footprinting (see *Protocol 10*) in 20 μl of 20 mM Hepes-NaOH (pH 7.9), 10% glycerol, 5 mM MgCl$_2$, 50 mM NaCl, 1 mM DTT. Incubate on ice for 1 h.

2. Prepare the cleavage reagent by mixing 20 μl of 1 mM MPE[a] with 20 μl of freshly prepared 1 mM Fe(NH$_4$)$_2$(SO4)$_2$.6H$_2$O and immediately diluting with:
   - 52 μl of water
   - 1 μl 1.0 M Tris-HCl (pH 7.6)
   - 2.5 μl 2.0 M NaCl
   - Add 5 μl of 100 mM DTT just prior to use.

3. Add 1 μl of 0.06% H$_2$O$_2$ and 5 μl of the cleavage reagent to the DNA−protein solution and incubate at room temperature for 1−5 min.

4. To stop the reaction, add 2.5 μl of 50 mM bathophenanthroline disulphonate followed by 100 μl of 1% SDS, 4 mM EDTA (pH 7.5), 0.2 M LiCl containing 0.2 mg/ml tRNA.

5. Phenol extract and ethanol precipitate the DNA as described in *Protocol 2*.

6. Analyse on a sequencing gel as described in *Protocol 9*.

[a] A stock solution of MPE in water can be stored at −20°C for several months.

6.3 Combined footprint – gel retardation method

The footprinting technique can be combined with the gel retardation method in a number of ways to make it more sensitive. Provided that the sequence-specific protein, once bound to its binding site, does not exchange rapidly between DNA molecules, one can digest the gel retardation incubation mixture with DNase I, then immediately load it on to the gel retardation electrophoresis gel. Then the DNA can be eluted from the gel retarded band and analysed on a sequencing gel to reveal the protected domain (see e.g. ref. 9). Alternatively, the gel retardation method can be carried out and after detecting the gel retarded band, the band can be excised from the gel and reacted with a chemical DNA cleaving reagent (22). The DNA is then eluted and analysed on a sequencing gel.

7. The exonuclease protection method

In this method, end-labelled DNA is incubated with protein and digested with exonuclease III which progressively digests each strand from the 3′ end until it encounters a protein bound to a specific sequence. Analysis of the denatured DNA on a sequencing gel then reveals this 'stop' as a sub-band (5). This gives one boundary of the binding site. The other boundary is determined by using a DNA fragment labelled on the other strand. The exonuclease protection method is similar to the DNase I footprinting technique but is more sensitive and has been used by a number of laboratories to detect DNA-binding proteins in unfractionated nuclear extracts. However, exonuclease III is not a very processive nuclease and spurious 'stops' occur during the digestion sequences which are not due to the binding of proteins to specific sequences but simply due to the general inhibition of the enzyme resulting from the non-specific binding of proteins. Thus it can be difficult to distinguish stops due to the binding of proteins to specific sequences. This problem can be minimized by using bacteriophage T4 DNA polymerase together with exonuclease III as described in *Protocol 12*, or by using bacteriophage lambda exonuclease which has been reported to be a more processive enzyme (23). As with other techniques described in this chapter, non-specific competitor DNA, poly(dI-dC)·poly(dI-dC) or pBR322, must be incorporated in the incubation mixtures, and the optimum concentration has to be determined empirically. To check that a sub-band is due to the binding of a protein to a specific sequence the unlabelled DNA fragment should be added to the incubation mixture to see if it competes for the binding of the putative sequence-specific protein (which should result in the loss of the sub-band). Another disadvantage of this method is that if there are multiple protein binding sites on a DNA fragment they may not all be detected. Also a protein bound to the middle of the DNA fragment may go undetected due to the instability of the double-helix when nearly half of each strand has been digested away.

Protocol 12. Exonuclease protection assay

1. Mix the following in 120 μl of 66 mM potassium acetate, 33 mM Tris-acetate (pH 7.9), 10 mM MgCl$_2$, 0.5 mM DTT, 0.1 mg/ml bovine serum albumin:
   - 5 fmols ($\sim$20 000 cpm) ^{32}P-5′-end-labelled DNA fragment
   - Poly(dI-dC)·poly(dI-dC)
   - 10−50 μl protein extract (5−10 mg/ml).

Protocol 12. *continued*

2. Incubate on ice for 1 h.

3. Digest at 37°C for 15 min with 65 units of exonuclease III and 10 units of T4 DNA polymerase.

4. Stop the reaction, extract the DNA and analyse it on a sequencing gel as described in *Protocol 9*.

8. Choice of method

A number of assays in addition to those detailed here have been described in the literature for detecting sequence-specific DNA-binding proteins. For example, the nitrocellulose filter-binding technique is similar to the gel retardation assay in that one is separating nucleoprotein complexes from uncomplexed DNA, but it is not as simple and reliable as the gel retardation assay and is not now the method of choice for attempting to detect minor components in crude extracts.

Faced with the task of identifying factors binding to a functional DNA sequence, it is recommended that an initial gel retardation analysis be carried out since this is simpler and quicker than the footprinting and exonuclease III protection techniques and there are fewer parameters to vary. Gel retardation with competitor DNA fragments will quickly ascertain whether any of the retarded bands are the result of the specific binding of proteins.

Having identified specific gel retarded bands, the binding sites are most easily determined by using the methylation interference technique provided there are G residues within the binding site that are important for binding. Note that for CCAAT-box binding factor, only G residues in the upper strand were found to be important for binding thus demonstrating the importance of analysing both strands. However, the methylation interference method is limited in the information it gives on the binding site since it analyses the role of only a few G residues. An alternative approach is the use of partially depurinated or depyrimidated DNA in interference assays to obtain more detailed information on the binding site of proteins (24).

Once identified, the protein(s) can be further purified by conventional techniques (25) such as HPLC ion-exchange chromatography, or DNA-cellulose chromatography (11) or, DNA affinity chromatography (26), assaying the fractions simply by the gel retardation method. The partially purified proteins can then be more readily used for footprinting than the original crude extracts.

Acknowledgements

The research described in this chapter was supported by grants from the Cancer Research Campaign and the Medical Research Council. My thanks to colleagues, in particular Dr M. Plumb and R. H. Nicolas for their contribution to this work.

References

1. Serfling, E., Jasin, M., and Schaffner, W. (1985). *Trends in Genetics*, **1**, 224.
2. Fried, M. and Crothers, D. M. (1981). *Nucleic Acids Res.*, **9**, 6505.
3. Garner, M. M. and Revzin, A. (1981). *Nucleic Acids Res.*, **9**, 3047.
4. Galas, D. J. and Schmitz, A. (1978). *Nucleic Acids Res.*, **5**, 3157.
5. Shalloway, D., Kleinberger, T., and Livingston, D. M. (1980). *Cell*, **20**, 411.
6. Luciw, P. A., Bishop, J. M., Varmus, H. E., and Capecchi, M. R. (1983). *Cell*, **33**, 705.
7. Laimins, L. A., Tsichlis, P., and Khoury, G. (1984). *Nucleic Acids Res.*, **12**, 6427.
8. Cullen, B. R., Raymond, K., and Ju, G. (1985). *Mol. Cell. Biol.*, **5**, 438.
9. Sealey, L. and Chalkley, R. (1987). *Mol. Cell. Biol.*, **7**, 787.
10. Goodwin, G. (1988). *J. Virol.*, **62**, 2186.
11. Plumb, M. A., Lobanenkov, V. V., Nicholas, R. H., Wright, C. A., Zavou, S., and Goodwin, G. (1986). *Nucleic Acids Res.*, **14**, 7675.
12. Swanstrom, R., DeLorbe, W. J., Bishop, J. M., and Varmus, H. E. (1981). *Proc. Nat. Acad. Sci. (USA)*, **78**, 1248.
13. Miller, J., McLachlan, A. D., and Klug, A. (1985). *EMBO J.*, **4**, 1609.
14. Svaren, J., Inagami, S., Levergren, E., and Chalkley, R. (1987). *Nucleic Acids Res.*, **15**, 8739.
15. Maniatis, T., Fritsch, E. F., and Sambrook, J. (1982). In *Molecular Cloning: A Laboratory Manual*, p. 466. Cold Spring Harbor Laboratory Press, NY.
16. Sienbenlist, U. and Gilbert, W. (1987). *Proc. Nat. Acad. Sci. (USA)*, **77**, 122.
17. Van Dyke, M. W. and Dervan, P. B. (1983). *Nucleic Acids Res.*, **11**, 5555.
18. Tullins, T. D. and Dombroski, B. A. (1986). *Proc. Nat. Acad. Sci. (USA)*, **83**, 5469.
19. Siebenlist, U., Simpson, R. B. and Gilbert, W. (1980). *Cell*, **20**, 269.
20. Becker, M. M. and Wang, J. C. (1984). *Nature*, **309**, 682.
21. Wang, Z. and Becker, M. M. (1988). *Proc. Nat. Acad. Sci. (USA)*, **85**, 654.
22. Kakkis, E. and Calame, K. (1987). *Proc. Nat. Acad. Sci. (USA)*, **84**, 7031.
23. Elbrecht, A., Tsai, S., Tsai, M.-J. and O'Malley, B. W. (1985). *DNA*, **4**, 233.
24. Brunelle, A. and Schleif, R. F. (1987). *Proc. Nat. Acad. Sci. (USA)*, **84**, 6673.
25. Harris, E. L. V. and Angal, S. (1989). *Protein Purification Methods: A Practical Approach*. IRL Press, Oxford.
26. Kadonaga, J. T. and Tjian, R. (1986). *Proc. Nat. Acad. Sci. (USA)*, **83**, 5889.

8

Electrophoresis of nucleosomes

ROBERT H. NICOLAS

1. Introduction

The basic structural unit of chromatin in the eukaryotic cell is the nucleosome. This is made up of about two turns of DNA (approx. 140 bp) wrapped around an octamer of two of each of the four core histones: H2A, H2B, H3, and H4. The DNA enters and leaves the nucleosome from the same side forming a binding site for the fifth histone, H1 (1). The DNA between nucleosomes is known as linker DNA and can vary in length from 5 to 80 bp. Nucleosomes can be liberated from nuclei or chromatin by treatment with nucleases and may be fractionated from oligonucleosomes on the basis of size, by sucrose gradient centrifugation and size exclusion chromatography. However, monomer nucleosomes prepared by these methods are found to be heterogeneous when the deoxyribonucleoprotein (DNP) complex is analysed by polyacrylamide gel electrophoresis (2).

This heterogeneity is a consequence of both the size of the DNA and the protein content of the nucleosome. The fact that nucleosomes are made up of both protein and DNA means that they lend themselves to analysis by two-dimensional electrophoresis. This enables one to separate nucleosomes by electrophoresis in a 5% polyacrylamide gel at low ionic strength in the first dimension, and then dissociate the protein from the DNA and analyse the DNA and proteins in two different second-dimensional gels. Results obtained from such analyses confirm the view that there are two major classes of monomer nucleosome, MN1 and MN2. MN1 nucleosomes, often known as the core particle, are made up of 145 bp of DNA plus a pair of each of the four core histones. The other particle, MN2, contains at least 160 bp of DNA and as well as the octamer of core histones, it contains histone H1 or H5 and is sometimes called the chromatosome. A number of non-histone proteins have also been shown to be associated with nucleosome particles, namely the HMG proteins. These are present only at about 5% of the level of the histones but they have been observed associated with nucleosomes by electrophoresis. *Table 1*, which is derived from a paper by Albright *et al.* (3), shows the complexity of monomer nucleosomes that can be analysed by two-dimensional electrophoresis. Heterogeneity in nucleosomes is also caused by post-synthetic modification of both the DNA and protein. The DNA can vary in its state of methylation while, in the case of the proteins, the states of phosphorylation, acetylation, ADP ribosylation and modification

249

Table 1. Composition of different electrophoretic forms of mononucleosomes.

Component	Mononucleosome class[a]			
	MI	MII	MII	MIV – V
DNA length (base-pairs)	140 – 175	140 – 184	160 – 210	185 – 220
H3, H2B, H2A, H4	+	+	+	+
M1, M2	+	+	+	+
uH2A (M3)	–	+	+	+
HMG14, 17	–	+	+	+
H1	–	–	+	+
HMG1, 2	–	–	+	+
M4,NF (H1°)	–	–	+	+
NHP	–	–	+	+

[a] MI to MV refer to different electrophoretic forms of mononucleosomes (3). M1, M4, NF refer to minor histones; NHP are non-identified non-histone proteins.

Table 2. Buffers required for the preparation of nucleosomes.

Solution	Composition
50 mM PMSF	50 mM phenylmethylsulphonyl fluoride in dried propan-2-ol (store at 4°C sealed in a dark bottle).
PBS	0.14 M NaCl, 0.01 M sodium phosphate (pH 7.5).
PBS-EDTA	PBS containing 10 mM EDTA (pH 7.5).
TMS	10 mM Tris-HCl (pH 7.5), 5 mM $MgCl_2$, 0.25 M sucrose.
TMS-Triton	TMS containing 0.5% (w/v) Triton X-100 (Pierce).
Digestion buffer	TMS containing 1 mM $CaCl_2$.
Micrococcal nuclease (Worthington)	10 000 U/ml in water (stored at – 20°C).
Stop buffer for salt-soluble nucleosomes	0.2 M NaCl, 10 mM EDTA, 20 mM Tris-HCl (pH 8.0).
5% (w/v) perchloric acid ($HClO_4$)	Diluted from concentrated stock solution.
Column buffer	Electrophoresis buffer (TAE or TBE buffer, see *Protocols 3* and *5*) containing 0.5 mM PMSF.
Sample loading buffer	20% (v/v) glycerol, 0.01% (w/v) bromophenol blue (BPB) in electrophoresis buffer.

with the ubiquitin polypeptide are all important in determining electrophoretic mobility. In fact core particles containing ubiquitinated H2A (uH2A) can be separated by electrophoresis from bulk nucleosomes (9).

Nucleosomes are soluble in low ionic strength buffers above pH 7.0 and are negatively charged. Most of the buffer systems used in the electrophoresis of DNA are applicable to nucleosomes. The original method described by Varshavsky *et al.*

250

(2) used 10 mM triethanolamine-HCl (TEA), 2 mM EDTA (pH 7.6). However, at very low ionic strength and in the absence of divalent metal ions, nucleosomes can unfold, and this may cause rearrangement artefacts. Goodwin *et al.* (4) included 40 mM NaCl to overcome this problem. As an alternative, the more concentrated buffer, TBE (89 mM Tris base, 89 mM boric acid, 2.5 mM EDTA, pH 8.3) can be used. In this buffer it has been shown that two molecules of HMG 14 and/or 17 bind co-operatively to the nucleosome (5), and this method is described in detail in this chapter. The method of Todd and Garrard (6) which gives high resolution is included also. Again it is a low ionic strength technique but the resolution is improved by use of a composite agarose—polyacrylamide gel.

Following the first-dimensional deoxynucleoprotein (DNP) gel electrophoresis, the proteins can be analysed by SDS polyacrylamide gel electrophoresis or acetic acid—urea gels in the second dimension. These systems are complementary, so both are described. In addition, methods for analysing DNA are included. DNA can be sized in a second-dimensional gel containing SDS. Alternatively, the DNA can be denatured and electrophoresed into a gel containing urea and then electroblotted onto activated paper or Nylon. This enables the DNA of the nucleosomes to be hybridized to specific gene probes and their sequences analysed. Since these methods depend on being able to prepare nucleosomes, a brief description of their isolation is given first.

2. Preparation of nucleosomes

The method of preparation of nucleosomes depends on the tissue being used. All the work described in this chapter was carried out with chicken red blood cells (RBC) using low speed centrifugation to isolate nuclei (7). This procedure is also appropriate for tissue culture cells but for most soft tissues it is better to use a high density sucrose method for preparing nuclei (8). Calcium-dependent nucleases can be inhibited by the addition of 0.5 mM EGTA to the buffers. If endogenous nucleases are very prevalent, then nuclei can be prepared in the absence of any divalent metal ions (10); in this case the nuclei are stabilized by including spermidine and spermine in the solutions. All steps should be carried out at 4°C unless otherwise indicated. Similarly, all solutions should contain 0.5 mM PMSF, this is added just prior to use.

The solutions required for the preparation of nuclei and nucleosomes are given in *Table 2*. The outline of the method for preparation of nuclei is given in *Protocol 1*. The method is very straightforward but after the cells have been lysed in TMS-Triton (*Table 2*) the nuclei are quite fragile and they have a tendency to clump. This is a sign that the nuclei are beginning to lyse. So after each centrifugation, suspend the nuclei gently with a loose-fitting Dounce homogenizer and discard any of the nuclei that have stuck hard to the bottom of the centrifuge tube. The measurement of nuclear DNA by its absorption at 260 nm when dissolved in NaOH solution is applicable only if the cell type contains only a small amount of nuclear RNA as in the case of chicken erythrocytes.

Protocol 1. Preparation of chick erythrocyte nuclei

1. Collect 20 ml of chicken blood in 80 ml of PBS−EDTA.

2. Filter through 4 layers of gauze and centrifuge for 10 min at 2000 *g*.

3. Suspend the erythrocytes (RBC) with the aid of a loose-fitting Dounce homogenizer in 100 ml of PBS and centrifuge for 10 min at 2000 *g*.

4. Suspend the RBC in 100 ml of TMS-Triton, leave for 5 min, resuspend and then centrifuge for 20 min at 4000 *g*.

5. Suspend nuclei in 100 ml TMS-Triton and centrifuge for 10 min at 2000 *g*.

6. If all the cells have lysed suspend in 100 ml of TMS and centrifuge for 10 min at 2000 *g*. If not repeat Step 5.

7. Resuspend the nuclei in 100 ml of TMS buffer and take an aliquot (100 μl) of the suspension, add 900 μl of 1 mM EDTA (pH 8.0) and 9 ml of 1 M NaOH and measure the optical density at 260 nm. Assume DNA at 1 mg/ml has an absorbance of 26.5 in a 1-cm-pathlength cell and calculate the yield of nuclei.

8. Pellet the nuclei and resuspend at 5 mg DNA/ml in digestion buffer.

Following micrococcal nuclease digestion of the nuclei, the digest can be fractionated into nucleosomes which are soluble in 0.1 M−0.15 M NaCl (mostly monomer nucleosomes devoid of histone H1 but containing HMG proteins) and those that remain insoluble at this ionic strength in the presence of EDTA (polynucleosome and H1-containing nucleosomes). *Protocol 2* details the procedure for isolating salt-soluble monomer nucleosomes. There are also a large number of other proteins and ribonucleoprotein complexes in the salt-soluble supernatant; these are best removed by Biogel A-5M gel filtration or by sucrose gradient rate-zonal centrifugation. The choice of buffer used for the gel filtration or centrifugation depends on the method of electrophoresis chosen. It is convenient to keep the buffer composition similar throughout the preparation. The partially purified monomers are then loaded on to the gel after the addition of glycerol and bromophenol blue; these latter two reagents can be dialysed into the sample. This has the advantage that it leads to a further two-fold concentration of the nucleosomes.

The method of preparation can be adapted in several ways, depending upon the particular application. The conditions described in *Protocol 2* normally render 4−5% of the DNA acid-soluble. Briefer digestions can increase the proportion of actively transcribed DNA liberated (11). The salt-insoluble mononucleosomes may be prepared by redissolving the pellet at Step 7 (*Protocol 2*) with 20 mM Tris-HCl, 2 mM EDTA (pH 8.0), passing the solubilized oligonucleosomes plus monomer nucleosomes down a Biogel A-5M column and collecting the monomer peak. Alternatively, a representative sample of bulk monomer nucleosomes can be prepared by digesting the nuclei to approximately 10% acid-soluble nucleotides, quenching the reaction on ice and pelleting the nuclei at 2000 *g* for 10 min. The supernatant will contain only a small proportion of the nucleosomes and may be discarded. The nuclei are then lysed by resuspension in 20 mM Tris-HCl, 2 mM EDTA (pH 8.0), stirred for 15 min on ice and then recentrifuged. The supernatant containing oligonucleosomes and monomer nucleosomes can then be fractionated by gel filtration or sucrose gradient centrifugation to isolate mononucleosomes.

Protocol 2. Preparation of salt-soluble nucleosomes

1. Warm nuclei to 37°C in digestion buffer.

2. Add micrococcal nuclease to a final concentration of 200 U/ml.

3. Incubate for 10 min at 37°C with occasional agitation.

4. Add an equal volume of ice-cold stop buffer with stirring and leave on ice for 10 min.

5. Take a 100 μl aliquot and add it to 4 ml of cold 5% (v/v) $HClO_4$. Vortex and centrifuge for 10 min at 4000 g.

6. Measure the absorbance of acid-soluble nucleotides at 260 nm and calculate the percentage digestion. The extinction coefficient is 33.9 for a 1 mg/ml solution of nucleotides in a 1-cm-pathlength cell.

7. Centrifuge the lysed nuclei at 70 000 g for 30 min and collect the salt-soluble monomer nucleosomes in the supernatant.

8. Dialyse the nucleosomes into column buffer overnight at 4°C and then spin out any aggregates at 70 000 g for 30 min.

9. Load the supernatant onto a Biogel A-5M gel filtration column at a flow rate of 0.3 ml/min. Up to 10 mg of DNA can be loaded on to a 90 × 2.5 cm column.

10. Collect the monomer nucleosome peak and concentrate the sample to about 0.5 mg/ml DNA using an Amicon ultrafiltration apparatus at 10 psi (0.7 Bar) with a PM10 membrane.

11. Add glycerol to 20% (v/v) glycerol and bromophenol blue to 0.01%.

---

3. One-dimensional electrophoresis of nucleosomes

3.1 Apparatus and materials

The materials and apparatus used for the work in this chapter are listed in *Table 3*. The sources of chemicals not listed are not critical, but one should use analytical grade reagents. A suitable electrophoresis apparatus such as that described by O'Farrell (13) can be constructed in most workshops but very similar types of apparatus are available commercially. An example of such an apparatus is shown in *Figure 1*. The safety lid has been removed for the purpose of clarity in this photograph. The reservoir buffers may be circulated through a cooling coil using a pump. The bottom chamber is deep to ensure efficient heat exchange between the gel plates and the buffer. Gels for the electrophoresis of nucleosomes should be run at 4°C and so this is best carried out in a cold room. The voltage and current conditions detailed in this chapter refer to a slab gel 14 × 14 cm. The gel plates and spacers may be sealed either with plastic tubing as shown in *Figure 1* or by using plastic adhesive tape or petroleum jelly.

Table 3. Electrophoresis apparatus and materials.

Apparatus

Power pack capable of producing 300 V and 100 mA (DC) current.

Slab gel electrophoresis apparatus capable of running two-dimensional gels with gel plates approximately 14 × 14 cm.

1, 2, and 3 mm spacers and appropriate sample well-formers (combs) with teeth 6 mm wide.

Deep plastic trays for staining and washing gels.

Orbital shaker.

Skin-graft knife blade for cutting gels (Eschmann).

UV transilluminator model TM20 (UV products).

Wratten No. 2E and No. 15 filters.

Electroblot apparatus, e.g. Pharmacia-LKB 2005 transphor electroblotting unit.

Plastic bag sealer (Calor).

Materials

Acrylamide and *N,N'*-methylene-bisacrylamide (bisacrylamide) (BDH Chemicals).

N,N,N',N'-tetramethylethylenediamine (TEMED) (Sigma Chemical Co.).

Sodium dodecyl sulphate (SDS) specially pure for electrophoresis (Fisons).

Urea (Gibco-BRL).

Mixed bed resin, MB1 (BDH Chemicals).

Ethidium bromide.

Coomassie blue R-250 now known as 'PAGE 83' (BDH Chemicals).

Bromophenol blue.

Pyronin Y (Sigma Chemical Co.).

Salmon sperm protamine sulphate (Sigma Chemical Co.).

Bovine serum albumin fraction V (Sigma Chemical Co.).

Hybond-N filters (Amersham International).

3.2 Electrophoresis in 5% polyacrylamide TBE gels

The analysis of nucleosomes on gels using TBE buffer was first described by Sandeen *et al.* (5); gels 2 mm or 3 mm thick are commonly used. A 2-mm-thick gel with a sample well-former (comb) that produces wells 6 mm wide is convenient when a number of samples are to be analysed by one-dimensional electrophoresis. However, if it is important to load more material on the gel for either two-dimensional analysis or preparative purposes, then a 3-mm-thick gel, with well-formers 2 cm wide should be used. The procedure is as described in *Protocols 3* and *4*.

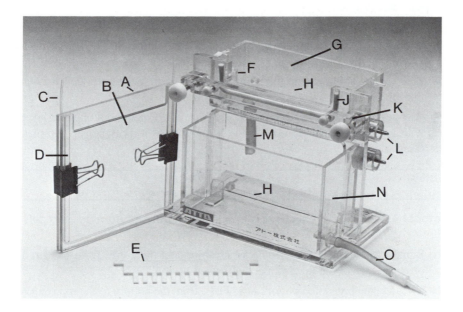

Figure 1. Electrophoresis apparatus. The electrophoresis apparatus shown here is manufactured by Atto. A, plain glass plate; B, notched, bevelled glass plate; C, plastic tubing to make seal; D, spacer; E, comb; F, overflow; G, top electrode reservoir; H, platinum electrodes; J, plastic seal between gel plates and top reservoir; K, gel holder held in place with a screw fitting; L, electrode terminals; M, buffer inlet from pump; O, buffer outlet to pump.

Protocol 3. Preparation of a polyacrylamide slab gel for nucleosome separations

Stock solutions

10 × TBE (pH 8.3): (0.89 M Tris base, 0.89 M boric acid, 25 mM EDTA). Dissolve 108 g of Tris, 55 g of boric acid and 9.3 g of disodium EDTA dihydrate in a final volume of a litre of distilled water. Check that the pH is close to pH 8.3 but do not adjust. Store at 25°C.

30% acrylamide: (29% acrylamide, 1% bisacrylamide). Dissolve 72.5 g of acrylamide and 2.5 g of bisacrylamide in 250 ml of water and deionize by stirring with 15 g of Amberlite MB1 for 30 min. Filter the solution and store at 4°C in the dark.

10% ammonium persulphate. Dissolve 1 g of ammonium persulphate crystals in 9 ml of water. This may be stored for 2−3 weeks at 4°C.

Procedure
1. For a 1 × TBE, 5% polyacrylamide gel mix, 7.5 ml 10 × TBE buffer, 12.5 ml 30% acrylamide stock solution, and 54.7 ml of water in a conical flask.

Protocol 3. *continued*

2. Stir under vacuum for 5 min to degas.

3. Add 50 μl of TEMED followed by 200 μl of 10% ammonium persulphate while stirring gently.

4. Pour the gel solution into the gel mould and place the sample well-former in the top. Make sure not to trap any air bubbles.

5. Any gel solution not covered by the sample well-former should be overlayered with 2 mm of 2-methylpropan-1-ol. Allow the gel to polymerize completely (approx. 2 h).

Protocol 4. Electrophoresis of nucleosomes on polyacrylamide slab gels in TBE

1. Remove the bottom spacer or plastic tubing and the comb from the gel and place the gel in the apparatus equilibrated at 4°C.

2. Circulate the precooled buffer, 1 × TBE, from the bottom to the top reservoir using a pump.

3. Pre-electrophorese the gel for 60 min at 100 V before loading the samples. The negative lead of the power pack should be connected to the top electrode (cathode).

4. After pre-electrophoresis, turn off the current, flush out the wells with buffer and load the samples using an automatic pipette. The samples are usually loaded in 1 × TBE, 20% glycerol, 0.01% BPB, but note that straighter electrophoretic bands are obtained if 4% Ficoll is used instead of glycerol. Place the tip of the pipettor containing the sample as close as possible to the bottom of the well and expel the sample from the tip slowly so that the sample forms a neat band at the bottom of the well.

5. Carry out electrophoresis for 500 V.h, either at 100 V for 5 h or at a proportionally lower voltage overnight.

6. After electrophoresis, stain the gel by placing it in a deep plastic tray with 500 ml of 0.5 μg/ml ethidium bromide (use a stock solution of 5 mg/ml in water) for 30 min. Results may be improved by destaining in 500 ml of 1 × TBE for 30 min before photography.

7. Place the gel on a UV transilluminator and photograph it using both a Wratten 2E UV filter and a Wratten 15 yellow filter. (See Section 2.8.2. of Chapter 2).

Figure 2 shows an example of such an electrophoretic separation. The monomer and dimer bands are marked. Complexes of DNP much larger than a dinucleosome are not resolved in this gel system. This gel shows that the salt-insoluble nucleosomes are depleted of core particles (MNI) and also that after limited digestions the majority

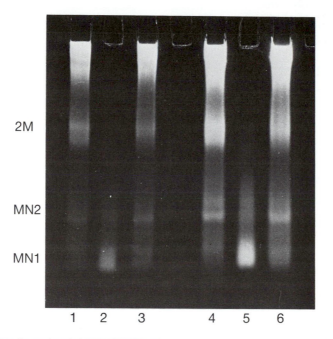

Figure 2. One-dimensional electrophoresis of nucleosomes in a 5% polyacrylamide gel in TBE buffer. Lanes 1 and 4, 10 and 30 μg (DNA) from salt-insoluble nucleosomes after a 2% acid-soluble digest of chicken erythrocyte (RBC) nuclei. Lanes 2 and 5, 3 and 9 μg (DNA) of salt-soluble monomeric nucleosomes from a 5% digest of chicken RBC nuclei. Lanes 3 and 6, 20 and 60 μg (DNA) of nucleosomes not salt-fractionated after a 2% acid-soluble digest of chicken RBC nuclei. NM1 and NM2 indicate the position of monomer nucleosomes, 2 M the dinucleosome bands.

of the chromatin is in the form of polynucleosomes. The salt-soluble fraction only contains mononucleosomes which separate into a major MNI band and minor, slower migrating species.

3.3 Electrophoresis in agarose – polyacrylamide TAE gels

The method of preparation of agarose—polyacrylamide composite gels is given in *Protocol 5*. Additional points to note are that to prevent the gel from slipping out from between the two glass plates, some 2% agarose should be set in the bottom of the glass plates before pouring the gel. This can be done by making the gel mould

---

Protocol 5. Preparation of an agarose—polyacrylamide gel

Stock solutions
10 × TAE (pH 8.5): (64 mM Tris base, 32 mM sodium acetate, 20 mM EDTA). Dissolve the Tris, sodium acetate, and EDTA in water. Adjust to pH 8.5 with acetic acid and make up to volume.

Protocol 5. *continued*

30% acrylamide stock solution (see *Protocol 3*)
10% ammonium persulphate (see *Protocol 3*)
agarose
glycerol (AR grade)
TEMED

Procedure

1. For an 0.5% agarose − 3.5% polyacrylamide gel: dissolve 0.5 g of agarose in 50 ml of water by heating to 100°C.

2. Mix 30 g of glycerol, 10 ml of 10 × TAE, 8.6 ml of 30% acrylamide solution and water to 50 ml, add 25 μl TEMED and 100 μl 10% ammonium persulphate.

3. Mix the two solutions (Steps 1 and 2) together and pour the gel immediately as described in the text.

---

without the bottom spacer and placing it in the gel apparatus. Then tip the mould back at 45° and pour molten 2% agarose into the apparatus so that it forms a wedge about 1 cm high at the bottom of the gel plates. The reservoirs contain 1 × TAE and the same buffer is used in the gel and in the 2% agarose plug. The sample buffer is 10% glycerol, 2 mM Tris-HCl, 2 mM EDTA (pH 8.0). Dialyse the samples into this buffer and centrifuge in a cooled microcentrifuge at 10 000 *g* for 10 min before loading on to the gel to remove particulate material. One of the reasons for the good resolution of this method is the low conductivity of the sample buffer which produces a band sharpening effect. However, as already stated, it may also cause rearrangement artefacts. The electrophoresis conditions are very similar to the previous method. Electrophoresis is carried out at 600 V.h at 4°C and the reservoir buffers are circulated. This gel system is used also as the first-dimensional gel in the two-dimensional analysis shown in *Figure 4* (see Section 4.3.1).

4. Two-dimensional analysis of nucleosomes

There are a number of two-dimensional systems that may be used but they all have features in common. The first-dimensional gel is electrophoresed and then cut into strips. These are equilibrated in a buffer that will dissociate the protein from the DNA and then either the DNA or the protein can be separated on the second-dimensional gel.

4.1 First-dimensional separation of nucleosomes

The best results are obtained by loading the nucleosome sample into a 2−4-cm-wide well of a 3-mm-thick gel. After electrophoresis as described in Section 3, leave the gel on one of the glass plates laid flat on a piece of graph paper. This makes it possible to orientate the gel while slicing the gel with a skin-graft knife blade or alternatively a scalpel blade. Cut out strips 5 mm or 10 mm wide from the centre of the lane. Do not cut out gel strips from the edge of the lane as the resolution may not be as good. To cut the gel strips using a skin-graft knife place the blade over the gel then place one end of it onto the glass plate just off the gel. Then slowly lower the other end of the blade. This will enable you to cut out a strip of gel with smooth straight sides. If a scalpel blade is used to slice the gel then it is necessary to use a ruler to make sure that you obtain an even strip of gel with straight sides. It is a good idea to make a nick at the bottom of each gel strip so that they can be orientated correctly. Once the gel strips are placed in equilibration buffer (see *Protocol 6*), stain the remainder of the gel with ethidium bromide. One can load up to 250 μg of DNA/cm^2, and a gel strip up to 1 cm wide may be loaded on to the second-dimensional gel. This means approximately 75 μg (DNA) of nucleosomes can be resolved on a second-dimensional gel 3-mm thick. In fact, if only the monomer nucleosome regions are going to be analysed, then more than one sample can be analysed on the second-dimensional gel.

4.2 Second-dimensional electrophoresis of DNA

4.2.1 Native DNA electrophoresis

Prepare a 5% polyacrylamide TBE gel containing 1% SDS. The method of preparation of a gel 3 mm thick is essentially the same as given in *Protocol 3*.

Protocol 6. Preparation of a second-dimensional polyacrylamide slab gel for the separation of native DNA

1. For 75 ml of gel solution, mix 7.5 ml of 10 × TBE buffer, 12.5 ml of 29% acrylamide, 1% bisacrylamide and 47.25 ml of water.

2. Degas the solution and then add 7.5 ml of 10% SDS, 50 μl TEMED and 200 μl 10% ammonium persulphate.

3. Pour the solution into the gel mould leaving a 1.5 cm space at the top. Overlayer the gel with 2-methylpropan-l-ol to a depth of 2 − 3 mm and allow to polymerize for 60 min.

4. Place the gel strips from the first-dimensional gel in test tubes with 10 ml of equilibration buffer (0.2 × TBE, 1% SDS, 20% glycerol, 0.01% bromophenol blue) and place on an orbital shaker for 30 min at 25°C. If the sample is not to be analysed immediately, it can be stored for several months at − 70°C after a fresh change of buffer. To analyse the sample, equilibrate for a further 30 min in a fresh change of buffer and then place the gel on a piece of sealing tissue and remove most of the buffer with a paper tissue. This gel strip is now ready to load on to the second-dimensional gel.

5. Wash the 2-methylpropan-l-ol off the top of the slab gel with water, followed by TBE buffer.

Protocol 6. *continued*

6. Place the gel and bottom reservoir buffer in the apparatus. The reservoir buffer is 1 × TBE containing 1% SDS.

7. Apply 1 ml of equilibration buffer to the top of the second-dimensional gel prior to loading the first-dimensional gel. Pick up the first-dimensional gel on the sealing tissue and push it onto the top surface of the second-dimensional gel using a plastic spatula. If one end of the gel is pushed on to the top of the second-dimensional first, then any air bubbles and buffer can be excluded from between the two gels. The gel strip usually fits snugly on top of the gel and between the two glass plates. If there is space, another strip from the first-dimensional gel may be loaded alongside the first one. If molecular weight markers need to be loaded, they should first be run into a DNA TBE-polyacrylamide gel for a short time (15 min), the gel cut up into short strips and then equilibrated as above. The gel piece is then loaded on to the second-dimensional gel in the same way as the first-dimensional gel strip. Loading DNA markers in this way ensures that they are electrophoresed under exactly the same conditions as the sample.

8. Carefully pour the 1 × TBE, 1% SDS electrophoresis buffer into the top reservoir. As the equilibration buffer is of lower conductivity than the remainder of the system, it gives a degree of band sharpening. However, to ensure straight bands there must be a uniform field strength, so any spaces between the gel strips should be filled by underlayering with equilibration buffer.

9. Electrophorese the gel at 100 V for 2.5 h at room temperature.

10. Since gels containing SDS cannot be stained directly with ethidium bromide, remove it by washing the gel three times with a litre of 50% methanol for 3 h with shaking. The gel will shrink but will rehydrate to its original size when stained in 0.5 μg/ml ethidium bromide in 0.1 × TBE.

This gel system is an excellent way of studying the complexity of DNA size in mono- and di-nucleosome populations. However, it should be remembered that the proteins associated with nucleosomes also move into the gel and in some cases co-migrate with DNA species. This can lead to quenching of the ethidium bromide fluorescence and, if the DNA is to be blotted, may interfere with subsequent analysis. This type of two-dimensional electrophoresis can also be used to analyse the product when the nucleosomes resolved by the first-dimensional gel electrophoresis are further digested by micrococcal nuclease or DNase I (3,12).

4.2.2 Denatured DNA electrophoresis

When the DNA from nucleosomes is to be electroblotted and then analysed by hybridization, it is convenient to denature and electrophorese the DNA into the second dimension in its denatured form. This analysis will also indicate if there are any single-stranded nicks in the DNA. It is also useful for analysing DNase I-digested nucleosomes (12).

Protocol 7. Separation of denatured nucleosomal DNA in the second dimension

Stock solutions
30% acrylamide (see *Protocol 3*)
10% ammonium persulphate (see *Protocol 3*)
10 × TBE (see *Protocol 3*)
urea (AR grade)
TEMED
SDS
Equilibration buffer: 9 M urea, 1% SDS, 1 × TBE
Reservoir buffer: 1 × TBE, 2% SDS

Procedure
1. For a 6% polyacrylamide−1 × TBE−7 M urea, 2% SDS gel, for 100 ml of gel, mix together in a conical flask:
   - 42 g urea
   - 20 ml 30% acrylamide stock solution
   - 10 ml 10 × TBE
   - 37 ml water
2. Degas the solution.
3. Add 2 g of SDS and dissolve.
4. Add 100 μl of TEMED and 200 μl of 10% ammonium persulphate, pour the gel and overlayer with 2-methylpropan-1-ol.
5. After electrophoresis of the first-dimensional gel (see Section 3), cut out the gel strips as described in Section 4.1 and place them each in 10 ml of urea equilibration buffer.
6. After equilibration in two changes of buffer for a total of 60 min as described in *Protocol 6*, transfer the tubes containing the gel strips to an 80°C water bath for 30 min.
7. Cool the tubes to room temperature and load the gel strips on to the urea-SDS second-dimensional gel as described in *Protocol 6*.
8. Electrophorese the gels at 100 V for 3 h. The bromophenol blue should run about half-way down the gel.

An example of such an analysis is shown in *Figure 3B*. In this experiment, mononucleosomes were treated with 0.6 M NaCl to remove any H1, H5, HMG14, and HMG17 proteins, then trimmed with micrococcal nuclease and the resulting core particles isolated. These were then reconstituted with pure protein HMG17 and the resulting nucleosomes separated on a 5% polyacrylamide gel with 1 × TBE buffer as the running buffer in the first dimension. Analysis of the denatured DNA shows that the nucleosomes with HMG17-bound (nucleosome 2) and HMG17-free

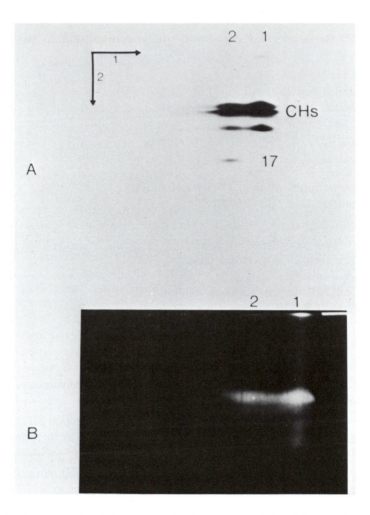

Figure 3. Two-dimensional electrophoresis of the proteins (A) and the denatured DNA (B) from core particles reconstituted with HMG17. Salt-stripped core particles were reconstituted with enough pure HMG17 to bind 25% of the core particles in 1 × TBE buffer. The resulting nucleosomes were fractionated by electrophoresis in a 5% polyacrylamide gel with 1 × TBE as buffer gel in the first dimension and their proteins analysed by acid – urea electrophoresis in the second dimension (A). This gel was stained with Coomassie blue R-250. Nucleosomes 1 and 2 contain the core histones (CHs) but only nucleosome 2 contains HMG17 (17). (B) DNA from the same first-dimensional gel strip was electrophoresed into a denaturing DNA gel containing SDS and urea. The positions of the two nucleosomes are indicated by 1 and 2. This gel was stained with ethidium bromide.

nucleosomes (nucleosome 1) contain the same size of DNA and that there has been a small number of nicks produced in the DNA during the isolation of the core particles.

4.3 Second-dimensional electrophoresis of proteins

The proteins from nucleosomes can be electrophoresed either into an SDS gel or into an acid—urea gel. The first method is essentially that described by O'Farrell (13) where the first-dimensional gel is equilibrated into a buffer containing SDS and then the proteins are resolved on a 15% polyacrylamide—SDS gel as described by Laemmli (14). O'Farrell used a polyacrylamide gradient gel in the second dimension. This is not usually necessary for the resolution of the histones and non-histone proteins present in nucleosomes.

The acid—urea method involves displacing the proteins with protamine sulphate (17), then electrophoresing the proteins on an acid—urea polyacrylamide gel (15) incorporating a stacking system (16). This method gives very good resolution of HMG14, HMG17, and histones and will detect histone modifications such as acetylation and phosphorylation.

The second-dimensional gels for both protein systems are 1 mm thick. However, since the first-dimensional gels are 3 mm thick, the gel plates used are modified from those shown in *Figure 1* to accommodate the 3-mm-thick first-dimensional gel being loaded on to a 1-mm-thick second-dimensional gel. The plates used are very similar to those described by O'Farrell (13) in his original paper. The glass used is 2 mm thick and the spacers 1 mm thick. The plain back plate is replaced by another notched plate with 'ears' 2.5 cm high, and a piece of plain glass is glued to the outside of this plate. When assembled with the inner plate, having a 1.5-cm-deep notch, it leaves a 3-mm-wide 1-cm-high slot at the top of the gel that will take the first-dimensional gel strip.

4.3.1 SDS gel electrophoresis

The method of preparation of a 15% polyacrylamide—SDS gel for the analysis of histones and non-histones is given in *Protocol 8* and the procedure is as follows.

Protocol 8. Preparation of SDS-polyacrylamide gels for separating proteins

1. For a 15% polyacrylamide separating gel (20 ml) mix 5 ml of water and 10 ml of 30% acrylamide solution in a conical flask and degas.

2. Add 5 ml separating gel buffer, 10 μl of TEMED, and 30 μl of 10% ammonium persulphate and pour a 1-mm-thick gel, leaving 2 cm from the top of the gel plates.

3. Overlayer with 2-methylpropan-1-ol and leave to polymerize for at least an hour.

4. Wash off 2-methylpropan-1-ol with water and store under separating gel buffer diluted 1 in 4.

5. Prepare the stacking gel by mixing 1.5 ml of 30% acrylamide solution and 6 ml of water in a conical flask and degas. Add 2.5 ml of stacking gel buffer, 10 μl of TEMED, and 30 μl of 10% ammonium persulphate.

Protocol 8. *continued*

6. Wash the buffer off the separating gel with water and then rinse with half the stacking gel mix plus catalysts. This step helps the two gels to adhere.

7. Pour on remainder of stacking gel to a level just higher than the bevel in the plates and overlayer with 2-methylpropan-1-ol, and leave to polymerize for an hour.

Protocol 9. Separation of nucleosomal proteins on a SDS-polyacrylamide gel

1. Cut the strips from the first-dimensional gel as described in Section 4.1 and incubate the strips in 20 ml of equilibration buffer (see *Table 4*) for 30 min.

2. Then put the gel strips in fresh equilibration buffer and incubate for another 30 min.

3. Wash off the top of the stacking gel with water and blot the top of the gel dry with tissue paper.

4. Place approximately 0.5 ml of molten agarose in equilibration buffer on top of the gel and place the first-dimensional gel strip on top of this making sure not to trap any air bubbles between the stacking gel and the gel strip.

5. Then add more agarose so that the first-dimensional gel is just covered. Allow this to set for a few minutes and then place the gel in the gel apparatus and fill the reservoirs with buffer.

6. Electrophorese the gel at 20 mA constant current for approximately 4 h until the bromophenol blue band reaches the bottom of the gel. The voltage increases from 120 V to 280 V during the run.

7. After electrophoresis stain the gels in 0.1% Coomassie blue R-250, 50% (v/v) methanol, 10% (v/v) acetic acid. To prepare the stain, dissolve the Coomassie blue R-250 in the methanol first and then add the acetic acid and water. Staining and destaining are carried out in a deep plastic tray placed on an orbital shaker. Stain the gel for at least 60 min with 240 ml of stain and destain with methanol/acetic acid/water (10/10/80, by vol.) with frequent changes.

8. The gel can then be photographed using a Wratten No. 15 yellow filter.

An example of the resulting SDS protein gel of chicken RBC salt-soluble nucleosomes fractionated on a 5% polyacrylamide gel (Method 2) is shown in *Figure 4*. These monomeric nucleosomes were not first purified by gel filtration chromatography or sucrose gradient centrifugation and so a large number of the proteins, especially those in the top left quarter of the gel, may be free proteins or proteins associated with ribonucleoprotein (RNP) particles. The first dimension was

Table 4. Stock solutions for the preparation of SDS – polyacrylamide gels.

30% acrylamide stock solution:

29.2% acrylamide	73.0 g
0.8% bisacrylamide	2.0 g

Deionize (see *Protocol 3*) and make up to 250 ml with water. Store at 4°C.

Separating gel buffer:

0.4% SDS	1.0 g
1.5 M Tris base	45.5 g

Adjust to pH 8.8 with 5 M HCl and make up to 250 ml. Store at 4°C.

Stacking gel buffer:

0.4% SDS	1.0 g
0.5 M Tris base	15.1 g

Adjust to pH 6.8 with 5 M HCl and make up to 250 ml. Store at 4°C.
10% ammonium persulphate (see *Protocol 3*).

Equilibration buffer

2.3% SDS	2.3 g
62.5 mM Tris base	0.76 g
10% glycerol	10.0 g
5% 2-mercaptoethanol	5.0 ml

Dissolve these components and adjust to pH 6.8 with 1 M HCl and make up to 10 ml.

Reservoir buffer

0.192 M glycine	43.2 g
0.025 M Tris base	9.0 g
0.1% SDS	3.0 g

Dissolve these components in a total of 3 litres of water, and do not adjust the pH of the buffer.

electrophoresed in the presence of DNase I and this may account for the band labelled 'X'. The 5% perchloric acid-soluble proteins (histones H1 and H5 together with the HMG proteins) were used as markers. These were loaded into a well, formed in the stacking gel on the right-hand side using the following procedure. The plastic well-former was not removed until after the bromophenol blue band had been electrophoresed into the stacking gel. The current and reservoir circulation pump were turned off, the well-former removed and the marker sample loaded. The current and pump were then turned on again. It is clear from the gel analysis that the salt-soluble MN2 nucleosomes contain HMG14 and HMG17 and the core histones whereas the MN1 nucleosomes are devoid of these HMGs. Bands of the proteins HMG1 and HMG2 are also seen but they are probably not associated with the nucleosomes and are probably aggregates or associated with RNP particles at low ionic strength. Note that the nucleosomal DNA may also be visualized on these gels after protein staining. To do this, first wash the gel in a litre of 50% (v/v) methanol for 60 min to remove the acetic acid. Then give the gel a further wash in 1 × TBE for 60 min before staining with ethidium bromide solution. The mononucleosomal DNA migrates just behind the core histones and in fact quenches the Coomassie staining of non-histone proteins in this region.

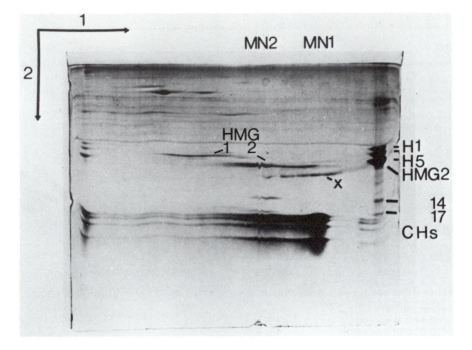

Figure 4. Two-dimensional electrophoresis of salt-soluble nucleosomes. Unfractionated salt-soluble mononucleosomes were electrophoresed in a Tris-acetate gel in the first dimension as indicated by arrow 1 and then the proteins analysed on a 15% polyacrylamide – SDS gel (arrow 2). The marker proteins (a perchloric acid extract of chicken RBC nuclei) were loaded in the well on the right-hand side of the gel. Band 'X' may be DNase I (see text) and CHs indicate core histones; the gel was stained with Coomassie blue R-250.

Table 5. Stock solutions for acetic polyacrylamide acid – urea polyacrylamide gels.

30% acrylamide (separating gel):

29.6% acrylamide	74.0 g
0.4% bisacrylamide	1.0 g

Deionize (see *Protocol 3*) and make up to 250 ml with water.

30% acrylamide stacking gel:
(29.2% acrylamide, 0.8% bisacrylamide) (see *Table 4*).

1.5 M potassium acetate (pH 4.0):
glacial acetic acid   9.0 g
Dilute with 50 ml of water and titrate to pH 4.0 with 4 M KOH.
Make up a final volume of 100 ml.
10% ammonium persulphate (see *Protocol 3*)

Electrophoresis buffer	0.9 M acetic acid
Equilibration buffer	9 M urea, 5% (v/v) 2-mercaptoethanol, 0.9 M acetic acid
1% Agarose	molten 1% agarose; make up in 0.375 M potassium acetate (pH 4.0)
Protamine sulphate	600 μg salmon sperm protamine sulphate is dissolved in 400 μl of equilibration buffer plus 0.01% pyronin Y.

4.3.2 Acetic acid – urea gel electrophoresis

The method of preparation of acetic acid–urea polyacrylamide gels is given in *Protocol 10*. The gel plates used are the same as described in the previous section; that is, they enable a first-dimensional 3-mm-thick gel strip up to 1 cm wide to be loaded on to and analysed in a 1-mm-thick second-dimensional gel.

Protocol 10. Preparation of acetic acid – urea polyacrylamide gels

1. For a 20% separating gel (30 ml) dissolve 4.5 g of urea in 20 ml of 30% acrylamide solution and 4.3 ml of water and degas.

2. Add 40 μl of TEMED, 1.87 ml of glacial acetic acid and 0.5 ml of 10% ammonium persulphate and pour the solution into the gel mould. Overlayer with 2-methylpropan-1-ol and allow to polymerize for 2 h.

3. Place the gel in the apparatus and pre-electrophorese for 4 h at 150 V. The reservoir buffer is 0.9 M acetic acid and is not circulated. The top electrode is the anode (+). During pre-electrophoresis the current drops from 30 mA to approximately 10 mA.

4. Next prepare the 7.5% polyacrylamide stacking gel (10 ml), dissolve 1.5 g of urea in 2.5 ml of 30% acrylamide solution, 2.5 ml of potassium acetate (pH 4.0), 100 μl of TEMED and 4.5 ml of water and degas.

5. Add 0.3 ml of 10% ammonium persulphate.

6. Pour off the acetic acid from the pre-electrophoresed separating gel, wash the top of this gel with 5 ml of the stacking gel solution and pour off. Then pour the remainder of the gel solution on to the top of the gel and overlayer with 2-methylpropan-1-ol. The stacking gel should polymerize up to the bevelled ledge of the notched glass plate; that is, 1.5 to 2 cm deep.

Protocol 11. Separation of nucleosomal proteins on an acid – urea polyacrylamide gel

1. Place the strip cut from the first-dimensional gel in a test tube and equilibrate the strip in 20 ml of acetic acid–urea equilibration buffer, gently shaking the tube for 30 min.

2. Repeat equilibration with another 20 ml of buffer, each time at room temperature.

3. Load the gel strip on top of the stacking gel and seal with 1% agarose in 0.375 M potassium acetate buffer as described in Section 4.3.1. Use just enough agarose to cover the DNP gel and give a completely flat level surface.

Protocol 11. *continued*

4. Once the agarose has set, fill the electrode chambers with the reservoir buffer (0.9 M acetic acid).

5. The first-dimensional gel and the agarose are then carefully overlayered with 400 μl of the protamine sulphate solution.

6. Carry out electrophoresis at 180 V for 7 h with the anode (+) as the top electrode and without circulating the reservoir buffers.

7. Stain the proteins with 0.1% Coomassie blue as described previously. Because the separating gel is 20% polyacrylamide, best results are obtained by staining overnight.

The electrophoresis causes the protamine to migrate through the first-dimensional gel strip and displace the proteins bound to the DNA. The DNA remains in the gel strip bound to the protamine. The proteins and the excess protamine migrate into the stacking gel and the bands 'stack' prior to being separated on the 20% acrylamide separating gel. After the marker dye pyronin Y has moved into the separating gel (approximately 2 h), the gel strip may be removed and the DNA in it analysed on a denaturing TBE−urea polyacrylamide gel (see Section 4.2.2).

The analysis of the proteins from HMG17-reconstituted core particles is shown in *Figure 3A*. The sample of core particles is the same as that described in Section 4.1 and the first-dimensional gel was electrophoresed in 1 × TBE buffer. One can see that the nucleosomes that contain HMG17 migrate more slowly than the core particles without HMG17 in the first-dimensional gel. Both core particles have a similar content of core histones and, as is borne out by the DNA analysis in *Figure 3B*, the size of the DNA in both cases is very similar. This method of protein analysis also gives good resolution of the histones H1 and H5 as well as the HMG proteins. If the nucleosomes are prepared in 10 mM sodium butyrate, an inhibitor of histone de-acetylase, acetylated forms of histone H4 can be detected migrating behind the non-acetylated form (18). It is clear that this method gives good resolution of most nuclear proteins associated with nucleosomes and has the added advantage that the DNA from the same strip can be analysed also.

5. DNA hybridization analysis after two-dimensional electrophoresis

So far in this chapter the techniques have been concerned with the separation of nucleosomes on polyacrylamide gels and analysing their DNA size and protein content. It is also of interest to probe the DNA for particular gene sequences to see whether the subclasses of nucleosomes are enriched in actively transcribed sequences or other sequences of interest. This involves hybridization of the mononucleosomal DNA with radioactively-labelled gene sequences. This can be carried out in a number of ways. The nucleosomal DNA can be eluted from the first-dimensional gel, purified and hybridized by dot blot hybridization analysis (19), or a polyacrylamide—agarose gel can be blotted by the Southern procedure (21), as described by Seale *et al.* (20). Alternatively, the DNA separated on a gel can be resolved in a second-dimensional DNA gel and then electroblotted on to diazo-paper (22,23).

General aspects of blotting methods for nucleic acids have been described in Sections 7.3 and 7.4 of Chapter 1 and Section 2.10 of Chapter 2. However, there are a number of problems associated with blotting and hybridizing nucleosomal DNA. The small size of the DNA means that the probe has to be labelled to high specific activity to produce a signal from a single copy gene. The small size also means that the nucleosomal DNA does not bind very strongly to nitrocellulose. This may be overcome by using diazo-paper which binds the DNA covalently. However, the low stability of diazo group and the time required for complete transfer of the DNA from gels using diffusion makes this method impracticable. Complete transfer can be obtained by electroblotting the DNA. The DNA cannot be denatured by treating a polyacrylamide gel with NaOH solution because the gel swells and subsequently breaks up during electroblotting. The DNA can be denatured by heating the gel to 100°C in transfer buffer or by denaturing the DNA in urea at 80°C and then electrophoresing it into a second-dimensional denaturing gel (see Section 4.2.2). However, all buffer components that contain amino groups must be removed from the gel prior to transfer since they will react with the diazo-paper and prevent the DNA binding. Protein in close proximity to the DNA being transferred may also interfere with binding. Another possible complication may be caused by RNA that is still present in the nucleosome preparation hybridizing to the probe. This RNA can be removed by treating the blot with NaOH solution after transfer, but this step is not possible when using nitrocellulose.

Table 6. Solutions required for hybridization analysis of nucleosomal DNA separated on a two-dimensional denaturing polyacrylamide gel.

Equilibration buffer (200 ml)	$1 \times$ TBE 1% SDS 9 M urea 0.01% bromophenol blue
Transfer buffer (5 litres)	$1 \times$ TBE
Stock 1 M phosphate buffer (1 litre)	1 mole NaH_2PO_4 (pH 7.2) 1 mM EDTA 7% SDS 1% bovine serum albumin (BSA)
Dissolve the reagents and filter the solution	
Hybridization wash Solution A (2 litres)	40 mM NaH_2PO_4 (pH 7.2) 1 mM EDTA 0.5% BSA 5% SDS
Hybridization wash Solution B (5 litres)	40 mM NaH_2PO_4 (pH 7.2) 1 mM EDTA 1% SDS

The method described by Goodwin *et al.* (23) overcomes these problems, but there is still room for improvement. Denaturing two-dimensional analysis of the core particle DNA shown in *Figure 3B* was electroblotted on to diazophenylthioester (DPT) paper (25) and hybridized with both a chicken β-globin cDNA, and ovalbumin probes. The result is shown in *Figure 5*. One can see signal from both of these single copy genes in the monomer DNA region of the gel. However, it is superimposed on a high background. Since this work was carried out, Nylon membranes such as Hybond-N have become readily available. Using Nylon membranes, Church and Gilbert (24) have shown that DNA of nucleosomal size can readily be detected after electroblotting directly from denaturing DNA sequencing gels. This method has the advantage of having a low noise-to-signal ratio, and the DNA can bind in the presence of buffer components containing amino groups. Once the DNA has been crosslinked to the membrane with UV light the membrane can be treated with alkali to remove RNA.

The method given in *Protocol 12* is that of Goodwin *et al.* (23) modified in the light of experience with the Church and Gilbert procedure (24). The solutions required are given in *Table 6*.

Protocol 12. Hybridization analysis of DNA separated on slab gels

1. Electrophorese the nucleosomes on a 5% polyacrylamide gel using 1 × TBE as the running buffer (*Protocol 4*).

2. Electrophorese the first-dimensional strip for protein analysis in the second dimension using the acetic acid−urea system (*Protocol 11*). This removes the proteins from the DNA and replaces them with protamine.

3. Remove the first-dimensional strip from the acid−urea gel after electrophoresis for 2 h, and continue electrophoresis for protein analysis.

4. Place the gel strip into 50 ml of TBE−urea−SDS equilibration buffer and shake gently with frequent changes of buffer. The bromophenol blue acts as an indicator. The equilibration buffer should be changed every time it begins to go yellow; in this way equilibration should be complete in 2 h.

5. Place the gel strip in 10 ml of fresh equilibration buffer and incubate it at 80°C for 30 min. Cool to room temperature.

6. Load the gel on to a denaturing DNA second-dimensional gel and electrophorese as described in *Protocol 7*. In this step the protamines are dissociated from the DNA and migrate well in front of the DNA on electrophoresis.

7. If the gel is not to be stained go to Step 8. To stain the gel, wash it in 50% methanol prior to staining (see Step 10 of *Protocol 6*).

8. Place a litre of 1 × TBE buffer in a developing tray and submerge the anode (+) side of the Pharmacia-LKB electroblot cassette.

9. Cut 4 sheets of Whatman 3MM paper and 1 sheet of Hybond-N to the size required to fit into the cassette.

10. Place the Dacron sponge and two sheets of Whatman paper on top of the

270

Protocol 12. *continued*

submerged cassette. Make sure not to trap any air bubbles. They can be removed by rolling a pipette over the papers.

11. Place the Hybond-N on top of the paper.

12. Place the gel on top of the Hybond-N. It is a good idea to cut off one corner of the gel to help with orientation.

13. Place two sheets of Whatman paper on top of the gel and then the cathode $(-)$ side of the cassette.

14. Assemble the complete cassette. The method used will depend on the apparatus. Place the closed cassette into the prefilled, precooled (4°C) transfer apparatus and apply a current of 0.5 A and 50 V for a minimum of 8 h. One should circulate a coolant through the heat exchangers to ensure the temperature does not rise above 4°C.

15. Disassemble the cassette making sure not to disturb the gel and membrane. With a soft pencil draw a line around the edge of the gel and then stain the gel.

16. Trim the membrane to within 2 mm of the pencil line and place between two sheets of Saran Wrap.

17. Place membrane DNA-side down on the UV transilluminator and irradiate for 3 min.

18. Incubate the membrane in 0.4 M NaOH at 45°C for 30 min to remove any RNA and then wash with a litre of 1 × TBE for 30 min at room temperature. This procedure may also be used to wash off the first probe when rehybridizing.

19. Seal in a plastic bag with 30 ml of hybridization buffer, minus the radioactive DNA or RNA probe and incubate for an hour at 65°C.

20. Remove the buffer and add the denatured probe in a minimum volume of hybridization buffer (5 ml). Exclude all air bubbles and reseal the bag.

21. Hybridize at 65°C for 24 h. The exact temperature and conditions for hybridization may have to be varied depending on the nature of the probe and DNA being analysed.

22. Preheat the hybridization wash solution to 65°C.

23. Immerse the membrane in a litre of hybridization wash Solution A (*Table 6*) and shake on an orbital shaker for 5 min at room temperature before repeating this with another litre of preheated hybridization Solution A.

24. Wash the membrane four times in a similar manner with hybridization wash Solution B (*Table 6*).

25. Wash the membrane at 65°C in a shaking water bath for 20 min with a further litre of Solution B.

26. Blot the membrane dry on Whatman paper and put it against Kodak XAR film at −70°C, using a Dupont intensifying screen for autoradiography.

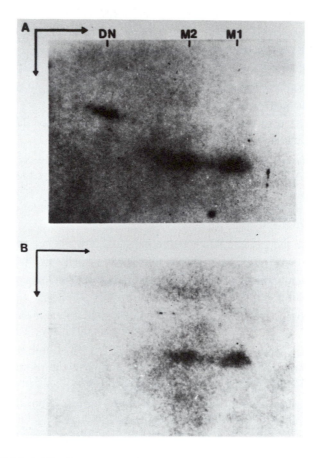

Figure 5. DNA hybridization analysis of core particles reconstituted with HMG17. The DNA from the two-dimensional DNA gel shown in *Figure 3B* was electroblotted on to DPT-paper. The blot was hybridized with 100 ng of nick-translated DNA probes labelled to 2×10^8 cpm per microgram. (A), a chicken β-globin cDNA containing recombinant plasmid (pHB1001). (B), a chicken ovalbumin cDNA containing plasmid (pCROV 2.1). After hybridization the blot was washed at 65°C and the final wash was on 15 mM NaCl, 1.5 mM sodium citrate (pH 7.0) (0.1 × SSC containing 0.1% SDS). M2 and M1 nucleosomes are respectively bound and unbound to HMG16. DN = dinucleosomes.

In the example shown in *Figures 3* and *5*, 75 μg (DNA) of salt extracted core particles, reconstituted with sufficient HMG17 to bind 25–30% of the nucleosomes, was electrophoresed in the first-dimensional 5% polyacrylamide gel. The proteins from this gel were analysed on an acetic acid–urea second-dimensional polyacrylamide gel (see *Figure 3A*), and the DNA from the same DNA gel strip electrophoresed into a denaturing DNA gel (see *Figure 3B*). After staining, the DNA gel was placed on a glass plate so as not to expose the DNA to high levels of UV light when being photographed. In this particular case the DNA was then electroblotted on to DPT-paper in phosphate buffer (23). If the DNA is to be blotted on to diazo-papers then the methanol washes in Step 7 are mandatory.

The hybridization analysis with the β-globin and ovalbumin probes show that the HMG17 containing M2 nucleosomes and the unbound M1 nucleosomes contain both the transcriptionally active β-globin and the inactive ovalbumin gene sequences. However, it is of interest to note that in this preparation of core particles there is a small proportion of dinucleosome (DN) that hybridizes only to the β-globin probe, indicating that it is enriched for actively transcribed sequences. One of the drawbacks of using DPT-paper is that 145 bp DNA gives a much lower signal than 320 bp DNA. Equal weights of mononucleosome DNA give only a third of the signal as dinucleosomal DNA. This problem is not as acute with the Hybond-N membranes. The advantage of both diazo-papers and Nylon membranes over nitrocellulose is that they are robust and the blot can be reprobed many times by washing off the previous probe with 0.4 M NaOH.

6. Applications

The isolation and analysis of nucleosomes has given a great deal of insight into the structure of chromatin. However, electrophoretic analysis of nucleosomes has shown that the composition of this basic unit of chromatin can be far more complex than 160 bp of DNA associated with five invariant histones. Nucleosomes may be associated with a variable number of protein and DNA modifications as well as being associated with a number of different non-histone proteins, and electrophoresis can separate some of these nucleosomal species. The techniques described here could be extended further by using specific antibodies for chromosomal proteins in Western blot analyses (29).

The investigation of the association of particular proteins with specific subclasses of electrophoretically separated nucleosomes is subject to a number of artefacts, the major ones being protein rearrangement and changes in modification states during the isolation of nuclei and the preparation and separation of nucleosomes. Chemical crosslinking techniques may be necessary to obviate rearrangement problems (26). Protein modifications such as acetylation and phosphorylation can be lost during nucleosome fractionation, but this can be prevented by an array of inhibitors such as sodium butyrate (to inhibit deactylase) and sodium bisulphite to inhibit phosphatases (28).

The gel systems for separating DNA described in this chapter are very similar to those used in gel retardation assays (ref. 27, and see Chapters 6 and 7). Thus some of the two-dimensional techniques described in this chapter can also prove useful in the analysis of specific DNA binding proteins (30). The same could also be true for the analysis of proteins binding to RNA.

Acknowledgements

The work described in this chapter was supported by grants from the Medical Research Council and the Cancer Research Campaign. I would like to thank Graham Goodwin, Carol Wright and Lesley Woodhead for their helpful discussions, and Marie Callahan for typing the manuscript.

References

1. Igo-Kemenes, T., Horz, W., and Zachau, H. G. (1982). *Annu. Rev. Biochem.*, **51**, 89−121.
2. Varshavsky, A. J., Bakayeva, T. G., and Georgiev, G. P. (1976). *Nucleic Acids Res.*, **3**, 477−92.
3. Albright, S. C., Wiseman, J. M., Lange, R. A., and Garrard, W. T. (1980). *J. Biol. Chem.*, **255**, 3673−84.
4. Goodwin, G. H., Wright, C. A., and Johns, E. W. (1981). *Nucleic Acids Res.*, **9**, 2761−75.
5. Sandeen, G., Wood, W. I., and Felsenfeld, G. (1980). *Nucleic Acids Res.*, **8**, 3757−78.
6. Todd, R. D. and Garrard, W. T. (1977). *J. Biol. Chem.*, **252**, 4729−38.
7. Plumb, M. A., Nicolas, R. H., Wright, C. A., and Goodwin, G. H. (1985). *Nucleic Acids Res.*, **13**, 4047−65.
8. Chaureau, J., Moule, Y., and Rouiller, C. H. (1956). *Exp. Cell Res.*, **11**, 317−24.
9. Varshavsky, A., Levinger, L., Sundin, O., Barsoum, J., Ozkaynak, E., Swerdlow, P., and Findley, D. (1983). *Cold Spring Harbor Symp. Quant. Biol.*, **47**, 511−28.
10. Hewish, D. R. and Burgoyne, L. A. (1973). *Biochem. Biophys. Res. Commun.*, **52**, 504−10.
11. Bloom, K. S. and Anderson, J. N. (1978). *Cell*, **15**, 141−50.
12. Todd, R. D. and Garrard, W. T. (1979). *J. Biol. Chem.*, **254**, 3074−83.
13. O'Farrell, P. H. (1975). *J. Biol. Chem.*, **250**, 4007−21.
14. Laemmli, U. K. (1980). *Nature*, **227**, 680−5.
15. Panyim, S. and Chalkley, R. (1969). *Arch. Biochem. Biophys.*, **130**, 337−46.
16. Spker, S. (1980). *Anal. Biochem.*, **198**, 263−5.
17. Shaw, B. R. and Richards, R. G. (1979). In *Chromatin Structure and Function* (ed. C. A. Nicoline) Part A, p. 125. Plenum, New York.
18. Riggs, M. G., Whittaker, R. G., Neumann, J. R., and Ingram, V. N. (1977). *Nature*, **268**, 462−4.
19. Nicolas, R. H., Wright, C. A., Cockerill, P. N., Wyke, J. A., and Goodwin, G. H. (1983). *Nucleic Acids Res.*, **11**, 753−72.
20. Seale, R. L., Annuiriato, A. T., and Smith, R. D. (1983). *Biochem.*, **22**, 5008−15.
21. Southern, E. M. (1975). *J. Mol. Biol.*, **98**, 503−17.
22. Levinger, L., Barsoum, J., and Varshavsky, A. (1981). *J. Mol. Biol.*, **146**, 287−304.
23. Goodwin, G. H., Nicolas, R. H., Cockerill, P. N., Zavou, S., and Wright, C. A. (1985). *Nucleic Acids Res.*, **13**, 3561−79.
24. Church, G. M. and Gilbert, W. (1984). *Proc. Nat. Acad. Sci. (USA)*, **81**, 1991−5.
25. Seed, B. (1982). *Nucleic Acids Res.*, **10**, 1799−810.
26. Schick, V. V., Belyavsky, A. V., Bavykin, S. G., and Mirzabekov, A. D. (1980). *J. Mol. Biol.*, **139**, 491−517.
27. Singh, H., Sen, R., Baltimore, D., and Sharp, P. A. (1986). *Nature*, **319**, 154−8.
28. D'Anna, J. A., Gurley, L. R., and Deaven, L. L. (1978). *Nucleic Acids Res.*, **5**, 3195−207.
29. Hames, B. D. (1990). In *Gel Electrophoresis of Proteins: A Practical Approach* Second edition (ed. B. D. Hames and D. Rickwood). IRL Press, Oxford.
30. Perkins, N. D., Nicolas, R. H., Plumb, M. A., and Goodwin, G. H. (1989) *Nucleic Acids Res.*, **17**, 1299−314.

9

Gel electrophoresis of ribonucleoproteins

ALBERT E. DAHLBERG and PAULA J. GRABOWSKI

1. Electrophoresis of ribosomes and polysomes

1.1 Introduction

The gel electrophoretic separation of bacterial polyribosomes (polysomes) and ribosomes was made possible by the development of composite gels containing both agarose and acrylamide by A. C. Peacock and S. L. Bunting (1). The addition of 0.5% agarose to low-percentage acrylamide gels forms a mechanically stable yet very porous gel. Gels composed of 1.5 – 3.0% polyacrylamide can be handled easily, and gels with even lower polyacrylamide concentrations can be managed if necessary. The fine resolution provided by these gels permits the separation of large macromolecules which may differ only in conformation or by the presence of a single ribosomal protein. The separation is on the basis of molecular sieving and the extent of separation is determined primarily by the structure (size and shape) of the particle rather than by its charge.

The structure of ribosomes and polysomes is very dependent on the surrounding ionic conditions, particularly the magnesium ion concentration. The inert matrix of agarose – acrylamide composite gels is particularly well-suited for permitting variations in the ionic conditions so as to permit the study of the ribosome structure. As an example, polysomes from bacteria, which migrate intact in a gel prepared in a buffer containing 1 – 10 mM magnesium ions, will dissociate and migrate as 30S and 50S ribosomal subunits if the magnesium ion concentration is reduced to 0.2 mM. This chapter describes the preparation of agarose – acrylamide composite gels as well as the different gel conditions which permit the optimum separation and characterization of bacterial polysomes and ribosomal subunits. These methods can also be used for separating eukaryotic ribosomes. However, eukaryotic polysomes, with the exception of those from yeast, tend to give smeared patterns (see Section 1.10).

1.2 Equipment

1.2.1 Electrophoresis apparatus

While agarose−acrylamide composite gels may be poured and electrophoresed in most types of electrophoresis apparatus, some are more easily adapted to the special requirements for pouring and running this type of gel than others. The slab gel apparatus of E.C. Apparatus Corporation is particularly suitable for the separation of ribonucleoproteins. One advantage of this vertical slab gel apparatus is the presence of cooling coils built into the Perspex (Plexiglass) plates on both sides of the gel mould. If this facility is not available, one can improvise by pouring the gel into a mould placed in a 20°C water bath (see Section 1.5). The dimensions of the E.C. slab gel are 12 cm × 17 cm and 3 mm thick. The sample well-former (comb) used gives eight wells 1 mm thick, 10 mm wide, and 10 mm deep. A 1-mm-thick well in the 3-mm-thick gel keeps the sample entirely within the gel matrix and appears to aid in the resolution of ribosomes and larger rRNAs in these composite gels. For larger sample volumes an adapter is available for the E.C. slab apparatus to increase the gel thickness to 6 mm, giving a sample well 4 mm thick.

1.2.2 Power supply

A unit that provides a high current as well as high voltage is preferable for gel electrophoresis of polysomes since the buffer used has a relatively high ionic strength and draws considerable current (120 V and about 340 mA). If such a power pack is not available, the voltage and current can be reduced as necessary and the time of electrophoresis increased accordingly.

1.3 Stock solutions

Generally, it is unnecessary to use the most expensive ultrapure reagents in composite gels; usually Analar quality reagents are sufficient. The stock solutions required are as follows.

20% acrylamide−bisacrylamide (19:1): Dissolve 190 g acrylamide and 10 g bisacrylamide in a litre of distilled water. This solution can be stored at 4°C.

6.4% DMAPN: Dilute 16 ml of concentrated 3-dimethylamino-proprionitrile to 250 ml with distilled water and store at 4°C.

1.6% ammonium persulphate: Dissolve 1.6 g of ammonium persulphate in 100 ml of distilled water. Prepare fresh at least every two weeks and store at 4°C.

Agarose. M.E. Sea Kem agarose from Marine Colloids Inc., or indeed almost any other brand suitable for electrophoresis can be used. Each gel requires 0.8 g of agarose (see *Protocol 2*). In addition, prepare 1 ml aliquots of 0.5% agarose in 60 mM KCl, 10 mM $MgCl_2$, in 25 mM Tris-HCl (pH 7.6) buffer for gelling polysome samples in the sample wells. Reflux the agarose in the buffer or heat in a microwave oven and store the agarose at 4°C in glass tubes until used. Melt the agarose by heating it in a boiling water bath just prior to use.

Tris-borate-EDTA (pH 8.3) buffer (1 × TBE): A tenfold concentrated stock solution contains 108 g Tris base, 55 g boric acid and 9.3 g disodium EDTA (dihydrate) per litre (Peacock's buffer; ref. 1).

1.0 M Tris-HCl (pH 7.6).[1] A range from pH 7.5 to pH 8.0 has been used without any noticeable difference in gel patterns.

1.0 M MgCl$_2$.[1]

2.0 M KCl.[1]

'Stains-all' (2). Prepare as a 0.1% stock solution in 100% formamide and dilute to 0.005% in 50% formamide for use as a stain (see Section 1.8).

1.4 Preparation of sample

The numerous methods available for preparing ribosomes and polysomes are described elsewhere (3,4) and will not be described in detail here. For bacteria such as *E. coli*, remove debris from the cell lysate by low-speed centrifugation (8000 *g* for 10 min at 5°C) and then the supernatant fraction containing polysomes and ribosomal subunits can be electrophoresed directly into the gel. Consistency in preparation of the sample is important for reproducible results since changes in the procedure, especially different buffer washes of the ribosomes, will affect the structure of ribosomes and, consequently, the electrophoretic pattern obtained.

1.5 Preparation of the gel apparatus

Before preparing the gel mixture prepare the apparatus for use as follows.

Protocol 1. Preparation of the gel apparatus

1. Circulate the coolant at 20°C through the cooling coils of the electrophoresis apparatus. If cooling is not available, place the gel mould in a 20°C water bath.

2. Place the sample well-former in a freezer (−20°C) and fill a stoppered test-tube with ice.

3. Check the gel apparatus to ensure that it is both clean and dry.

4. Place a piece of sponge or a small roll of fibreglass screening (2 mm holes) at the bottom of the apparatus (furthest from the sample wells) to serve as a gel support.

1.6 Preparation of agarose – acrylamide composite gels

All composite gels contain 0.5% agarose. Thus, gels designated as 2%/0.5% and 3%/0.5% contain 0.5% agarose plus 2% and 3% acrylamide respectively (with 5% crosslinking).

---

[1] Acetate is substituted for chloride ion in these solutions for gel electrophoresis of yeast polysomes.

Protocol 2. Preparation of an agarose-polyacrylamide composite gel

1. For a final gel volume of 160 ml, dissolve 0.8 g of agarose in 100 ml of distilled water by refluxing for 10 min or heating in a microwave oven for 2−4 min (caution! use a low power setting).

2. Afterwards cool the solution to 50−55°C and re-adjust the volume to 100 ml.

3. Next, add 20% acrylamide-bisacrylamide solution to give the desired gel concentration (e.g. 8 ml gives 1% acrylamide in the final volume of 160 ml, so add 18 ml for a 2.25%/0.5% gel). Add 10 ml of 6.4% DMAPN to each gel mixture.

4. Add the appropriate buffer stock solution along with distilled water to give a total volume of 155 ml. Specific buffers are described in *Table 1*, but, as an example, for the gel shown in *Figure 1*, 4.0 ml of 1.0 M Tris-HCl (pH 7.6), 3.2 ml of 1.0 M KCl and 1.6 ml of 1.0 M $MgCl_2$ plus 18 ml distilled water were added.

5. After cooling the solution slowly to 35°C under tepid water, add 5 ml of 1.6% ammonium persulphate and swirl gently to mix.

6. Pour the gel mixture immediately into the gel mould, taking care to avoid air bubbles in the gel. Immediately, put the cold sample well-former into the gel. If using the E.C. electrophoresis cell, place a stoppered tube of ice in front of the sample well-former to aid cooling of the gel in the area not cooled by the cooling coils. It is essential to work carefully but quickly, since the agarose sets rapidly at 20°C. The reason for pouring the gel solution into a unit constantly cooled at 20°C is to permit the formation of a thin coat of agarose around the gel slab. This gives the gel added stability. At this temperature, the agarose sets before the acrylamide polymerizes. The latter is complete after about an hour, after which time the gel is ready for pre-electrophoresis.

In preparation for the pre-electrophoresis, remove any extra gel around the sample well-former. Then add the reservoir buffer (same as the gel buffer) before removing the sample well-former. This prevents air bubbles from entering the area between the gel plate and the gel. It is important to remove the former gently. Ribosomes have a net negative charge over a wide range of pH and therefore will migrate from the negative (cathode) to the positive pole (anode) in an electric field. Thus, place the negative electrode at the top and the positive electrode at the bottom of the gel. Pre-electrophorese the gel for 60 min at 50−100 V with continual recirculation of the reservoir buffer while gradually reducing the temperature of the circulating coolant around the gel from 20°C to 4°C. After pre-electrophoresis, discard the gel buffer and replace with fresh buffer after loading the sample.

1.7 Sample loading and electrophoresis

For optimum resolution of polysome bands, mix the sample with an equal volume of warm (50°C) 0.5% agarose in buffer (see Section 1.3) and allow the sample to

Table 1. Recipes of gel mixtures used for separating polysomes and ribosomes

Fig. no.	Gel	Buffer	Stock solutions for polyacrylamide gels									Time of electrophoresis	Voltage (V)	Temp. (°C)
			Agarose (g)	Water (ml)	20% Acrylamide-bisacrylamide solution	6.4% DMAPN (ml)	10 × TBE buffer (ml)	1 M Tris-HCl (ml)	3 M KCl (ml)	1 M MgCl$_2$ (ml)	1.6% Ammonium persulphate (ml)			
1	2.25%/0.5%	25 mM Tris-HCl 60 mM KCl 10 mM MgCl$_2$ (pH 7.6)	0.8	118	18	10	—	4	3.2	1.6	5	6 h (buffer change every 2 h)	120	4
2	2.5%/0.5%	25 mM Tris-acetate 60 mM potassium acetate 10 mM magnesium acetate (pH 7.6)	0.8	116	20	10	—	4[a]	3.2[a]	1.6[a]	5	6 h (buffer change every 2 h)	120	10
3	2.25%/0.5%	25 mM Tris-HCl 6 mM KCl 2 mM MgCl$_2$ (pH 7.6)	0.8	122	18	10	—	4	0.32	0.32	5	4 h (buffer change every 2 h)	200	4
4	2.75%/0.5%	25 mM Tris-HCl 0.2 mM MgCl$_2$ (pH 7.6)	0.8	119	22	10	—	4	—	0.032	5	6 h (buffer change every 2 h)	200	4
5	3%/0.5%	Tris-borate-EDTA (TBE) (pH 8.3)	0.8	105	24	10	16	—	—	—	5	6 h (no buffer change)	200	4
6	First dimension as *Figure 3* gel. Second dimension: 2 h, 200 V, 3°C in 1 × TBE buffer.													

[a] In this case chloride was replaced by acetate.

279

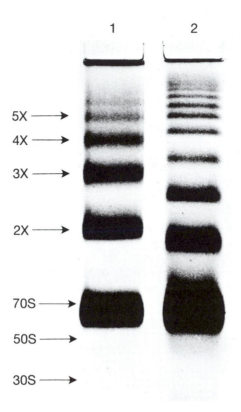

Figure 1. Electrophoresis of bacterial polysomes. Lane 1, polysomes from an exponentially growing culture of *E. coli*; lane 2, polysomes from a culture treated with streptomycin. The bands 2×, 3×, etc., represent polysomes of sizes disome, trisome, etc. The gel electrophoretic conditions used are given in *Table 1*.

set in place in a sample well from which the buffer has been removed. Buffer can be removed from the wells either using strips of filter paper or by suction with a Carlsberg or Lang-Levy micropipette. Samples not applied as a gel in the well give a streaked but otherwise identical pattern. About $0.2-0.4$ A_{260} unit of polysomes in $20-50$ μl can be added to each slot, along with an equal volume of 0.5% agarose solution.

Gelling of the sample in the well is not necessary for the analysis of ribosomal subunits in gels containing a low concentration (0.2 mM $MgCl_2$) of magnesium ions (see Section 1.9, *Figure 4*) or EDTA (see *Figure 5*). In these cases, the addition of sucrose (10% final concentration) is sufficient for layering the samples into the wells after the reservoir buffer has been added to the electrophoresis apparatus. Bromophenol blue (0.1%) can be added to the samples as a dye marker as it aids in visualizing the samples during application.

Electrophorese the samples towards the anode at a constant voltage of 120−200 V (use a voltage gradient of approximately 5−10 V/cm) at 4°C. Recirculate the buffer constantly through both reservoirs; let it overflow from the top to the bottom reservoir and return it to the top using a pump; replace the buffer with fresh buffer every two hours. Specific electrophoresis conditions are listed in *Table 1*.

1.8 Staining of gels

Immediately after the electrophoretic run stain the gels in Stains-all solution. Place the gel in a photographic tray containing 10 ml of 0.1% stock solution of Stains-all (see Section 1.3), 90 ml of 100% formamide and 100 ml of distilled water, cover the tray with aluminium foil and stain overnight. Destain the gels in running tap water while protecting the gels from direct light because of the photosensitivity of this stain. Gel strips may be scanned at 570 nm or photographed on a fluorescent viewing box, using a yellow or green filter.

The stained gels can be kept for several weeks either in distilled water at room temperature in covered trays (in the dark) or wrapped first in plastic film and then in aluminium foil without loss of resolution. These gels can also be dried easily onto filter paper with or without heat and suction (but in the dark) for a permanent record (see Section 6.5 of Chapter 1). Other stains can also be used to stain ribosomes and polysomes including methylene blue and Coomassie brilliant blue (2).

1.9 Two-dimensional gel electrophoresis

Polysomes and ribosomal subunits separated in a slab gel can be dissociated into rRNA and ribosomal proteins by soaking the gel in a solution of SDS and EDTA prior to electrophoresis in the second dimension. This permits identification of the rRNA composition of particles separated in the first-dimensional separation (2). The procedure used is as follows.

Protocol 3. Second-dimensional separation of ribosomal subunits

1. After a first-dimensional separation of polysomes in a well situated to one side of the gel, soak the gel with stirring, in a litre of 1 × TBE buffer (see Section 1.3) at 4°C for 60 min.

2. Add SDS to a final concentration of 0.2% and continue stirring for a further hour.

3. Blot the gel dry and place it in a dry electrophoretic cell at right angles to the original orientation.

4. Fix the gel in place with 1.25% agarose in 1 × TBE buffer and electrophorese for 2 h at 200 V and 4°C. Ribosomal RNA can be used as a marker in both dimensions in one of the other sample wells.

5. Wash the gel in running tap water at 20°C for 6 h to remove SDS before staining the gel with Stains-all. If SDS is omitted from this procedure, then ribosomal subunits, not rRNAs, are separated by the second-dimensional electrophoretic separation.

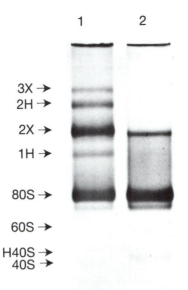

Figure 2. Electrophoresis of yeast polysomes and 'halfmers'. Lane 1, sample taken from the disome – trisome region of a sucrose gradient separating polysomes of cycloheximide-treated yeast cells. 1H and 2H represent monosomes and disomes to which are attached a 40S initiation complex and hence are termed 'halfmers'; lane 2, the same sample treated with RNase A to hydrolyse the mRNA and release the 40S initiation complex (H40S) from the remainder of the polysomes. The gel electrophoretic conditions used are given in *Table 1*.

1.10 Applications of different composite gels

The gels shown in *Figures 1 – 6* demonstrate specific applications of different composite gels for the analysis of polysomes and ribosomal subunits. Variations in buffers and acrylamide concentration (with corresponding adjustments in time and current) produce considerable differences in the separation of the macromolecules; the specific electrophoresis conditions are listed in *Table 1*.

1.10.1 Electrophoresis of bacterial polysomes

Polysomes electrophoresed into an agarose−acrylamide composite gel separate according to the size of the particle, with subunits moving fastest and the largest polysomes moving slowest (*Figure 1*, lane 1) (2,5). All polysomes penetrate the gel and polysomes as large as decasomes (10 ribosomes associated with a molecule of mRNA) are resolved as discrete bands. Since the size of polysomes is determined, in part, by the structural arrangement of the ribosomes on the mRNA, an alteration in this arrangement can change polysome mobility in the gel.

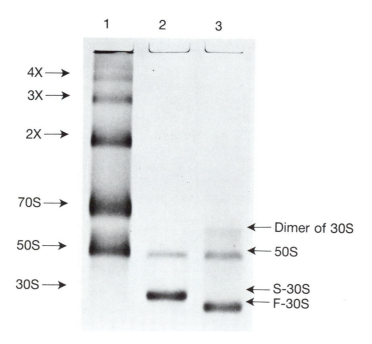

Figure 3. Separation of bacterial polysomes and ribosomal subunits. Lane 1, cell lysate containing polysomes; lane 2, ribosomal subunits from a sucrose gradient; lane 3, sample of lane 2 plus poly(U) to remove protein S1 from S-30S subunits. The gel electrophoretic conditions used are given in *Table 1*.

Streptomycin causes ribosomes to stack together on mRNA, making the polysomes more resistant to RNase degradation. These polysomes also migrate faster in gels than untreated polysomes of comparable size (6) (*Figure 1*, lane 2). One of the authors has found a similar effect with polysomes from cells starved of amino acids and with polysomes assembled *in vitro* using natural mRNA. The gel and buffer used for *Figure 1* (see *Table 1*) provide the optimum resolution of polysome bands. If a shorter time of electrophoresis is desired and some loss of fine detail of larger polysomes is acceptable, the potassium and magnesium ion concentrations in the buffer can be reduced to 15 mM and 5 mM, respectively (see also *Figure 3*).

1.10.2 Electrophoresis of eukaryotic polysomes from the yeast *Saccharomyces cerevisiae*

The substitution of acetate for chloride ions in the gel buffer improves the resolution of yeast polysomes (*Figure 2*). In this gel, the acrylamide concentration is raised to 2.5% for greater resolution of the smaller polysomes. Also shown is the separation of 'halfmer' structures; polysomes containing a 40S initiation complex produced by blocking the binding of the 60S subunit to the 40S initiation complex using cycloheximide (7).

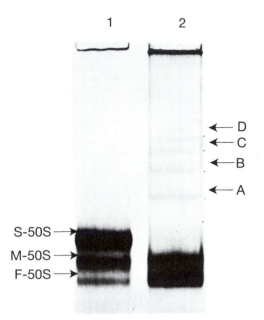

Figure 4. Separation of multiple forms of bacterial ribosomal subunits. Lane 1, 50S ribosomal subunits washed in 0.5 M NH$_4$Cl and separated into three forms in a gel containing 0.2 mM MgCl$_2$; lane 2, sample of lane 1 incubated with IgG antibody to protein L7/L12 just prior to electrophoresis. Bands A, B, C, and D represent antibody–subunit complexes (see text). The gel electrophoretic conditions used are given in *Table 1*.

1.10.3 Rapid electrophoretic separation of ribosomal subunits and polysomes

In the gel shown in *Figure 3*, the potassium and magnesium ion concentrations used are 6 mM and 2 mM, respectively, permitting the utilization of a higher voltage and a shorter time of electrophoresis (8). Due to the gentle nature of the electrophoretic separation (e.g. no pressure effect as with ultracentrifugation) polysomes remain intact even in 1 mM MgCl$_2$ (2). The larger polysomes are not as well resolved as in *Figure 1*, but small polysomes and native subunits are separated, including the resolution (lanes 2 and 3) of 30S subunits differing only by the presence (S-30S) or absence (F-30S) of a single ribosomal protein, S1 (8,9).

1.10.4 Electrophoretic separation of multiple forms of bacterial ribosomal subunits

Reduction of the magnesium ion concentration to 0.2 mM causes polysomes to dissociate into subunits but to retain considerably more structure than if dissociated in EDTA. The gel shown in *Figure 4* shows the separation of three forms of 50S ribosomal subunits from *E. coli*. For optimum separation of the three forms, the

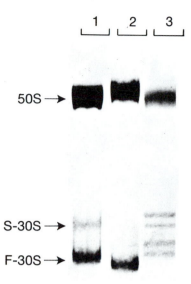

Figure 5. Electrophoretic separation of different ribosomal subunits in an EDTA-containing buffer. Lane 1, 70S ribosomes dissociated into 30S and 50S subunits in EDTA; lane 2, sample of lane 1 treated with colicin E3; lane 3, ribosomes from *E. coli* mutant defective in processing the 5'-end of 16S rRNA, containing normal subunits and slower migrating precursor particles. The gel electrophoretic conditions used are given in *Table 1*.

acrylamide concentration is raised to 2.7%. The three bands shown in lane 1 represent native subunits (S-50S), subunits lacking four copies of the ribosomal protein L7/L12 (M-50S), and subunits lacking proteins L7/L2 and additional proteins as a result of a salt wash (F-50S) (10). Upon incubation of the subunits with antibody to protein L7/L2, stable complexes are formed which survive electrophoresis and migrate slower than the subunits. The antibody molecules bind only to subunits containing proteins L7/L12 (S-50S) and are resolved in the gel as four new bands, A, B, C, and D (lane 2), representing 1, 2, 3, and 4 antibody molecules bound per ribosomal subunit. This technique of gel immunoelectrophoresis has the potential to be very useful for the analysis of a variety of nucleoprotein particles.

1.10.5 Separation of ribosomal subunits in TBE-buffered gels

The considerable buffering capacity of Peacock's TBE buffer (1) permits an extended time of electrophoresis (15 h at 200 V) without recirculation or change of buffer. The presence of EDTA unfolds the ribosomal subunits extensively but does permit the separation of eukaryotic ribosomal subunits. Just as previous gels have shown the separation of subunits differing in protein composition, the conditions used for the gel shown in *Figure 5* allow the separation of subunits with small differences in rRNA length. These differences can also be resolved by using the electrophoresis conditions used for the gels shown in *Figure 4*.

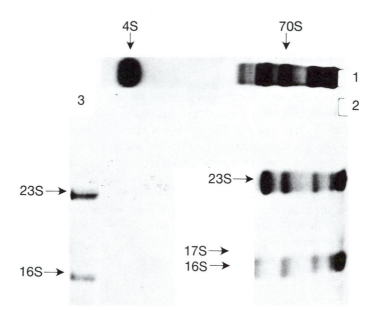

Figure 6. Two-dimensional gel analysis of polysome and ribosomal subunits. A lysate of *E. coli* containing polysomes was electrophoresed in lanes 1 and 2 (direction: right to left). The gel strip containing lane 2 was soaked in SDS, EDTA-containing buffer, rotated through 90°, fixed to the top of a fresh gel with agarose and electrophoresed in the second dimension (top to bottom). Lane 3 contained rRNA as a marker for the second dimension. The gel electrophoretic conditions used are given in *Table 1*.

Treatment of 70S ribosomes with colicin E3 cleaves 16S rRNA in the 30S subunit at a site 49 nucleotides from the 3′ end (11,12). During electrophoresis, the RNA fragment along with ribosomal protein S1, is separated from the unfolded subunit (lane 2) which migrates faster than the untreated 30S subunits (lane 1) (8). Ribosomal subunits from the *E. coli* BUMMER mutant strain include both normal 30S subunits (S-30S and F-30S) and subunits containing incompletely processed, precursor 16.3S rRNA (60 nucleotides longer at the 5′ end of the 16S rRNA) (13). These particles contain all 21 ribosomal proteins but migrate slower than normal subunits (lane 3). Rat-liver ribosomal subunits, although not shown in *Figure 5*, migrate much slower than bacterial subunits due to their larger rRNAs and more ribosomal proteins.

1.10.6 Two-dimensional gel analysis of polysomes and ribosomal subunits

The two-dimensional separation of the type shown in *Figure 6* permits an analysis of the composition of ribosomal subunits and polysomes. Note the identification of precursor rRNA (17S) associated with the 26S particle migrating faster than nascent 30S subunits. The samples can be re-electrophoresed within the same gel or into a new gel of higher acrylamide concentration using any of the many different buffer conditions. See Section 4 of Chapter 8 for descriptions of the types of protocol that can be used for two-dimensional separations.

1.11 Possible problems and their remedies

1.11.1 Irregularities in the gel

A variety of factors can contribute to irregularities in polymerization of the gel. It is recommended that care be taken to follow the protocols exactly. The volume of gel solution should be kept constant. The volume of agarose solution after microwave heating must be checked and adjusted for any loss of water. The formation of agarose clumps on the inside of flasks should be avoided by using only *gradual* cooling to 35°C. It is also important that the gel mixture is poured into a gel mould cooled at 20°C to permit gelling of agarose on the plates before polymerization of the acrylamide. For the same reason, a cold sample well-former should be used. Gels such as those shown in *Figures 1* and *2* can heat up during electrophoresis unless properly cooled. A temperature increase of even 10−20°C may affect the ribosome conformation (and mobility in the gel) as shown for naked rRNA (14,15).

1.11.2 Distortion of band patterns

Streaking of polysome bands, particularly at the margin of each lane, is reduced by gelling the sample into the well. Distortion of bands can also occur if the gel is overloaded, if the sample contains too much salt, or if the surfaces of the gel plates are not clean, causing separation of the gel from one or both of the plates. These problems are less severe when using a 3-mm-thick gel with a 1-mm-thick well since the sample is surrounded by gel which can compensate, somewhat, for overloading and salt effects as well as preventing the sample from coming into contact with the gel plates.

1.11.3 Streaked gel patterns with most eukaryotic polysomes

The only eukaryotic polysomes easily and distinctly separated as polysomes in composite gels are from yeast. The presence of agarose increases a streaked gel pattern with mammalian polysomes. It is thought that the association of membrane proteins with polysomes producing large aggregates may create considerable heterogeneity in particle size and account for the streaked pattern. Eukaryotic ribosomes from all sources can be separated as unfolded subunits in TBE-buffered gels. One of the authors has found that brief treatment of *Dictyostelium* (cellular slime mould) polysomes with proteinase K will produce a sharper polysome band pattern. High salt washes and treatment with non-ionic detergents to remove non-ribosomal proteins may also aid in the resolution of polysomes from higher eukaryotic cells but no successful method has yet been developed.

2. Electrophoresis of spliceosomes

2.1 Introduction

An analogy has been made between the spliceosome and the ribosome as they are both ribonucleoprotein complexes and have enzymatic activity (16). The similarities also extend to their abilities to be separated by gel electrophoresis. Pikielny and Rosbash (17) were the first to report the use of the composite gel system to study successfully the interaction between snRNPs and pre-mRNA during splicing. They showed that the gels permit electrophoretic separation and analysis of three splicing complexes formed with pre-mRNA during *in vitro* splicing using cell-free extracts from the yeast *Saccharomyces cerevisiae*. The time course of the appearance of the complexes and their RNA composition permitted them to predict the pathway of splicing complex assembly.

2.2 Gel conditions for the separation of spliceosomes

Pikielny and Rosbash separated yeast spliceosomes by electrophoresis in a 3% acrylamide−0.5% agarose composite gel containing 0.5 × TBE buffer. Other laboratories have successfully separated mammalian spliceosomes by gel electrophoresis (19,20). However, the same problem occurs with electrophoresis of spliceosomes as is seen with electrophoresis of mammalian ribosomes (Section 1.11.3); when agarose is present in the gel there is aggregation and streaking of samples in composite gels unless EDTA is added to the gel buffer. Consequently, the separation of yeast and mammalian spliceosomes in a composite gel of 3%/0.5% (17,18) or 6%/0.5% (20) has only been reported using EDTA (0.5 × TBE buffer in ref. 20), and the separation of mammalian spliceosomes in a Tris-glycine buffer without EDTA can only be achieved by electrophoresis in an acrylamide gel without agarose (19). The ratio of acrylamide to bisacrylamide is increased considerably in spliceosome gels to as much as 80:1 in 4% acrylamide gels (19) and 250:1 in 6%/0.5% gels (20). The temperature of the gel during electrophoresis (5°C or 15°C) does not appear to be critical for separation of the spliceosome complexes. The equipment, solutions, and preparation of the gels as described in Section 1 for the electrophoresis of ribosomes may be used successfully for the separation of spliceosomes. However, other conditions, as described in the reports mentioned above (refs. 17−20), also give good separations.

Acknowledgements

Many of the techniques described in this chapter were initially developed at N.I.H. in the laboratory of the late Dr Andrew C. Peacock with the assistance of the late Dr Sylvia L. Bunting. They are credited with the discovery of acrylamide gel electrophoresis of RNA, the Tris-borate-EDTA buffer, acrylamide−agarose composite gels, and Stains-all. Their absence as scientific colleagues and good friends is deeply felt.

References

1. Peacock, A. C. and Dingman, C. W. (1968). *Biochemistry*, **7**, 668.
2. Dahlberg, A. E., Dingman, C. W., and Peacock, A. C. (1969). *J. Mol. Biol.*, **41**, 139.
3. Lindahl, L. and Forchhammer, J. (1969). *J. Mol. Biol.*, **43**, 593.

4. Hobden, A. N. and Cundliffe, E. (1980). *Biochem. J.*, **190**, 765.
5. Talens, A., Van Diggelen, O. P., Bongers, M., Popa, L. M., and Bosch, L. (1973). *Eur. J. Biochem.*, **37**, 121.
6. Dahlberg, A. E., Lund, E., and Kjeldgaard, N. O. (1973). *J. Mol. Biol.*, **78**, 627.
7. Helser, L., Baan, R. A., and Dahlberg, A. E. (1981). *Mol. Cell Biol.*, **1**, 51.
8. Dahlberg, A. E. (1974). *J. Biol. Chem.*, **249**, 7673.
9. Szer, W. and Leffler, A. (1974). *Proc. Nat. Acad. Sci. (USA)*, **71**, 3611.
10. Tokomatsu, H., Strycharz, W., and Dahlberg, A. E. (1981). *J. Mol. Biol.*, **152**, 397.
11. Bowman, C. M., Dahlberg, J. E., Ikemura, T., Konishky, J., and Nomura, M. (1971). *Proc. Nat. Acad. Sci. (USA)*, **68**, 964.
12. Senior, B. W. and Holland, I. B. (1971). *Proc. Nat. Acad. Sci. (USA)*, **68**, 959.
13. Dahlberg, A. E., Dahlberg, J. E., Lund, E., Tokimatsu, H., Rabson, A. B., Calvert, P. C., Reynolds, F., and Zahalak, M. (1978). *Proc. Nat. Acad. Sci. (USA)*, **75**, 3598.
14. Dahlberg, A. E. and Peacock, A. C. (1971). *J. Mol. Biol.*, **55**, 61.
15. Dahlberg, A. E. and Peacock, A. C. (1971). *J. Mol. Biol.*, **60**, 409.
16. Sharp, P. (1981). *Cell*, **23**, 643.
17. Pikielny, C. and Rosbash, M. (1986). *Cell*, **45**, 869.
18. Pikielny, C., Rymond, B., and Rosbash, M. (1986). *Nature*, **324**, 341.
19. Konarska, M. and Sharp, P. (1987). *Cell*, **49**, 763.
20. Christofori, G., Frendewey, D., and Keller, W. (1987). *EMBO J.*, **6**, 1747.

A1

Nucleic acid molecular weight markers

STEPHEN J. MINTER, PAUL G. SEALEY, and RAKESH ANAND

1. DNA size markers

1.1 Plasmid pBR322

Plasmid pBR322 is extremely convenient for use as a molecular weight marker. By using a relatively small range of restriction endonucleases a wide range of molecular weight markers can be generated. The linear 4362-bp molecule, useful as a marker for agarose gels, can be obtained using any one of the following restriction enzymes, each of which cleaves at a single site in the plasmid:

*Aat*II, *Ava*I, *Bal*I, *Bam*HI, *Cla*I, *Eco*RI, *Hin*dIII, *Nde*I, *Nru*I, *Pst*I, *Pvu*II, *Rru*I, *Sal*I, *Sph*I, *Tth*I, *Xma*III, *Xor*II.

Other enzymes which have multiple restriction sites, together with the sizes of the fragments obtained, are listed in *Table 1*.

Table 1. Sizes of restriction fragments of pBR322[a,b,c]

BvuI	HincII	AccI	XmnI	EcoK	AhaIII	BglI	RsaI	NarI	NaeI	ThnII
4348	3256	2767	2430	2377	3651	2319	2117	3571	3481	3196
14	1106	1595	1932	1985	692	1809	1565	657	367	1127
					19	234	680	113	354	32
								21	160	7

MstI	HgiDI	BstNI	TaqI	AvaII	DdeI	HgiAI	HgiCI	HinfI	HaeII	HgaI
2134	2699	1857	1444	1746	1652	1161	2027	1631	1876	867
1095	657	1060	1307	1433	542	826	1123	517	622	731
1035	490	928	475	303	540	604	439	506	439	633
98	382	383	368	279	465	587	294	396	430	578
	113	121	315	249	426	498	218	344	370	415
	21	13	312	222	409	310	113	298	227	314
			141	88	166	291	84	221	181	245
				42	162	85	43	220	83	239
							21	154	60	158
								75	53	150
									21	32

NciI	MboII	FokI	HphI	Sau96	AluI	ScrFI	Sau3A	SfaNI	HaeIII	TacI
724	790	1171	1106	1461	910	696	1374	1052	587	581
699	755	853	853	616	659	592	665	424	540	493
696	753	659	576	352	655	525	358	395	504	452
632	592	649	415	279	521	363	341	375	458	372
363	492	287	387	274	403	351	317	248	434	355
351	271	188	282	249	281	347	272	243	267	341
328	254	181	227	222	257	328	258	234	234	332
308	196	141	221	191	226	308	207	223	213	330
226	109	78	207	189	136	218	105	220	192	145
35	78	62	45	179	100	199	91	192	184	129
	72	48	34	124	63	184	78	192	124	129
		45	9	88	57	121	75	164	123	122
				79	49	42	46	134	104	115
				42	19	40	36	96	89	103
				17	15	35	31	88	80	97
					11	13	27	78	64	69
							18	63	57	66
							17	39	51	61
							15	37	21	27
							12	25	18	26
							11	12	11	10
							8	11	7	5
								9		2

[a] The enzymes are listed in order of increasing number of fragments generated.

[b] The fragment sizes (in base pairs) do not include any single-stranded extensions which may be generated by the restriction endonuclease.

[c] Data derived using the pBR322 sequence (1) on file in the EMBO DNA Sequence Library.

HpaII	MnlI	HhaI	Fnu4H
622	591	393	480
527	400	348	328
404	334	337	299
309	262	332	277
242	247	270	229
238	218	259	206
217	206	206	189
201	206	190	165
190	206	174	155
180	204	153	150
160	185	152	143
160	179	151	129
147	166	141	123
147	156	132	119
122	150	131	118
110	96	109	116
90	88	103	107
76	81	100	97
67	77	93	95
34	61	83	85
34	60	75	83
26	58	67	81
26	38	62	79
15	36	60	71
9	30	53	65
9	27	40	57
		36	53
		33	51
		30	46
		28	45
		21	34
			27
			18
			14
			7
			3
			3
			3
			3
			3
			3
			3

1.2 Plasmid pAT153

pBR322 and pAT153 are closely-related plasmids. pAT153 was derived from pBR322 by removing the B (622 bp) and G (83 bp) *Hae*II restriction fragments of pBR322 (2). Linear molecules of pAT153 are 3657 bp long; they can be obtained by restriction with any one of the following enzymes:

*Aat*II, *Acc*I, *Ava*I, *Bal*I, *Bam*HI, *Cla*I, *Eco*K, *Eco*RV, *Hin*dIII, *Nru*I, *Pst*I, *Rru*I, *Sal*I, *Sph*I, *Xma*III, *Xmn*I, *Xor*II.

Other enzymes which have multiple restriction sites, together with the sizes of the fragments obtained, are listed in *Table 2*.

Table 2. Sizes of restriction fragments of pAT153[a]

BvuI	RsaI	HincII	TthII	AhaIII	BglI	NarI	NaeI	MstI	HgiDI	BstNI	DdeI
3643	3486	2551	3618	2946	1809	2866	2776	1429	1994	1857	1652
14	165	1106	32	692	1614	657	367	1095	657	928	540
		7		19	234	113	354	1035	490	383	464
						21	160	98	382	355	426
									113	121	409
									21	13	166

FokI	TaqI	AvaII	HgiAI	NciI	HaeII	HinfI	HgiCI	HgaI	HphI	MboII	AluI
1584	1444	1433	1161	724	1876	1631	1322	867	986	790	659
853	602	1320	619	696	439	517	1117	731	853	755	655
659	475	303	604	665	430	396	439	578	415	592	622
287	368	249	587	632	370	298	294	501	387	492	521
181	315	222	310	363	227	221	218	314	282	271	403
48	312	88	291	351	181	220	113	245	227	254	257
45	141	42	85	226	60	154	84	239	221	196	226
					53	145	43	150	207	109	136
					21	75	21	32	45	78	100
									34	72	63
										48	15

[a] See footnotes to *Table 1*.

294

Sau96	SfaNI	ScrFI	TacI	Sau3A	MnlI	HaeIII	HpaII	HhaI	Fnu4H
1224	1052	696	581	876	591	587	622	393	480
616	437	592	493	665	400	458	492	337	328
352	424	525	452	358	247	434	404	332	277
274	395	363	355	341	218	339	242	270	229
249	375	351	332	317	206	267	238	259	206
222	248	313	330	272	206	234	217	206	189
191	234	218	182	258	206	213	201	190	165
179	192	199	145	105	204	192	190	174	155
124	164	184	129	91	200	184	160	153	150
88	88	121	129	78	179	124	160	152	143
79	25	42	122	75	166	123	147	151	129
42	12	40	115	46	156	104	122	132	119
17	11	13	97	36	150	89	110	131	118
			66	31	96	80	90	109	113
			61	27	88	64	76	100	107
			27	18	81	57	67	93	95
			26	17	77	51	34	75	85
			10	15	61	21	26	67	83
			5	12	60	18	26	62	79
				11	38	11	15	60	71
				8	27	7	9	53	65
							9	40	57
								36	51
								33	45
								28	34
								21	27
									18
									14
									7
									3 × 6

295

1.3 Bacteriophage lambda (λ)

Bacteriophage λ is the most extensively-studied phage of *Escherichia coli*. The DNA of strain λ cI *indl ts* 857 Sam7 has been sequenced (3). The native molecule is circular and double-stranded. Heating at 68°C for 10 min, in a buffer containing a low concentration of salt, followed by rapid cooling causes separation of the cohesive ends at the 'cos' site. This generates a linear molecule with single-stranded 'sticky' ends 12 nucleotides long. *Table 3* lists the sizes of fragments generated by a variety of restriction endonucleases.

Table 3. Sizes of restriction fragments of phage λ DNA (strain cl *ts* 857)[a,b]

XhoI	XbaI	SalI	KpnI	SstI	AvrII	SmaI	PvuI
33 498	24 508	32 745	29 942	24 776	24 322	19 399	14 321
15 004	23 994	15 258	17 057	22 621	24 106	12 220	12 712
		499	1503	1105	74	8612	11 936
						8271	9533

SstII	EcoRI	BamHI	BglII	HindIII	AvaI	HpaI	PvuII
20 323	21 226	16 841	22 010	23 130	14 677	8666	21 088
18 780	7421	7233	13 286	9416	8614	6911	4421
8113	5804	6770	9688	6557	6888	5414	4268
1076	5643	6527	2392	4361	4720	4535	4194
210	4878	5626	651	2322	4716	4491	3916
	3530	5505	415	2027	3730	4347	3638
			60	564	1881	3408	2296
				125	1674	3384	1708
					1602	3042	636
						2240	579
						734	532
						441	468
						410	343
						251	211
						228	141
							64

[a] See footnotes *a* and *b* to *Table 1*.
[b] Data derived from the sequence of λ cl *ts* 867 on file in the EMBO DNA Sequence Library.

1.4 Simian virus 40 (SV40)

The SV40 genome is a double-stranded circular DNA 5243 bp long. It can be linearized using *Acc*I, *Bam*HI, *Eco*RI, *Eco*RV, *Hae*II, *Kpn*I, *Msp*I (*Hpa*II), *Nae*I or *Taq*I. *Table 4* lists the sizes of restriction fragments generated by other enzymes which have multiple cleavage sites.

Table 4. Sizes of restriction fragments from SV40[a,b,c]

HhaI	NdeI	PstI	XhoII	PvuII	HphI	HpaI	AvaII
4753	4225	4027	3007	2007	3091	2147	1580
490	1018	1216	1566	1719	1856	2009	1525
		760	760	1446	160	1067	995
					136	20	682
							430
							31

HindIII	HincII	MboII	HinfI	HaeI	AhaIII	RsaI	MboII
1768	1980	1347	1845	1739	1753	1605	1350
1169	1538	1264	1085	1661	739	708	756
1116	1067	945	766	383	565	675	687
526	369	610	543	348	430	551	645
447	240	396	525	329	411	497	409
215	29	384	237	300	364	351	395
	20	234	109	227	318	294	383
		60	83	179	315	226	375
			24	33	141	153	69
			24	30	136	111	65
				14	71	57	31
						15	30
							14
							13
							11
							10

HaeIII	AluI	
1661	775	46
765	483	41
540	329	38
373	288	30
329	275	29
325	253	28
300	253	27
299	243	12
227	224	10
179	223	8
49	177	7
45	157	
41	154	
33	153	
30	146	
29	144	
14	123	
9	75	
6	54	
	53	
	50	
	49	

[a] See footnotes a and b to Table 1.
[b] From sequence data for SV40 DNA reported in references 4 and 5.
[c] There are no cleavage sites for AvaI, BglII, ClaI, FndDII, PvuI, SalI, SmaI, XbaI, XhoI or XmaII.

1.5 Coliphage M13 mp7

This is a widely-used cloning and sequencing vector which consists of a circular DNA molecule, 7238 bp long (RF form). It can be linearized by *Acy*I, *Ava*I, *Ava*II, *Avr*I, *Bgl*I, *Bgl*II, *Mst*II, *Nae*I, *Nar*I, *Pvu*I, *Pst*I, *Sau*I, *Sua*I. *Table 5* lists the sizes of fragments generated from M13 mp7 DNA by enzymes which have multiple restriction sites.

Table 5. Sizes of restriction fragments for phage M13 mp7[a,b]

AccI	SalI	HincII	BamHI	EcoRI	GdiII	PvuII	HgiAI	XmnI	ClaI
7226	7226	7226	7214	7196	6993	6835	6516	4949	4343
12	12	12	24	42	245	310	722	2289	2895

AsuI	CauII	NciI	XhoI	HaeI	HgiCI	HaeII	EcoRII	SfaNI	HgaI
6578	4313	4313	4020	2836	4428	3514	3975	2625	1638
446	4162	2336	2535	2813	2022	2520	1809	1684	1517
190	558	558	659	1325	324	434	952	966	1089
24	31	31	24	264	322	433	179	871	1075
					130	329	139	716	846
					12	8	127	363	758
							57	13	315

Sau3A	BbvI	KpnI	TaqI	HaeIII	Fnu4HI	HphI	RsaI	FndDII	HpaII
4020	1826	1694	1018	2527	1826	2212	1345	1290	1596
1696	1739	1241	971	1623	1739	1595	1334	889	829
507	1154	1226	927	849	891	1013	1004	772	818
434	665	792	703	341	629	589	742	681	651
332	611	666	639	311	611	489	604	646	545
129	436	332	612	309	436	271	522	541	543
96	420	318	579	245	420	241	322	496	472
24	387	301	564	214	387	194	258	495	454
		196	441	169	164	144	201	361	357
		196	381	158	45	135	190	353	183
		163	239	117	27	127	163	304	176
		113	152	106	24	76	143	190	156
			12	102	22	39	107	63	130
				98	14	39	102	57	123
				69	3	33	93	54	79
						26	65	24	60
						9	27	20	30
						6	16	2	18
									18

[a] See footnotes *a* and *b* of *Table 1*.
[b] There are no cleavage sites for *Bst*EII, *Hind*III, *Hpa*I, *Kpn*I, *Mst*I, *Sma*I, *Xho*I or *Xma*I.

HhaI	AluI	HinfI	DdeI	EcoRI*	MnlI	
967	1446	1288	998	1402	423	31
732	1330	771	866	665	421	30
725	600	571	672	567	391	28
714	555	486	652	508	384	27
683	484	413	600	416	351	22
495	336	348	563	406	340	21
434	331	345	399	323	318	19
340	313	328	381	283	313	18
329	220	324	376	272	304	16
312	204	274	303	247	286	15
293	201	261	294	213	269	15
272	180	253	272	209	263	15
244	159	234	160	176	243	15
190	151	232	153	174	234	15
115	140	212	128	152	226	15
92	111	209	72	142	217	15
74	111	160	63	109	158	15
65	104	137	46	109	143	12
56	93	96	42	102	130	8
28	72	80	39	99	119	6
26	63	63	28	96	117	5
22	39	46	27	88	111	4
13	39	45	24	76	105	4
9	27	22	15	69	102	
8	26	21	15	63	98	
	24	19	15	55	90	
			15	53	86	
			15	41	76	
			14	40	74	
				36	69	
				33	66	
				12	63	
				1	51	
				1	49	
					48	
					48	
					43	
					38	

1.6 Bacteriophage ΦX174

This is a bacteriophage of *E. coli* with a single-stranded DNA molecule. The RF form is double-stranded and 5386 bp long. It can be linearized using *Ava*I, *Ava*II, *Mst*II, *Pst*I, *Sac*II or *Xho*I. The restriction fragment sizes for enzymes with multiple restriction sites are given in *Table 6*.

Table 6. Sizes of restriction fragments of phage φX174[a,b,c]

*Aha*III	*Nar*I	*Acc*I	*Eco*RII	*Xmn*I	*Hpa*II	*Hae*I
4307	3429	3034	2767	4126	2748	2712
1079	1957	2352	2619	974	1697	872
				286	374	738
					348	603
					219	389
						72

*Hae*II	*Hph*I	*Taq*I	*Rsa*I	*Hae*III	*Mbo*II	*Hinc*II
2314	1638	2914	1560	1353	1103	1057
1560	1116	1175	964	1078	1066	770
786	791	404	645	872	857	612
269	777	327	525	605	812	495
185	777	321	472	310	396	397
125	110	141	392	281	394	392
93	77	87	247	271	324	345
54	43	54	197	232	224	341
	39	33	157	194	118	335
		20	138	118	89	291
			89	72	3	210
						162
						79

*Dde*I	*Fnu*DII	*Hha*I
1012	1050	1553
998	870	640
927	718	614
542	695	532
486	530	305
393	490	300
303	259	269
302	176	201
186	156	192
165	127	145
38	114	143
18	103	123
10	79	101
6	19	93
		84
		54
		35
		2

[a] See footnotes *a* and *b* to *Table 1*.
[b] Data from reference 6.
[c] *Hinf*I gives 21 fragments and *Alu*I gives 24 fragments. There are no cleavage sites for *Bam*HI, *Eco*RI, *Hind*III, *Nae*I, *Pvu*I, *Pvu*II, *Sal*I, *Sma*I, *Xba*I or *Xho*II.

1.7 Coliphage fd

This is a male-specific coliphage closely related to the phage M13. The virion is a single-stranded circular DNA molecule 6408 nucleotides long. The fd DNA can be linearized using either *Acc*I or *Hinc*II. The restriction fragment sizes for enzymes with multiple cleavage sites are given in *Table 7*.

Table 7. Sizes of restriction fragments for phage fd DNA[a,b,c]

*Bam*HI	*Hae*II	*Mbo*II	*Hae*III	*Taq*I	*Msp*I(*Hpa*II)	*Alu*I
3425	3550	4349	2528	2019	1596	1446
2983	2033	666	1633	850	829	1330
	817	384	849	703	819	705
	8	332	352	652	652	554
		318	311	579	648	484
		196	309	441	501	366
		163	154	381	454	314
			106	357	381	257
			103	287	156	220
			69	139	129	204
					123	201
					60	166
					42	111
					12	29
					6	27
						24

[a] See footnotes *a* and *b* of *Table 1*.
[b] Data from reference 7.
[c] *Hinf*I gives 24 fragments and *Mnl*I gives 51 fragments. There are no cut sites for *Bcl*I, *Eco*RI, *Hin*dIII or *Kpn*I.

301

1.8 Yeast chromosomes

Yeast chromosomes from the organism *Saccharomyces cerevisiae* (X2180-1B) can be used as molecular weight markers for pulsed field electrophoresis separations. Note that chromosome size is strain specific and so the chromosomal DNA of other strains may vary from the sizes given in *Table 8*.

Table 8. Sizes of yeast chromosomal DNA

Band no.	Chromosome	Size (kb)[a]
1.	I	245
2	VI	280
3	III	360
4	IX	450
5	V & VIII	590
6	XI	680
7	X	750
8	XIV	790
9	II	820
10	XIII	940
11	XVI	970
12	VII & XV	1115
13	IV	1370
14[b]	XII	1750
15[b]	XII	1850

[a] Bands 1 to 13 were sized using an oligomer ladder of λcI857Sam7 DNA.
[b] Chromosome XII contains almost all the genes for rRNA and in this strain exhibits some instability, migrating as two bands. The sizes of these bands have been extrapolated from the mobility of the smaller sized bands.

2. RNA size markers

RNA can be electrophoresed on non-denaturing agarose or polyacrylamide gels, or on denaturing gels, either agarose containing methyl mercuric hydroxide or polyacrylamide gels containing 98% formamide or 8 M urea. Hence the apparent molecular weight will depend upon the conditions of electrophoresis. In addition to the RNA molecular weight markers listed in *Table 9*, restriction endonuclease fragments of DNA (see *Tables 1−8*) may also be useful as size markers in certain situations. Some companies also sell mixtures of RNA markers, both labelled and unlabelled, ready for use.

Table 9. Molecular weight markers for gel electrophoresis of RNA

RNA species	Molecular weight[a]	Number of nucleotides	Reference
Myosin heavy chain mRNA (chicken)	2.02×10^6	6500	11[c]
28S rRNA (HeLa)[d]	1.90×10^6	6333	8
25S rRNA (*Aspergillus*)[d]	1.24×10^6	4000	17
23S rRNA (*E. coli*)[d]	1.07×10^6	3566	9
18S rRNA (HeLa)[d]	0.71×10^6	2366	8
17S rRNA (*Aspergillus*)[d]	0.62×10^6	2000	17
16S rRNA (*E. coli*)[d]	0.53×10^6	1776	10
A2 crystallin mRNA (calf lens)	0.45×10^6	1460	12[c]
Immunoglobulin light chain mRNA (mouse)	0.39×10^6	1250	13[c]
β-globin mRNA (mouse)	0.24×10^6	783	14
β-globin mRNA (rabbit)	0.22×10^6	710	15[c]
α-globin mRNA (mouse)	0.22×10^6	696	14
α-globin mRNA (rabbit)	0.20×10^6	630	15[c]
Histone H4 mRNA (sea urchin)	0.13×10^6	410	16[b]
5.8S RNA (*Aspergillus*)	4.89×10^4	158	17
5S RNA (*E. coli*)	3.72×10^4	120	18
4S RNA (*Aspergillus*)	2.63×10^4	85	17

[a] Molecular weights are approximate only and based upon average 'molecular weight' of 310 for each nucleotide residue.
[b] Non-denaturing gel system.
[c] Denaturing (formamide) gel system.
[d] A more extensive list of sizes of rRNAs is given in ref. 19.

References

1. Sutcliffe, J. G. (1979). *Cold Spring Harbor Symp. Quant. Biol.*, **43**, 77.
2. Twigg, A. J. and Sherrat, D. (1980). *Nature*, **283**, 216.
3. Sanger, F., Coulson, A. R., Hong, G. F., Hill, D. F., and Petersen, G. B. (1982). *J. Mol. Biol.*, **162**, 729.
4. Fiers, W., Contreras, R., Haegeman, G., Rogiers, R., Van de Voorde, A., Van Heuverswyn, H., Van Herreweghe, J., Volckaert, G., and Ysebaert, M. (1978). *Nature*, **273**, 113.
5. Van Heuverswyn, H. and Fiers, W. (1979). *Eur. J. Biochem.*, **100**, 50.
6. Sanger, F., Coulson, A. R., Friedmann, T., Air, G. M., Barrell, B. G., Brown, N. L., Fiddes, J. C., Hutchison, C. A., Slocombe, P. M., and Smith, M. (1978). *J. Mol. Biol.*, **125**, 225.
7. Beck, F., Sommer, R., Auerswald, E. A., Kurz, Ch., Zink, B., Osterburg, G., and Schaller, .H (1978). *Nucleic Acids Res.*, **5**, 4495.
8. McConkey, E. and Hopkins, J. (1969). *J. Mol. Biol.*, **39**, 545.
9. Stanley, W. M. and Bock, R. M. (1965). *Biochemistry*, **4**, 1302.
10. Pearce, T. C., Rowe, A. J., and Turnock, G. (1975). *J. Mol. Biol.*, **97**, 193.
11. Mondal, H., Sutton, A., Chen, V. J., and Sarkar, S. (1974). *Biochem. Biophys. Res. Commun.*, **56**, 988.
12. Berns, A., Jansson, P., and Bloemendal, H. (1974). *Biochem. Biophys. Res. Commun.*, **59**, 1157.
13. Stravnezer, J., Huang, R. C. C., Stravnezer. E., and Bishop, J. M. (1974). *J. Mol. Biol.*, **88**, 43.
14. Morrison, M. R. and Lingrel, J. B. (1976). *Biochim. Biophys. Acta*, **447**, 104.
15. Hamlyn, P. H. and Gould, H. J. (1973). *J. Mol. Biol.*, **94**, 101.
16. Grunstein, M. and Schedl, P. (1976). *J. Mol. Biol.*, **104**, 323.
17. Scazzochio,C., personal communication.
18. Brownlee, G., Sanger, F., and Barrell, B. G. (1968). *J. Mol. Biol.*, **34**, 379.
19. Loening, U. E. (1968). *J. Mol. Biol.*, **38**, 355.

Suppliers of specialist items for electrophoresis

B. DAVID HAMES and DAVID RICKWOOD

Many of the larger companies have subsidiaries in other countries whilst most of the smaller companies market their products through agents. The name of a local supplier is most easily obtained by writing to the relevant address listed here.

Acronym Pvt Ltd, Boronia 3155, Victoria, Australia.

Aldrich Chemical Company, The Old Brickyard, New Road, Gillingham, Dorset SP8 4JL, UK; 940 West Saint Paul Avenue, Milwaukee, WI 53233, USA.

Amersham International, Lincoln Place, Green End, Aylesbury, Bucks HP20 2TP, UK; 2636 South Clearbook Drive, Arlington Heights, IL 60005, USA.

Amicon Corporation, Upper Mill, Stonehouse, Gloucs GL0 2BJ, UK; 17 Cherry Hill Drive, Danvers, MA 01923, USA.

Anderman & Co. Ltd, 145 London Road, Kingston-upon-Thames, Surrey KT2 6NH, UK.

Applied Biosystems, 850, Lincoln Drive Centre, Foster City, CA 94404, USA.

Atto Corp., 2−3 Hongo 7 Chome, Bunkyo-Ku, Tokyo 113.

W & R Balston Ltd, Springfield Mill, Maidstone, Kent, UK.

BDH Chemicals Ltd, Broom Road, Poole, Dorset BH12 4NN, UK.

Bio-Rad Laboratories, Caxton Way, Watford Business Park, Watford, Herts WD1 9RP, UK; 1414 Harbor Way South, Richmond, CA 93804, USA.

Bethesda Research Laboratories (BRL), *see* Gibco-BRL.

Biotech Instruments Ltd, 183 Camford Way, Luton, Beds LU3 4AN, UK.

Boehringer Mannheim Biochemicals, Boehringer Mannheim House, Bell Lane, Lewes, East Sussex BN7 1LG, UK; PO Box 50816, Indianapolis, IN 46250−0816, USA; PO Box 310120, 6800 Mannheim 31, FRG.

Buchler Scientific Instruments, 1327 Sixteenth Street, Fort Lee, NJ 07024, USA.

Calbiochem Corporation, Novabiochem, UK; PO Box 12087, San Diego, CA 92112−1480, USA.

Cambridge Research Biochemicals, Button End, Harston, Cambridge CB2 5NX, UK; Suite 202, 10 East Merrick Road, Valley Stream, NY 11580, USA.

CP Laboratories, PO Box 22, Bishops Stortford, Herts CM23 3DH, UK.

Difco Laboratories Ltd, PO Box 14B, Central Avenue, East Moseley, Surrey KT8 0SE, UK.

Dupont Company Biotechnology Systems, Barley Mill Plaza, Chandler Mill Building, Wilmington, DE 19898, USA.

E-C Apparatus Corporation, 3831 Tyrone Boulevard, North St, Petersburg, FL 33709, USA.

Fisher Scientific, 52 Fadem Road, Springfield, NJ 07081, USA.

Fluka Chemical Corp., 980 South Second Street, Ronkonkoma, NY 11779, USA; Industriestrasse 25, CH-9470, Buchs, Switzerland.

FMC Corp., PO Box 308, 5 Maple Street, Rockland, ME 04841, USA.

Fuji Photo Film Co. Ltd, Fuji Photo Film (UK) Ltd, Cresta House, 125 Finchley Road, London NW3 6JH, UK; Chemical Products Dept. 26−30, Nishiazabu, 2-chome, Minato-ku, Tokyo 106, Japan.

Gelman Sciences Inc., 12 Peter Road, Lancing, Sussex, UK; 600 South Wagner Road, Ann Arbor, MI 48106, USA.

Gibco-BRL, PO Box 35, 3 Washington Road, Sandyford Industrial Estate, Paisley, Renfrewshire PA3 4EP, UK; 8400 Helgerman Court, Gaithersburg, MD 20877, USA.

V.A.Howe & Co. Ltd, 12−14 St Ann's Crescent, London SW18 2LS, UK.

Hamilton, V.A.Howe & Co. Ltd, UK; PO Box 10030, Reno, NV 89510, USA.

Hamilton Bonaduz AG, PO Box 26, CH-7402 Bonaduz, Switzerland.

Hanimex (UK), Faraday Road, Dorcan, Swindon, Wiltshire SN3 5HW, UK.

Hoefer Scientific Instruments, PO Box 351 Newcastle-under-Lyme, Staffs ST5 0TW, UK; 65 Minnesota Street, PO Box 77387, San Francisco, CA 94107, USA.

IBI International Biotechnologies, P.O. Box 9558, New Haven, CT 06535, USA.

ICN Biomedicals Ltd, Freepress House, Castle Street, High Wycombe, Bucks HP13 6RN, UK.

Ilford Ltd, Town Lane, Mobberley, Knutsford, Cheshire WA16 7HA, UK.

Instrumentation Specialities Co. (ISCO), Life Science Laboratories Ltd, UK; PO Box 5347, Lincoln, NE 68505, USA.

Jannsen-Pharmaceutical, ICN Biomedicals Ltd, UK; Jannsen Biotech NV, Lammerdries 55, B-2430 Olen, Belgium.

Joyce-Loebl Ltd, 48 Princes Way, Team Valley, Gateshead NE11 0UJ, UK.

Kodak, Kodak Ltd, PO Box 33, Swallowdale Lane, Hemel Hempstead, Herts HP2 7EU, UK; Eastman Kodak Co., 343 State Street, Rochester, NY 14650, USA.

Life Science Laboratories, Sarum Road, Luton, Beds LU3 2RA,UK.

LKB, *see* Pharmacia-LKB Biotechnology.

Machery-Nagel GmbH & Co. KG, Neumann-Neander Str., P.O. Box 307, D5160 Dueren, FRG.

Marine Colloids Division, FMC Corporation, PO Box 308, 5 Maple Street, Rockland, MA 04841, USA.

E.Merck, Frankfurter Strasse 250, D-6100 Darmstadt 1, FRG.

Miles Research Products Division, Miles Laboratories Ltd, ICN Biomedicals Ltd, UK; PO Box 70, Elkhart, IN 46514, USA.

Millipore Corporation, 11−15 Peterborough Road, Harrow, Middlesex HA1 2YH, UK; 80 Ashby Road, Bedford, MA 01730, USA.

Nalgene Labware, 75 Panorama Creek Drive, P.O. Box 365, Rochester, NY 14602, USA.

National Technical Information Service (NTIS), Springfield, VA 22161, USA.

New England Biolabs, Inc., CP Laboratories, UK; 32 Tozer Road, Beverly, MA 01915−5510, USA; Postfach 2750, 6231 Schwalbach bei Frankfurt, FRG.

New England Nuclear (NEN) Research Products, NEN Chemicals GmBH, Postfach 401240, 6072 Dreieich, FRG; NEN Corporation, 549 Albany Street, Boston, MA 02118, USA.

Novabiochem, 3 Heathcoat Building, Highfields Science Park, University Boulevard, Nottingham NG7 2QJ, UK.

Oxoid Ltd, Wade Road, Basingstoke, Hants RG24 0PW, UK.

Pierce, Pierce Europe BV, PO Box 1512, 3260 BA Oud-Beijerland, The Netherlands; PO Box 117, Rockford, IL 61105, USA.

Procon, DK-7171, Uldum, Denmark.

Pharmacia-LKB Biotechnology, Pharmacia House, Midsummer Boulevard, Milton Keynes MK9 3HP, UK; 800 Centennial Avenue, Piscataway, NJ 08854, USA; S-75182 Uppsala, Sweden.

Photodyne Inc., 16700 W. Victor Rd., New Berlin, WI 53151, USA.

PL Biochemical, *see* Pharmacia-LKB Biotechnology.

Polaroid, Ashley Road, St Albans, Herts AL1 5PR, UK.

Polysciences, 24 Low Farm Place, Moulton Park, Northampton NN3 1HY, UK; Paul Valley Industrial Park, Warrington, PA 18976, USA.

Sartorius GmbH, Weender Landstrasse 94/108, PO Box 3243, D-3400 Göttingen, FRG.

Schleicher & Schuell, Inc., Anderman & Co. Ltd, UK.

Schleicher & Schüll GmbH, 10 Optical Avenue, Keene, NH 03431,USA; D-3354 Dassel, FRG.

Schwartz Mann Division, Becton Dickinson Corp., Mountain View, CA 940213, USA.

Seikagaku Kogyo Co. Ltd, Tokyo.

Serva Fine Biochemicals, Cambridge Bioscience, 42 Devonshire Road, Cambridge CB1 2BL, UK; PO Box A, 18 Villa Place, Garden City Park, NY 11040, USA.

Serva Fenbiochemica GmbH, PO Box 105260, Carl-Benz-Strasse 7, D-6900 Heidelberg, FRG.

Shandon Southern, 93−96 Chadwick Road, Astmoor Industrial Estate, Runcorn, Cheshire WA7 1PR, UK.

Sigma Chemical Corporation, Fancy Road, Poole, Dorset BH17 7NH, UK; PO Box 14508, 3500 DeKalb St, St Louis, MO 63178, USA.

Sterilin Ltd, Sterilin House, Clockhouse Lane, Feltham TW14 8QS, Middlesex, UK.

Tribotics Ltd, 24/27 Crawley Hill Industrial Estate, Oxford OX8 5TJ, UK.

Union Carbide, Union Carbide Ltd, Peter House, Oxford Street, Manchester 1, UK; Union Carbide Corporation, 6733 West 65th Street, Chicago, IL 60632, USA.

Universal Scientific Ltd, 9 The Broadway, Woodford Green, Essex 1G8 0H2, UK.

UV Products Inc., 5100 Walnut Grove Ave., P.O. Box 1501, San Gabriel, CA 91778, USA. Science Park, Milton Road, Cambridge CB4 4BN, UK.

Watson Products, 1068 North Allen Avenue, Pasadena, CA 91104, USA.

Watson-Marlow Ltd, Falmouth, Cornwall TR11 4RU, UK.

Wellcome Reagents Ltd, 303 Hithergreen Lane, London SE13 6TL, UK.

Whatman Biosystems Ltd, Springfield Mill, Maidstone, Kent ME14 2LE, UK; 9 Bridewell Place, Clifton, NJ 07014, USA.

Index

Index

scanning gels 9,204
scanning gels 9
SDS gels of proteins 263
slab gels
  agarose 34−6
  apparatus 23,52,126,194
  polyacrylamide 24,127,164
slicing gels 10
solubilization of gels 16
spliceosomes 287−8
stains-all 277,281
staining methods
  DNA 65,204,276
  ribosomes 281
  RNA 10,36,162
submarine gels 33−6,54
suppliers 305−8

temperature effects 16,44,64,127,163,190,208
tube gels
  analysis of gels 9
  apparatus 4,6
  polyacrylamide recipes 5
  preparation 5,211

two-dimensional separations
  DNA 182
  nucleosomes 258
  ribosomes 281
  RNA 153,166,174
  spliceosomes 287

urea gels 19,60,161
UV light
  fixing DNA to filters 74
  footprinting 237
  fragmentation of DNA 73,119
  visualization of nucleic acids 28,43,136,
    165,204

virus fingerprinting 171

yeast
  chromosome markers 107,109,302
  ribosomes 283,287

zinc finger proteins 231